STEM CELL RESEARCH TRENDS

STEM CELL RESEARCH TRENDS

JOSSE R. BRAGGINA
EDITOR

Nova Biomedical Books
New York

For permission to use material from this book please contact us:
Telephone 631-231-7269; Fax 631-231-8175
Web Site: http://www.novapublishers.com

LIBRARY OF CONGRESS CATALOGING-IN-PUBLICATION DATA

Stem cell research trends / editor, Josse R. Braggina.
p. ; cm.
Includes bibliographical references and index.
ISBN-13: 978-1-60021-622-0 (hardcover)
ISBN-10: 1-60021-622-6 (hardcover)
1. Stem cells. 2. Stem cells--Transplantation. I. Braggina, Josse R.
[DNLM: 1. Stem Cells--physiology. 2. Biomedical Research--trends. 3. Stem Cell Transplantation. QU 325 S824 2007]
QH588.S83S7423 2007
616'.02774--dc22 2007006567

Published by Nova Science Publishers, Inc. ✦ New York

CONTENTS

PREFACE

Among the many applications of stem cell research are nervous system diseases, diabetes, heart disease, autoimmune diseases as well as Parkinson's disease, end-stage kidney disease, liver failure, cancer, spinal cord injury, multiple sclerosis, Parkinson's disease, and Alzheimer's disease. Stem cells are self-renewing, unspecialized cells that can give rise to multiple types all of specialized cells of the body. Stem cell research also involves complex ethical and legal considerations since they involve adult, fetal tissue and embryonic sources. This new book presents important research from around the globe.

Chapter I - Tissue stem cells are multipotent and maybe clonogenic. These characters are similar to those of the earliest progenitor cells of carcinomas. In the gastrointestinal tract, epithelial stem cells exist in the region of the isthmus/neck in the stomach, in the small intestine in the crypt base, just superior to the Paneth cells, and towards the bottom of the crypt in the colon, giving rise to all epithelial cell lineages in all these tissues. Recent reports have revealed that many supposed alterations found in cancer cells are also present in morphologically-normal stem cells. In this context, carcinomas can perhaps be explained as stem cell diseases. To understand the clonal expansion of a single, normal or mutated stem cell brings new insights for not only into the biology of the gastrointestinal tract but also in tumorigenesis in the intestine.

In this review, the authors will describe; (1) general aspects of gastrointestinal tract biology, (2) the development of the gastrointestinal tract and molecular factors regulating gut embryogenesis, (3) the proposal that stem cells yield all of the gastrointestinal epithelial cell lineages; the 'Unitarian Theory', and (4) the way in which stem cells, or cancer cells, clonally expand, especially focusing on the two basic models; the 'top-down' theory and the 'bottom-up' concepts. In addition (5) the authors will assess the mechanisms which maintain the crypt stem cell populations, (6) putative stem cell markers, (7) the stem cell niche, thought to be provided by subepithelial myofibroblasts, (8) the plasticity of the adult bone marrow stem cells in relation to gastrointestinal epithelial cells and subepithelial myofibroblasts, and finally (9) molecular mechanisms regulating epithelial proliferation and differentiation.

Chapter II - Optimized islet cell transplantation therapies hold the prospect of providing a curative treatment for diabetes. However, the scarcity of donors prevents the development of these therapies on a large scale. One possible way to reduce the shortage of islets for transplantation would be to produce large amounts of β-cells *in vitro*. Although this approach still remains theoretical, recent advances in the knowledge of embryonic stem cells biology have raised the prospect that these cells may be used for the production of pancreatic β-cells. The experience acquired during the past few years indicates the need to recreate *in vitro* the steps which occur *in vivo*. Stem cells must first be induced to become definitive endoderm, then to differentiate into pancreatic endoderm, and finally to reach the stage of differentiated mature β-cells. Thus, our capacity to drive this process *in vitro* will depend on a thorough understanding of the molecular mechanisms controlling the development of the pancreas. It is now well established, that all the pancreatic cell types are derived from a small pool of apparently identical stem cells, which originate from the early gut endoderm. After a phase of intense proliferation, these progenitor cells differentiate into the pancreatic exocrine and endocrine cell types. Considerable progress has been made in the identification of transcription factors that are required for the differentiation or maintenance of the pancreatic progenitors into the different pancreatic cell types. However less is known about the signalling pathways and the transcription factors controlling the generation of the pancreatic precursors in the primitive endoderm. Also, only very recently have we started to uncover the mechanisms controlling the maintenance of the pancreatic progenitors in an undifferentiated state. This review will focus on both the signalling pathways and transcription factors, which determine the specification of the pancreatic progenitors, as well as on the mechanisms controlling the maintenance of their self-renewal capacity.

Chapter III - Current concepts of gastrointestinal tumor formation favor a multi-step model with the acquisition of several targeted mutations in oncogenes and tumor suppressor genes that finally lead to the acquisition of different "hallmarks of cancer" like self-sufficiency in growth or anti-growth signals, evasion of apoptosis, immortality and angiogenesis as has been highlighted by Hanahan and Weinberg in their outstanding review [1]. Recent evidences indicate that targeted genetic events alone (e.g. point mutations in oncogenes) are not sufficient to provide tumor cells with these growth advantages but that the process of carcinogenesis closely resembles and involves processes of trans- and dedifferentiation as have been observed during embryonic development. Comparative analyses of embryogenesis and tumorigenesis, especially of the gastrointestinal tract, revealed immense analogies of morphological patterning as well as of expression of differentiation markers. Both processes of development and carcinogenesis as well as integral processes of regeneration like reparation after acute or chronic inflammation are characterized by pattern maintenance which is essentially maintained by different gradients of markers of embryonic differentiation. So called epigenetic modulations of chromatin and DNA elements lead to changes in phenotypic properties that can also favor the malignant conversion of progenitor cells. Furthermore, the role of cell-cell and cell-matrix contacts has gathered new attention as these signaling pathways strongly influence survival and differentiation of embryonic and tumorigenic cells and is one of the key features of the still to be defined (cancer) stem cell niche, especially in the gastrointestinal tract. The so far unclear factors that contribute to the maintenance and longevity of tumor stem cells may also contribute to relapsing tumor diseases after successful first-line treatment and the outbreak of disseminated micro metastases. This novel concept yields great impact for the development of new diagnostic and therapeutic approaches, as highly chemo- and radioresistant long-living tumor progenitor or stem cells will enter the focus of future cancer medicine and research.

Chapter IV - An essential prerequisite for the application of adult and embryonic stem cells in clinical therapy is the development of efficient protocols for directing their differentiation or commitment to specific lineages. It must be noted that there probably exist a subtle balance somewhere between the undifferentiated and terminally-differentiated state that would be optimal for transplantation/transfusion therapy, which is likely to differ for different cell/tissue/organ models of regeneration. For example, differentiated/committed stem cell progenies may express specific cell-surface markers or extracellular matrix that could facilitate their rapid engraftment and integration within the recipient tissue or organ. Moreover, the commitment/differentiation of stem cells prior to clinical therapy, would also avoid spontaneous differentiation into undesired lineages that can compromise tissue/organ regeneration i.e. teratoma and fibroblastic scar tissue formation. On the other hand, differentiated/committed stem cell progenies have reduced proliferative capacity and possess complex nutritional requirements that could compromise their survival within the 'hostile' pathological environment at the transplantation/transfusion site. There are two major approaches for directing lineage-specific commitment/differentiation of stem cells. Firstly, a 'milieu-based' approach would involve exposing stem cells to a cocktail of exogenous chemicals, cytokines, growth factors or extracellular matrix substratum, over an extended duration of in vitro culture. Nevertheless, the non-specific pleiotropic effects exerted by various cytokines, growth factors, and extracellular matrix would make this a relatively inefficient approach. Moreover, a "milieu-based" approach is likely to require extended durations of in vitro culture, which might delay autologous transplantation of adult stem cells to the patient and might alter their immunogenicity through prolonged exposure to xenogenic proteins within the culture milieu. Secondly, a 'gene-modulatory' approach would involve transfection of stem cells with either recombinant DNA or iRNA, to directly influence gene regulatory networks implicated in cellular fate and lineage determination. However, overwhelming safety and ethical concerns are likely to preclude the application of genetically modified stem cells in human clinical therapy for the foreseeable near future. To avoid permanent genetic modification, new technology platforms have been developed for the modulation of gene expression without recombinant DNA. This includes the use of protein transduction domain (PTD) – fusion transcription factors, PTD-PNA (peptide nucleic acid) analogs, as well as immunoliposome-mediated delivery of proteins or iRNA directly into the cytosol. Nevertheless, because such molecules have a limited active half-life in the cytosol and are obviously not incorporated into the genetic code of the cell, these would only exert a transient modulatory effect on gene expression.

Chapter V - Parathyroid hormone (PTH) is known as a regulator of calcium exchange; it also stimulates and activates proliferation of spindle-shaped N-cadherin positive osteoblasts which are able to control hematopoietic stem cells (HSC) in the niche. Because PTH could be an attractive drug for expansion of long-term repopulating HSC *in vivo*, the influence of this hormone on hematopoietic and stromal stem cells was studied *in vivo* and *in vitro*. Limiting dilution analysis *in vivo* indicated a significant increase in the frequency of competitive repopulation units (CRU) in the bone marrow of PTH treated mice. The calculated frequency of CRU was dependent upon the time of the recipient analysis; it increased from 2.5- to 4-fold between 10 and 16 months since reconstitution. The LTC-IC frequency increased 5-fold and CAFC 28-35 2.5-fold, whereas the number of differentiated progenitors CFU-S and

CFU-GM, as well as terminally differentiated mature hematopoietic cells, did not change after PTH treatment. Possible explanation of such disproportion between early and late hematopoietic progenitors lies in discriminate PTH influences on osteoblastic and vascular niches specialized for quiescent and differentiating hematopoietic progenitors, respectively. Seeding efficiency (F-24) in the bone marrow stromal microenvironment differed significantly between long-term repopulating cells (CAFC 28-35) that did not change after PTH treatment and short-term repopulating cells (CFU-S) that decreased dramatically in PTH treated animals. On the contrary, the spleen microenvironment was not sensitive to long-term PTH exposure. Such functional discrepancies in stromal microenvironments could be explained by different impacts of PTH on cell components of niches specialized for long-term and short-term hematopoietic precursor cells. Moreover, it is not clear whether PTH also affects mesenchymal stem cells (MSC). MSCs from mice bone marrow are able to rebuild the hematopoietic microenvironment de novo after transplantation of a bone marrow plug under the renal capsule of a syngeneic animal. In ectopic hematopoietic foci formed 6 weeks later, the stromal cells present belong to the donor of the bone marrow; the hematopoietic cells are of recipient origin. Using this method, it was shown that 4 weeks of PTH treatment has no influence on MSC either during de novo foci formation or in the case of transplantation of pretreated with PTH bone marrow. Therefore PTH does not affect MSC directly, but affects its more differentiated progeny – osteoblasts and probably cells of vascular niches. PTH stimulates osteoblasts in completely formed niches, as was shown by *in vivo* ectopic foci formation de novo, as well as the experiments *in vitro*. In long-term bone marrow culture, the most pronounced stimulating effect of PTH addition on hematopoietic precursors was revealed after 10 weeks of cultivation. Simultaneously, the expression level of genes specific for osteogenic differentiation increased. The expression level of Notch-1 ligand Jagged-1 also increased steadily and significantly after 7 weeks in culture. The obtained data suggest PTH is a possible tool for specific expansion of early hematopoietic precursors only.

Chapter VI - In the field of medicine, stem cell research is now the most exciting opportunity for the treatment of some diseases that until now have been incurable. The discovery of the regenerative properties of stem cells, both embryonic and adult, has been one of the main advances in Biology in the past years. This fact has contributed to the increase in the number of projects of human stem cell line derivation.

In the process of obtaining stem cell lines for research or clinical application purposes, there are still some problems that should be resolved. Apart from the difficulties of cell line generation, the problems with the culture media and the possibility of tumoral degeneration, one of the main complications of stem cell cultures is the possibility of contamination. The main sources for this to occur are the stem cells themselves, the laboratory workers and the environment in which these cell lines are being handled.

Due to the large quantity of microorganisms that could contaminate these cultures (e.g. bacterias, fungi, viruses and prions), all the stem cell research centres should incorporate an exhaustive microbiological and environmental program in order to both detect the possible pathogens and to avoid the transmission to the recipients.

Any microbial contamination during the manufacturing process presents a serious hazard to recipients. In order to establish the sterility of stem cell cultures, there are several methods to study the possible presence of pathogens both in cell cultures and in the environment of the manufacturing cell factories.

Until now, the methods used for this purpose are mainly standard protocols based on the culture in liquid or solid growth media and incubated at different temperatures, reflecting conditions for human pathogen culture and environmental microorganisms. Other techniques used are PCR and RT-PCR methods, electron microscopy and in vitro inoculation and in vivo animal inoculation. However, some of these tests are long standing techniques and can take a number of weeks and can prove difficult to implement in the scheduling for release testing of time critical products. In this sense, the technologies based on the use of hybridization chips, using microarrays of immobilized oligonucleotides or antigens/antibodies/genes for microorganisms can provide a rapid and useful methodology for the identification of contaminants.

This review will discuss the methodology that could be used in stem cell research centres to assure the quality and biosafety of cell lines and biotechnological products and avoid the transmission of pathogens like bacteria, virus and prion particles.

Chapter VII - Traumatic brain injury (TBI) usually occurs as a result of a direct mechanical insult to the brain, and induces degeneration and death in the central nervous system (CNS). CNS disorder can be caused by widespread neuronal and axonal degeneration induced by TBI. It was originally thought that recovery from such injuries was severely limited because neuronal loss and degeneration in the adult brain were irreversible in the mammalian nervous system. However, recent studies have indicated that the mammalian nervous system has the potential to replenish populations of damaged and/or destroyed neurons via proliferation of neural stem cells (NSCs). NSCs have been identified in adult mammals and have the potential to differentiate into either glial or neural phenotypes. However the proliferation, differentiation and neurogenesis of NSCs at damaged regions after TBI remain unclear. The authors investigated the chronology of the occurrence and differentiation of NSCs around cerebral cortical damaged areas after traumatic brain injury in the rat.

NSCs, isolated and cultured from damaged cerebral cortical brain tissue after rat TBI, were confirmed to have the potential to differentiate into neurons and glial cells, although it was unclear if NSCs migrated to the damaged area from the subventricular zone (SVZ) or the subependymal zone (SEZ) at the dentate gyrus (DG)-hilus interface, and if astrocytes showed blastogenesis in the reactivity after injury. Moreover, *in vivo*, there were immature neurons among the gliosis and glial scars around the damaged area at 7days, in decreasing stages of NSCs, after rat TBI. Additionally, the authors confirmed that some immature neurons matured and survived among the glial scars at 30days after injury. Their data suggested that promotion of maturation and differentiation of newly formed immature neurons around a damaged area may improve brain dysfunction induced by glial scars after brain injury.

Chapter VIII - Bone marrow stromal cells (MSCs) are multipotent stem cells that differentiate into cells of the mesodermal lineage like bone, cartilage, fat, and muscle, and though adult, they are able to transdifferentiate. The authors aimed to evaluate their transdifferentiation potential into myelinating “Schwann cell-like” cells to offer new therapeutic strategies for a wide range of diseases and injuries of the nervous system.

Schwann cells are of special interest not only as central players in peripheral nerve regeneration, where they act as pathfinders for the outgrowing fibers, but also as therapeutic option in multiple sclerosis. They are not attacked by the immune cells and they are able to break down devastated myelin and to clear debris by phagocytosis, an important prerequisite for remyelination. Human Schwann cells can be obtained from nerve biopsies for autologous transplantation without the need for immunosuppression. However, the method has inevitable disadvantages, to mention only e few: limitations of nerve material, decreased yield of cells due to their restricted mitotic activity, and sacrificing one or more functioning nerves. Hence, alternative cell systems are desirable and stem cells may be such an alternative cell source.

The authors transformed cultivated rat MSCs into Schwann cell-like cells by using different cytokine cocktails. Treated MSCs changed morphologically into cells resembling typical spindle shaped Schwann cells with enhanced expression of $p75^{NTR}$, Krox-20, CD104 and S100ß proteins and decreased expression of BMP receptor-1A. As final proof of successful transdifferentiation the functionality, i.e. the myelinating capacity was checked. Therefore, Schwann cells and transdifferentiated or untreated MSCs cultured from male rats were grafted into autologous muscle conduits bridging a 2cm-gap in a female sciatic nerve. PCR of the SRY gene and S100 immunoreactivity of pre-labeled cells confirmed their presence in the grafts. After three and six weeks, regeneration was monitored clinically, histologically and morphometrically. Autologous nerves and cell-free muscle grafts were used as controls.

Revascularization studies suggested that transdifferentiated MSCs facilitated neo-angiogenesis and did not negatively influence macrophage recruitment. Autologous nerve grafts demonstrated the best results in all regenerative parameters. An appropriate regeneration was noted in the Schwann cell-groups and, albeit with restrictions, in the transdifferentiated MSC-groups, while regeneration in the MSC-group and in the cell-free group was impaired.

Although the results must be interpreted with caution, the authors want to speculate that the transdifferentiation technique provides a tool to manipulate adult stem cells for cell-based approaches in regenerative medicine of demyelinating diseases.

Chapter IX - Bone marrow stroma is the source of mesenchymal stem cells. Several research articles dealing with the use of bone marrow stromal stem cells (BSCs) for regenerating bone or cartilage have been published. However, there has been no research on the effect of BSC transplantation on wound healing. The BSCs secrete collagen and several cytokines essential for wound healing. However, it is unclear as to what extent these materials are produced and it has not yet been studied whether they are superior to fibroblasts (which have been used conventionally) in promoting the wound healing process. It has been hypothesized that BSCs and fibroblasts would differ in their ability to promote wound healing, and furthermore that BSCs should have greater activity than fibroblasts. If so, then the current cell therapy method using fibroblasts to stimulate the healing of chronic wounds could be replaced by a therapy which uses BSCs and has a superior effect. The purpose of this chapter is to compare the wound healing activity of BSCs with that of fibroblasts *in vitro* and *in vivo*. Cultured human BSCs and dermal fibroblasts taken from same patients were tested. *In vitro* study was focused on cell proliferation, synthesis of collagen, and production of three growth factors – basic fibroblast growth factor (bFGF), vascular endothelial growth factor (VEGF), and transforming growth factor beta (TGF-β) – which are considered to be important in chronic wound healing. *In vivo* study was performed to compare collagen synthesis, epithelization, and angiogenic activity of BSCs with those of fibroblasts using rat wound models. In the *in vitro* study the authors did not observe great differences in cell proliferation and TFG-β secretion. In contrast, the amount of collagen synthesis and the levels of bFGF and VEGF were much higher in the BSC group than in the fibroblast group. In particular, the VEGF level of the BSC group was 12 times higher than that of the fibroblast group. In the *in vivo* study, great differences were noted in collagen synthesis, epithelization, and angiogenesis among the three groups. The BSC group showed the best results, followed by the fibroblast group and then the no cell group. These results suggest that the potential of BSCs in wound healing acceleration may be superior to that of fibroblasts and that BSCs may be possibly used as a replacement for fibroblasts, which are currently being used for wound healing.

Chapter X - Rheumatoid arthritis is a chronic inflammatory disease, and angiogenesis observed in synovial tissue involves various factors. Based on the hypothesis that angiogenesis inhibition suppresses arthritis, studies have shown anti-arthritic effects in animal models of arthritis after the administration of angiogenesis inhibitors. Endostatin is a fragment of the C-terminal non-collagenous domain of type XVIII collagen, and has a potent angiogenesis-inhibiting effect. The administration of endostatin to a mouse model of arthritis inhibited angiogenesis, resulting in the suppression of arthritis. No side effects were observed during the period of drug administration. As a therapeutic agent for rheumatoid arthritis, endostatin includes the advantages of having few side effects, an action mechanism different from those of conventional therapeutic agents for rheumatoid arthritis, an angiogenesis-inhibiting effect through a broad-spectrum mechanism, and little potential for inducing tolerance; therefore, it may be a promising therapeutic agent for rheumatoid arthritis.

Chapter XI - Stem cells are a major stake in biology, biotechnology and therapy. The ability of certain cells to regenerate tissues and functions has a great theoretical importance in Developmental Biology and constitutes a great hope to treat debiliting conditions resulting from aging or injuries. The development of efficient methods to amplify, differentiate and graft stem cells is a major challenge for today's biotechnology industry. In the recent past, major progresses have arisen from the field of developmental biology. A clear example is the use of the sequence of signals in the embryo to drive the differentiation of stem cells into preselected cell types in vitro. Recent results indicate the importance of a tissue-level organization in the maintenance and differentiation of the pools of stem cells in vivo. In the present article, we describe how spatially organized stem cells are responsible for the elongation of the spinal cord. We also show that a limited number of constraints are responsible for the orientation of cell divisions of neuroepithelial stem cell, a parameter critical for the maintenance of stemness in these cells. Taken together, these results suggest a lineage of stem cells from ES cells to neural differentiation that may be significant for the embryo. In addition, these results suggest that future strategies for the manipulation of stem cell in vitro will require to change growth methods into three-dimensional matrices and reconstitution of apico-basal organization of the cells. This should help fine tune the properties of the stem cells and provide a better spatial control of the sequence of events that take place in the embryo.

Chapter XII - *Objective:* Neural stem cells have been shown to participate in the repair of experimental CNS disorders due to their self-renewal and multi-potency. In a study, the authors attempt to explore the feasibility for the therapy of spinal cord injury (SCI) by combining neural stem cell (NSC) with lentivirus.

Methods: Following the construction of the genetic engineering NSC modified by Lentivirus to secrete both neurotrophic factor-3 (NT-3) and green fluorescence protein (GFP), hemisection of spinal cord at the level of T_{10} was produced in 48dult Wistar rats that were randomly divided into four groups (n=16), namely three treated groups and one control group. The treated groups were dealed with NSC, genetic engineering NSC respectively. Then used fluorescence microscope to detect the transgenic expression in vitro and in vivo, migration of the grafted cells in vivo and used the method of BBB to assess the function recovery.

Results: The transplanted cells could survive for long time in vivo and migrate for long distances; the stably transgenic expression could be detected in vivo; the hind-limb function of the injured rats, especially for the rats that had been dealed with genetic engineering NSC, had obviously improved.

Conclusion: The genetic engineering NSC modified by lentivirus to deliver NT-3, acting as a source of neurotrophic factors and function cell in vivo, have the potential to participate in spinal cord repair.

Chapter XIII - Embryonic stem cells (ESCs) are self-renewing pluripotent cells that generate all the cell types of the body. ESCs hold the potential to cure a broad range of diseases and injuries, from diabetes, heart diseases, muscle damages, to neurological diseases and injuries. Because ESCs are derived from embryos, their use for clinical research and therapy is the source of debates and controversies. The stringent conditions require for maintaining them in culture, and their risk to form tumors upon grafting limit their therapeutic application. In this chapter, the authors will review the therapeutic potential of ESCs. They will discuss the ethical and political debates and controversies, and the limitations associated with the use of ESCs for therapy.

In: Stem Cell Research Trends
Editor: Josse R. Braggina, pp. 1-2
ISBN: 978-1-60021-622-0

Expert Commentary A

Embryonic Stem Cells from Non-Viable Tissues: An Alternative Source of Stem Cells for Therapy?

Philippe Taupin*
National Neuroscience Institute
National University of Singapore
11 Jalan Tan Tock Seng, 308433 Singapore

Embryonic stem cells (ESCs) are self-renewing pluripotent cells. ESCs generate cells from the three germ layers of the individual; neurectoderm, mesoderm and endoderm. As such, they carry the hope to cure a broad range of diseases and injuries, from diabetes, strokes, to heart, neuro-muscular and neurological diseases and injuries. ESCs are derived from embryos, particularly from the inner cell mass of blastocystes. Because it involves the destruction of embryos, the derivation and use of human ESCs for clinical research and therapy are the source of ethical and political debates, controversies, and strict regulations [1].

To circumvent these ethical and political issues, investigators are devising strategies to derive ESCs without the destruction of embryos. Meissner and Jaenisch used a variation of somatic cell nuclear transfer, in which a gene crucial for trophectoderm development -the gene CDX2- is mutated, to derive ESCs from mice. This mutation produces embryos unable to implant into the uterus [2]. Lanza and coworkers derived ESCs from single-blastomere biopsies. Single-blastomere biopsy is a technique similar to pre-implantation genetic diagnosis used in fertility clinics [3, 4]. These studies did not resolve the ethical and political issues over the destruction of embryos to derive ESCs [1, 5]. For more discussions over the limitations, debates and controversies over these strategies to derive ESCs without the destruction of embryos, please, refer to chapter entitled "Therapeutic potential of embryonic stem cells".

*.Tel. (62) 6357 7533. Fax (62) 6256 9178. Email: obgpjt@nus.edu.sg

Zhang et al. reported the derivation of ESCs from developmentally arrested embryos. Developmentally arrested embryos are embryos produced during in vitro fertilization process that are believed to be non-viable, as their development in vitro is altered. This study suggests that developmentally arrested embryos provide an alternative source of tissue to derive stem cells for therapy [6]. But does it solve the ethical and political issues over the derivation of ESCs without the destruction of embryos? And does it provide a potential source of tissue for therapy?

On the one hand, it is argued that developmentally arrested embryos are dead and their donation assimilated to organ donation of post-mortem tissue [7]. On the other hand, it is not known whether such embryos would have developed normally after implantation. Therefore, the isolation of ESCs, without the destruction of embryos and without ethical and political controversies and debates, remains an unresolved issue.

With regard to the developmental potential of stem cells isolated from non-viable embryos, on the one hand, progenitor and stem cells can be isolated and cultured in vitro from post-mortem tissues [8]. On the other hand, studies from cloned embryos revealed the importance of developmental cues for establishing ESC lines [9]. Therefore, stem cells derived from non-viable tissues, as opposed to post-mortem tissues, may have altered developmental potential, which would affect their therapeutic capacity. Studies remain to be performed to further evaluate the developmental and therapeutic properties of ESCs derived from non-viable embryos.

REFERENCES

[1] Taupin P. (2006) Derivation of embryonic stem cells for cellular therapy: challenges and new strategies. *Med. Sci. Monit.* 12, RA75-8.

[2] Meissner A, Jaenisch R. (2006) Generation of nuclear transfer-derived pluripotent ES cells from cloned Cdx2-deficient blastocysts. *Nature.* 439, 212-5.

[3] Chung Y, Klimanskaya I, Becker S, Marh J, Lu SJ, Johnson J, Meisner L, Lanza R. (2006) Embryonic and extraembryonic stem cell lines derived from single mouse blastomeres. *Nature.* 439, 216-9.

[4] Klimanskaya I, Chung Y, Becker S, Lu SJ, Lanza R. (2006) Human embryonic stem cell lines derived from single blastomeres. *Nature*. 444, 481-5. Erratum in: (2006) *Nature*. 444, 512.

[5] Weissman IL. (2006) Medicine: politic stem cells. *Nature.* 439, 145-7.

[6] Zhang X, Stojkovic P, Przyborski S, Cooke M, Armstrong L, Lako M, Stojkovic M. (2006) Derivation of human embryonic stem cells from developing and arrested embryos. *Stem Cells*. 24, 2669-76.

[7] Landry DW, Zucker HA.(2004) Embryonic death and the creation of human embryonic stem cells. *J. Clin. Invest.* 114, 1184-6.

[8] Palmer TD, Schwartz PH, Taupin P, Kaspar B, Stein SA, Gage FH. (2001) Cell culture. Progenitor cells from human brain after death. *Nature*. 411, 42-3.

[9] Shiels PG, Kind AJ, Campbell KH, Waddington D, Wilmut I, Colman A, Schnieke AE. (1999) Analysis of telomere lengths in cloned sheep. *Nature.* 399, 316-7.

In: Stem Cell Research Trends
Editor: Josse R. Braggina, pp. 3-4
ISBN: 978-1-60021-622-0

Expert Commentary B

THYMIDINE ANALOGS AND BRDU LABELING FOR STUDYING NEUROGENESIS: BEWARE OF THE LIMITATIONS AND PITFALLS

Philippe Taupin*
National Neuroscience Institute
National University of Singapore
11 Jalan Tan Tock Seng, 308433 Singapore

Tritiated ([3H]) thymidine and bromodeoxyuridine (BrdU) immunohistochemistry have been used to study cell proliferation, and neurogenesis in the developing and adult central nervous system [1]. Strategies have been devised to determine the time of origin, migration, lineage and fate of neuronal cells using thymidine analogs [2]. These paradigms have limitations for studying cell proliferation and neurogenesis. Particularly, the quantification of the number of proliferating cells using thymidine analogs is relative and depends on parameters, like the labeling paradigm, cell cycle of the dividing cells, the survival and number of time the cells divide in presence of thymidine analogs. Besides these limitations, the use of thymidine analogs for studying cell proliferation and neurogenesis is subject to limitations and pitfalls [3].

Thymidine analogs are not markers of cell division or proliferation; they are markers of DNA synthesis. As such they incorporate DNA, not only during cell division, but also in cells undergoing DNA repair, abortive cell cycle re-entry and gene duplication [4]. Thymidine analogs can also be transferred from the genome of labeled dying cells to newly generated cells, an artifact that may affect the temporal resolution of birth of the dividing population of cells [5].

The use of [3H]-thymidine, a radiolabeled substrate, and the time-consuming process involved in autoradiography has contributed to the advent of BrdU-immunohistochemistry. However, BrdU-labeling presents some limitations for studying cell proliferation and neurogenesis. BrdU-labeling is not stoechiometric. The use of primary antibodies directed

*.Tel. (62) 6357 7533. Fax (62) 6256 9178. Email: obgpjt@nus.edu.sg

against BrdU on single-stranded DNA requires denaturing the DNA. Such treatment can affect the integrity of the tissue and antigenicity recognition, thereby limiting morphological and phenotype identification of newly generated cells [6]. BrdU is also a toxic substance that has side effects. The integration of bromine into the DNA alters its stability, increases the risk of sister-chromatid exchanges, mutations, DNA double-strand breaks, induces cell death and the formation of teratomas. BrdU lengthens the cell cycle and has mitogenic, transcriptional and translational effects, on cells that incorporate it [7]. Additionally, there are debates about the bioavailability of BrdU for labeling dividing cells and the saturating concentration of BrdU for labeling most S-phase cells [8].

In all, thymidine analogs have proven most useful in studying cell proliferation and neurogenesis in the developing and adult nervous system. However, their uses are not without limitations and pitfalls. These must be taking into account when analyzing and interpreting studies involving thymidine analogs, and particularly BrdU. Appropriate controls must be performed and alternative strategies considered [9].

References

[1] Taupin P, Gage FH. (2002) Adult neurogenesis and neural stem cells of the central nervous system in mammals. *J. Neurosci. Res.* 69, 745-9.

[2] Miller MW, Nowakowski RS. (1988) Use of bromodeoxyuridine-immunohistochemistry to examine the proliferation, migration and time of origin of cells in the central nervous system. *Brain Res.* 457, 44-52.

[3] Nowakowski RS, Hayes NL. (2001) Stem cells: the promises and pitfalls *Neuropsychopharmacology* 25, 799-804.

[4] Nowakowski RS. Hayes NL. (2000) New neurons: extraordinary evidence or extraordinary conclusion? *Science* 288, 771.

[5] Burns TC, Ortiz-Gonzalez XR, Gutierrez-Perez M, Keene CD, Sharda R, Demorest ZL, Jiang Y, Nelson-Holte M, Soriano M, Nakagawa Y, Luquin MR, Garcia-Verdugo JM, Prosper F, Low WC, Verfaillie CM. (2006) Thymidine Analogs are Transferred from Pre-Labeled Donor to Host Cells in the Central Nervous System After Transplantation: A Word of Caution. *Stem Cells* 24, 1121-7.

[6] Moran R, Darzynkiewicz Z, Staiano-Coico L, Melamed MR. (1985) Detection of 5-bromodeoxyuridine (BrdUrd) incorporation by monoclonal antibodies: role of the DNA denaturation step. *J. Histochem. Cytochem.* 33, 821-7.

[7] Morris SM. (1991) The genetic toxicology of 5-bromodeoxyuridine in mammalian cells. *Mutat. Res.* 258, 161-88.

[8] Eadie BD, Redila VA, Christie BR. (2005) Voluntary exercise alters the cytoarchitecture of the adult dentate gyrus by increasing cellular proliferation, dendritic complexity, and spine density. *J. Comp. Neurol.* 486, 39-47.

[9] Taupin P. (2006) BrdU immunohistochemistry for studying adult neurogenesis: Paradigms, pitfalls, limitations, and validation.*Brain Res. Brain Res. Rev*. In press.

In: Stem Cell Research Trends
Editor: Josse R. Braggina, pp. 5-7
ISBN: 978-1-60021-622-0

Expert Commentary C

Less than Perfect HLA Matching in Human Embryonic Stem Cell Banking – Can Combined Transplantation/Transfusion with Autologous Adult Stem Cells Provide a Solution?

Boon Chin Heng, Hua Liu and Tong Cao*
Stem Cell Program, Faculty of Dentistry,
National University of Singapore,
5 Lower Kent Ridge Road, 119074 Singapore

Keywords: adult, embryonic, immunology stem cells, synergy, transfusion, transplantation.

The recent scandal involving the fabrication of scientific data by Hwang and co-workers [1] highlighted the formidable technical challenges, as well as ethical problems (i.e. donor oocyte procurement) faced with therapeutic cloning for the derivation of immunocompatible patient-specific human embryonic stem cell (hESC) lines. Banking of numerous hESC lines derived from donated surplus in vitro fertilized (IVF) embryos for subsequent human leukocyte antigen (HLA) -matching with patients has been proposed as an alternative solution [2]. Nevertheless, a major deficiency of this approach is the near-impossibility of finding a completely perfect match for each and every individual patient, given the tremendous diversity and almost infinite permutation of HLA gene loci within the human population. At best, a close but still less than perfect HLA match can be achieved for individual patients, which would mean that there still exist a likelihood of immunorejection

* Correspondence and reprint requests: Dr. Tong Cao, Stem Cell Program, Faculty of Dentistry, National University of Singapore. 5 Lower Kent Ridge Road, 119074 Singapore; e-mail : dencaot@nus.edu.sg. Tel : +65-6516-4630. Fax : +65-6774-5701.

upon transplantation/transfusion. This would in turn necessitate life-long dependency on immunosuppressive drugs together with their adverse side-effects [3].

A novel strategy may be to combine the use of closely-matched allogenic hESC together with autologous adult stem cells in regenerative medicine. Numerous studies have demonstrated that bone marrow-derived mesenchymal stem cells (MSC) possess immunosuppressive properties [4-6]. Hence, it is possible that autologous MSC may mitigate immunological rejection against co-transplanted or co-transfused hESC derivatives within the patient. Once adequate tissue/organ regeneration has been achieved and the immunosuppressive effects of the autologous MSC have worn off (i.e., through differentiation), then the engrafted human embryonic stem cell progenies may gradually be killed off by the recipient's immune system while being replaced by proliferating autologous cells (i.e., co-transplanted or co-transfused MSC). Another possibility is that differentiated lineages may still retain some immuno-modulatory properties of their MSC precursors, as demonstrated by a recent study from our research group [7]. Hence, there may be long-term tolerance induction of the closely-matched allogenic hESC derivatives through the immunomodulatory action of the co-transplanted or co-transfused autologous MSC. Also, the concentrated localized immunomodulatory effect of co-transplanted autologous MSC at one particular transplantation site is preferable to the holistic adverse side-effects of immunosuppressive drug administration.

Of course, the pertinent question that arises is why not exclusively utilize autologous adult stem cells in transplantation/transfusion therapy? This is because autologous adult stem cells possess numerous shortcomings which would limit their usefulness in regenerative medicine. For a start, most adult stem cell sub-populations are relatively scarce, difficult to extract from the human body, and possess limited proliferative capacity and multi-lineage differentiation potential [8, 9]. Additionally, there have also been numerous reports on the age-related decline in the quality of adult stem cell sub-populations, which would limit their therapeutic usefulness for older patients [10, 11]. Another shortcoming of utilizing autologous adult stem cells may be the requirement for prolonged durations of ex vivo culture, either for proliferation in cell numbers, or for differentiation and lineage-commitment. This would obviously delay urgent life-saving treatment to the patient.

Hence, combined transplantation or transfusion with closely-matched hESC derivatives can somehow make up for the shortfall in the total numbers of viable adult stem cells that can be extracted from older patients; because a certain threshold number of transplanted/transfused cells are probably required to achieve optimal efficacy of tissue/organ regeneration. Moreover, the putative 'nascent' state of the closely-matched hESC derivatives can also compensate for the age-related decline in the proliferative capacity and regenerative potential of autologous adult stem cells derived from older patients [10, 11]. Additionally, differentiated or lineage-committed hESC progenies may provide paracrine signaling and cellular contacts, resulting in a more conducive environment for directing co-transplanted autologous adult stem cells into a required lineage at the transplantation site; since the natural milieu may already be compromised by the pathological state of the diseased/damaged tissue or organ requiring transplantation therapy. It must be noted that because the closely-matched hESC are from a donated allogenic source, an extended period of time is available to differentiate them into a particular lineage before transplantation. By contrast, such an option is rarely available in the case of autologous adult stem cells, because of the urgency of life-saving treatment for the patient. In this manner, differentiated progenies of closely-matched hESC can possibly be utilized as "cellular catalysts" to promote the appropriate differentiation or lineage-commitment of co-transplanted autologous adult stem cells in situ, which could in turn enhance their subsequent engraftment and integration within recipient tissues/organs.

Hence, there are indeed many exciting possibilities in combining and synergizing the complementary therapeutic advantages of closely-matched allogenic hESC with autologous adult stem cells. This would certainly warrant further investigation.

REFERENCES

[1] Rusnak AJ, Chudley AE. Stem cell research: cloning, therapy and scientific fraud. *Clin. Genet.* 2006 Oct;70(4):302-305.

[2] Taylor CJ, Bolton EM, Pocock S, Sharples LD, Pedersen RA, Bradley JA. Banking on human embryonic stem cells: estimating the number of donor cell lines needed for HLA matching. *Lancet.* 2005 Dec 10;366(9502):2019-25.

[3] Crane E, List A: Immunomodulatory drugs. *Cancer Invest* 23: 625–634, 2005.

[4] Le Blanc K: Immunomodulatory effects of fetal and adult mesenchymal stem cells. *Cytotherapy* 5: 485–489, 2003.

[5] Djouad F, Plence P, Bony C, et al: Immunosuppressive effect of mesenchymal stem cells favors tumor growth in allogeneic animals. *Blood* 102: 3837–3844, 2003.

[6] Inoue S, Popp FC, Koehl GE, Piso P, Schlitt HJ, Geissler EK, Dahlke MH. Immunomodulatory effects of mesenchymal stem cells in a rat organ transplant model. *Transplantation.* 2006 Jun 15;81(11):1589-95

[7] Liu H, Kemeny DM, Heng BC, Ouyang HW, Melendez AJ, Cao T. The Immunogenicity and Immunomodulatory Function of Osteogenic Cells Differentiated from Mesenchymal Stem Cells. *Journal of Immunology.* 2006 Mar 1;176(5): 2864-71.

[8] Verfaillie CM. Adult stem cells: Assessing the case for pluripotency. Trends Cell Biol. 2002; 12:502–508.

[9] Hedrick MH, Daniels EJ. The use of adult stem cells in regenerative medicine. *Clin. Plast. Surg.* 2003; 30:499–505.

[10] Quarto R, Thomas D, Liang CT. Bone progenitor cell deficits and the age associated decline in bone repair capacity. *Calcif. Tissue Int.* 1995; 56:123–129.

[11] D'Ippolito G, Schiller PC, Ricordi C, Roos BA, Howard GA. Age-related osteogenic potential of mesenchymal stromal stem cells from human vertebral bone marrow. *J. Bone Miner Res.* 1999; 14:1115–1122.

In: Stem Cell Research Trends
Editor: Josse R. Braggina, pp. 9-72
ISBN: 978-1-60021-622-0

Chapter I

GUT STEM CELLS: MULTIPOTENT, CLONOGENIC AND THE ORIGIN OF GASTROINTESTINAL CANCER?

Shigeki Bamba[1,2], William R. Otto[1] and Nicholas A. Wright[1,3]
[1]Histopathology Unit, Cancer Research UK,
Lincoln's Inn Fields, London, UK
[2]Department of Gastroenterology,
Shiga University of Medical Science, Japan
[3]Histopathology, Barts and The London,
Queen Mary's School of Medicine and Dentistry, London, UK

ABSTRACT

Tissue stem cells are multipotent and maybe clonogenic. These characters are similar to those of the earliest progenitor cells of carcinomas. In the gastrointestinal tract, epithelial stem cells exist in the region of the isthmus/neck in the stomach, in the small intestine in the crypt base, just superior to the Paneth cells, and towards the bottom of the crypt in the colon, giving rise to all epithelial cell lineages in all these tissues. Recent reports have revealed that many supposed alterations found in cancer cells are also present in morphologically-normal stem cells. In this context, carcinomas can perhaps be explained as stem cell diseases. To understand the clonal expansion of a single, normal or mutated stem cell brings new insights for not only into the biology of the gastrointestinal tract but also in tumorigenesis in the intestine.

In this review, we will describe; (1) general aspects of gastrointestinal tract biology, (2) the development of the gastrointestinal tract and molecular factors regulating gut embryogenesis, (3) the proposal that stem cells yield all of the gastrointestinal epithelial cell lineages; the 'Unitarian Theory', and (4) the way in which stem cells, or cancer cells, clonally expand, especially focusing on the two basic models; the 'top-down' theory and the 'bottom-up' concepts. In addition (5) we will assess the mechanisms which maintain the crypt stem cell populations, (6) putative stem cell markers, (7) the stem cell niche, thought to be provided by subepithelial myofibroblasts, (8) the plasticity of the adult bone marrow stem cells in relation to gastrointestinal epithelial cells and subepithelial myofibroblasts, and finally (9) molecular mechanisms regulating epithelial proliferation and differentiation.

INTRODUCTION

The human gastrointestinal system is approximately 500 cm in length *in vivo* and has about 200 times the surface area of skin [1], making the largest contact to the external environment in the human body. Hence it performs a barrier function to protect our bodies from various kinds of noxious substances. To maintain this barrier function, turnover of the epithelial cell lineages within the gastrointestinal tract occurs every two to seven days under normal homeostasis, and this time is reduced after damage. Multipotent intestinal stem cells play a central role in this process. The small intestinal epithelium consists of four main cells; columnar cells, goblet cells, endocrine cells, and Paneth cells. All these mature cells arise from intestinal stem cells, which, with the exception of the Paneth cells, migrate upward along the crypt-villus axis towards the tips of the villi, gradually differentiating as go, and eventually become apoptotic or are shed from the tip.

Recent studies have revealed that Delta/Notch signalling plays a critical role in cell fate decisions during the maturation of gut epithelial cells, and Wnt signalling seems to have a critical involvement in the maintenance of gut stem cell properties, possibly acting on stem cells to regulate asymmetric division [2, 3].

Mutant stem cells appear to expand in a clonal manner by cell division and subsequent crypt fission. Then, how do these cells spread in the gut? A recent report by Shih *et al* (2001) [4] proposed a 'top-down' concept, where mutated cells located within the intercryptal zones on the mucosal surface spread downwards into the adjacent crypts. In contrast to this is the 'bottom-up' hypothesis; the mutated stem cell in the crypt base fills the crypt by cell division and expands locally by crypt fission. Our recent studies suggest that clonal expansion in early adenomas is a property of mutated stem cells, indicating that the 'bottom-up' proposal is more likely in the earlier stages of carcinogenesis.

The plasticity of adult bone marrow stem cells is a controversial topic. Some reports have revealed that adult bone marrow cells can transdifferentiate and repopulate the gastrointestinal epithelium and also the supportive mesenchymal cells in both humans and the mice. Such reports cannot exclude the possibility of cell fusion events in the gastrointestinal tract as being responsible, but this might not be a critical point for clinical medicine, since fusion might also help regeneration and repair. Genetic modifications or the assistance of growth factors or cytokines may be required to encourage stem cell engraftment.

We will review the above topics expanding on the role of stem cells in gastrointestinal tumorigenesis, and also covering other topical matters.

GENERAL ORGANIZATION

The gastrointestinal tract is divided into several major anatomical regions, each with its own unique physiological functions and structure; oesophagus, stomach, duodenum, small intestine, colon and rectum. The various regions share a basic structure [5] (Figure 1). There is a mucosal lining beneath which is a thin muscular layer known as the muscularis mucosae. The mucosa serves many of the digestive and absorptive functions of the gastrointestinal tract. Beneath the muscularis mucosae is the submucosa. The next layer is the muscularis propria, which is responsible for propelling the gastrointestinal contents forward throughout the gut.

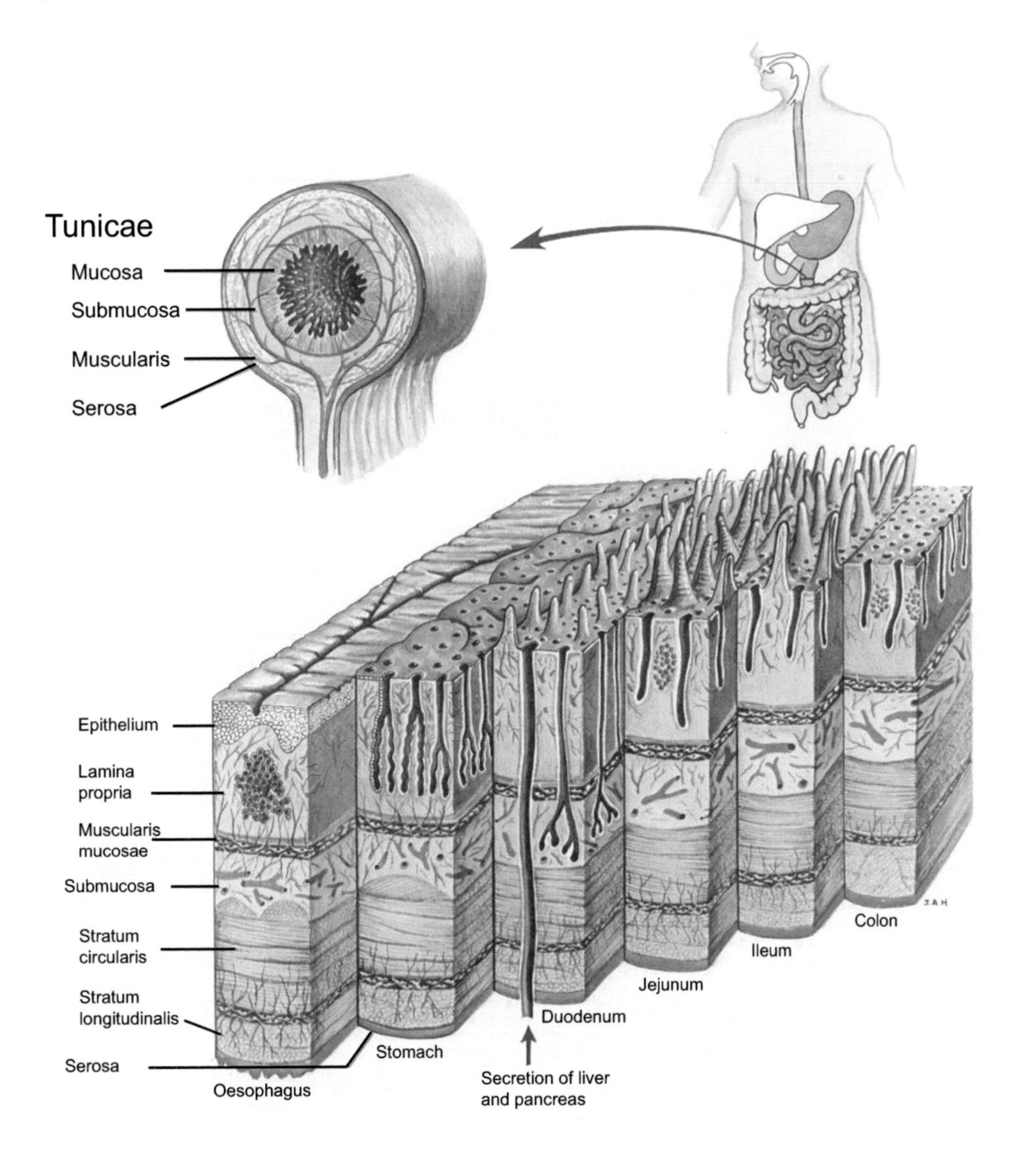

Figure 1. General histological diagram of the gastrointestinal tract. The general arrangement of the alimentary canal, its mural tunicae and (below) the general histological arrangement (from [5] with permission).

(1) Epithelial Cells

The intestinal epithelium lining the gastrointestinal tract has a well-defined architecture. The simple columnar epithelium rests on a basement membrane and is folded to form a number of invaginations, or crypts, each of which contain about 2000 cells in the human colon [6-9]. In the small intestine, several crypts surround the base of finger-like luminal projections called villi. This arrangement generates a large surface area, allowing efficient absorption of nutrients from the intestinal lumen [10]. Epithelial cells produced in the lower part of the crypt migrate up the crypt onto an adjacent villus in coherent columns (Figure 2), where they perform their differentiated absorptive function before being shed, intact, into the lumen. In the colon, cells migrate to the intercrypt table at the top of the colonic crypt.

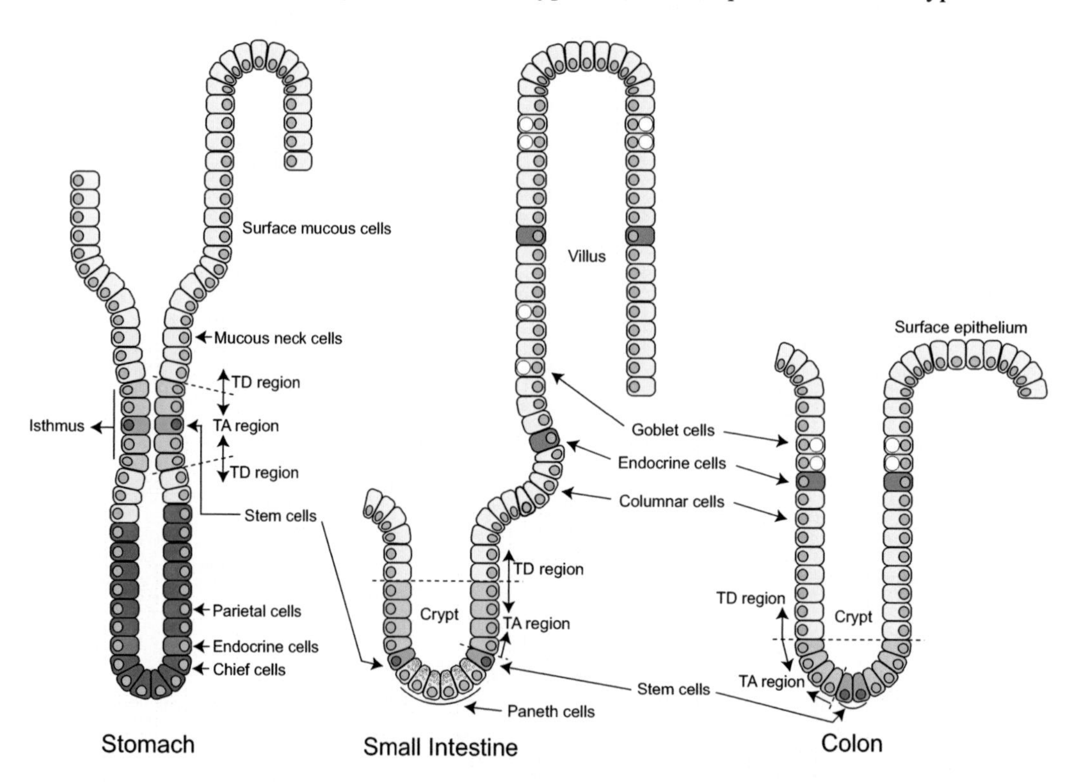

Figure 2. Gastrointestinal epithelium. The proliferative compartments of gastrointestinal tract are located at the isthmus/neck in the stomach, at positions 4th to 5th superior to the Paneth cell in the small intestine, and towards the crypt base in the colon. Stem cells produce transit amplifying (TA) progeny, and these cells move bi-directionally in the stomach, predominantly upwards in the small intestine and colon and decycle in the terminally-differentiated (TD) regions.

In both the small intestine and colon, cells differentiate into three main functional cell types common to both as they migrate: the predominant villus columnar cell (the enterocyte or colonocyte), the mucus-secreting goblet cell and the peptide hormone secreting endocrine cells. Endocrine, 'enteroendocrine', or 'neuroendocrine' cells are an abundant cell population distributed throughout the intestinal epithelium, which secrete peptide hormones in an endocrine or paracrine manner from their dense-core neurosecretory granules. Addtionally in the small intestine and the ascending colon, a number of cells migrate down to the base of the crypt to become the fourth cell type, the Paneth cells. Paneth cells contain large apical secretory granules and express a number of proteins including lysozyme, tumour necrosis factor (TNF)-α [11] and the antibacterial β-defensins, or cryptdins (3~4 kDa molecular weight peptides) [12], which contribute to innate immunity by sensing bacteria and bacterial antigens and secreting microbicidal peptides. Other less common cell lineages are also present, such as caveolated cells, which are characterized by a microvillous tuft protruding into the glandular lumen, and long narrow convoluted caveoli that open between the microvilli, and M- (membranous- or microfold-) cells which are present where the epithelium covers masses of lymphoid tissue in the intestinal wall, and are thought to transfer antigens from the lumen of the intestine to the underlying tissues, acting as a sampling system to enable the lymphoid tissue to produce appropriate antibodies for secretion.

In the stomach, epithelial cells are organized in vertical tubular structures consisting of an apical pit region, an isthmus just below the pit, the neck region and a gland region forming the lower part of the vertical unit [13]. A single-cell-thick epithelium covers the whole surface and lines the pit, isthmus, neck and gland. These gastric units contain various cells; mucus-producing surface mucous cells, mucous neck cells, acid-producing parietal cells, pepsinogen-producing chief cells, and endocrine cells. The progenitor cells of the gastric unit are believed to exist the isthmus/neck region of the tubular unit, and give rise to all gastric epithelial cell types that migrate either up or down from this point [14]. One type, called the foveolar/pit cell lineage, migrates up towards the luminal surface and differentiates into surface mucous cells. Other cell lineages move downward, slowly differentiating into mucous neck, parietal, chief, and endocrine cells. These bi-directional processes are known as foveolar and glandular differentiation, respectively.

(2) Subepithelial Myofibroblasts

The intestinal crypts are enclosed within a fenestrated sheath of intestinal subepithelial myofibroblasts (ISEMFs). These cells are present immediately under the basement membrane in the normal intestinal mucosa, juxtaposed to the bottom of the epithelial cells. The ISEMFs are specialized mesenchymal cells that exhibit the ultrastructural features of both fibroblasts and smooth muscle cells and can be characterized by positive immunoreactivity for α-smooth muscle actin (α-SMA) and vimentin, but negative immunoreactivity for desmin. These cells play a central role in the regulation of a number of epithelial cell functions, such as proliferation and differentiation, and extracellular matrix (ECM) metabolism affecting the growth of the basement membrane [15,16], and also intestinal inflammation via the secretion of proinflammatory cytokines, such as IL-6 together with a variety of chemokines [17,18]. In particular, ISEMFs are believed to mediate the epithelial-mesenchymal interactions by secreting growth factors and cytokines, such as platelet-derived growth factor (PDGF) (required for formation of villi [19]), hepatocyte growth factor (HGF) [20], transforming growth factor (TGF)-β [21], keratinocyte growth factor (KGF) [22], basic-fibroblast growth factor (b-FGF, or FGF-2) [23], and Interleukin (IL)-11 [24].

DEVELOPMENT OF THE GUT

(1) Morphologic Gut Development

From the view point of developmental biology, the gastrointestinal tract is a fusion of multiple different cell lineages from different germinal layers. The epithelium of the foregut, midgut and hindgut is derived from the embryonic yolk sac, which is of endodermal origin. The vascular, lymphatic and haematopoietic system, including haematopoietic stem cells, are derived from mesodermal germ layer, while the peripheral nerves which control the gastrointestinal tract are of ectodermal origin.

Morphologic gastrointestinal tract development starts from two ventral invaginations, one at the anterior (anterior intestinal portal, AIP) and the other at the posterior (caudal intestinal portal, CIP) end of the embryo. These invaginations elongate in the endodermal layer to form the foregut, the midgut and the hindgut. The midgut communicates with the yolk sac via the vitello-intestinal duct.

In humans, intestinal epithelial development occurs at a relatively earlier stage than in most laboratory animals (Figure 3) [25]. At 8 weeks of embryogenesis, the visceral endoderm appears uniform and presents stratified cell layers. Interestingly, the functional structures of the small intestine, the villi, develop earlier than crypts from 9 weeks. Subsequently from 14-16 weeks, invagination of the intervillous epithelium in the adjacent mesenchyme starts to form the crypts which contain proliferative but poorly differentiated cells. The classical intestinal unit, the crypt-villus axis, is formed by 16 weeks in the small intestine. In the colon, a villus structure is transient and disappears after 30 weeks to form a flat surface epithelium (SE) with intervening crypts.

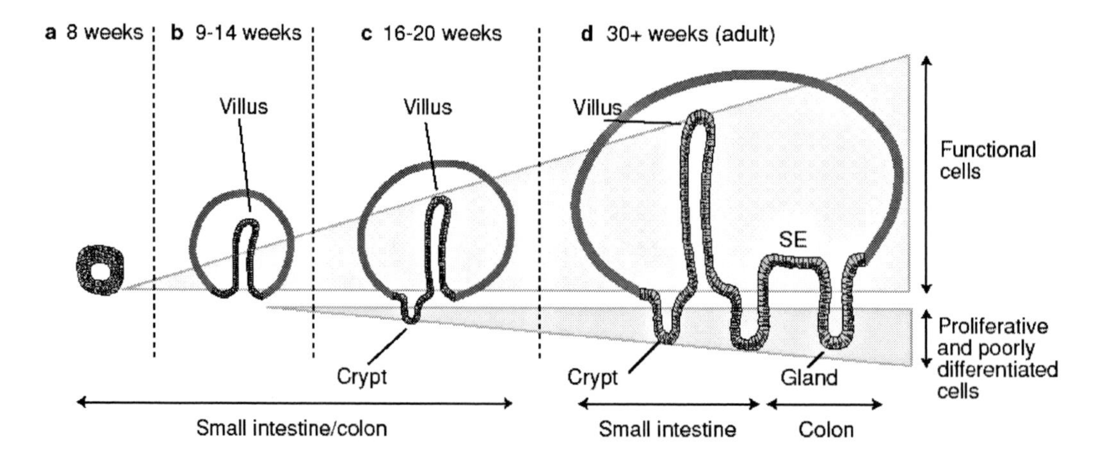

Figure 3. Development of the human small intestine and colon. (a) At 8 weeks, the epithelium appears stratified and undifferentiated. (b) From 9 weeks, villi start to develop in the small intestine and colon. (c) Subsequently, from 14-16 weeks, invagination of the intervillous epithelium in the adjacent mesenchyme starts to form the crypts. (d) In the colon, the villus structure is transient and disappears after 30 weeks to form the surface epithelium (SE) (Reprinted from Interactions between laminin and epithelial cells in intestinal health and disease, Teller IC and Beaulieu JF, Expert Reviews in Molecular Medicine, 28:1-18, (2001) with permission from Cambridge University Press).

(2) Molecular Factors Regulating Gut Development

The gut epithelium is a constitutively developing tissue. Signals between the adjacent mesoderm and epithelial cells are required for gut development, differentiation, and apoptosis. Recent reports have demonstrated the critical molecular pathways in these processes, such as the Hox and Sox transcription factor, the hedgehog (Hh), and the bone morphogenetic protein (BMP) pathways.

(i) Sox Transcription Factors

Sox17, which is a high-mobility group (HMG) transcription factor gene related to the sex-determining factor gene SRY [26], controls early endoderm formation in *Xenopus* [27, 28]. Murine Sox17 knockouts demonstrate that Sox17 is essential for embryonic cells to acquire an endodermal cell fate and that disruption of the Sox17 gene displayed complete deletion of the mid- and hindgut endoderm but does not affect the formation of the foregut endoderm [29]. Several Sox gene expressions are expressed in the gut endoderm - Sox2 [30], Sox7 [31, 32], Sox9 [33], Sox17 [29,34] and Sox18 [35] - suggesting that Sox genes may be important in the gut development. Moreover, the functions of Sox genes in intestinal epithelium are indicated by several expression studies. In the normal gastrointestinal tract, Sox17 mRNA was preferentially expressed in the oesophagus, stomach and small intestine rather than in colon and rectum. On the other hand, Sox18 mRNA was relatively highly expressed in stomach and jejunum but both Sox mRNAs were almost undetectable or less expressed in human cancer cell lines than the normal gastrointestinal tract [34, 35].

(ii) Hox Transcription Factors

Hox genes, which are homeobox-containing transcription factors, are principally expressed in mesoderm and play an important role in patterning the gut along the anterior-posterior axis, influencing both the gross morphology of the gut and later the epithelial-mesenchymal interactions responsible for normal gut epithelial differentiation [36-38]. In the gut, Hox genes are expressed in a spatially and temporally-specific manner in the posterior mesoderm, from the post umbilical portion of the midgut through to the hindgut [36-39]. Hoxa13 and Hoxd13 are co-expressed in the distal-most hindgut mesoderm and uniquely throughout the hind endoderm [36,40]. In mice, Hoxa13$^{+/-}$/Hoxd13$^{-/-}$ mutants have gastrointestinal anomalies, showing dilatation of the rectum with abnormalities of the epithelial layer, which lacked rectal glands, while the smooth muscle layer, which appeared abnormally distant from the mucosa and interrupted in some areas. One (out of 14 analyzed) newborn mice showed an almost complete absence of the terminal part of the rectum [41], suggesting their importance in the early morphogenesis of posterior trunk structures. Moreover, de Santa Barbara *et al* (2002) investigating the role of Hoxa13 in the posterior endoderm using avian system, showed that a Hoxa13 mutant protein, which behaves in a dominant negative fashion, was specifically expressed in the early developing chick posterior endoderm [37]. This resulted in decreased wild-type protein, and the chicks developed with dramatic malformations in the gut and genitourinary system with atresia of the hindgut anterior to the cloaca, cystic mesonephric maldevelopment, and atresia of the distal Müllerian ducts. Hoxa8 has also been found to be expressed in the small and large intestinal endoderm; however, no functional studies have been made [42, 43].

(iii) Hedgehog (Hh) Pathway

The Hedgehog (Hh) pathway is known to play an important role in gut development [44-46]. The Hh family includes three members in most vertebrates: Sonic hedgehog (Shh), Indian hedgehog (Ihh), and Desert hedgehog. All hedgehog proteins can bind two common homologous receptors: Patched (Ptc)-1 and Ptc-2. In the unstimulated state, these receptors downregulate the activity of a seven transmembrane receptor Smoothened (Smo) [47]. Once Hh binds to Ptc, the suppression of Smo is relieved and pathway activation through the Gli family of transcription factors ensues. Both Shh and Ihh are important endodermal signals in gut tube differentiation.

Shh is an important factor, implicated in the first phase of endoderm to mesoderm signalling in the gut [38,48]. Shh is expressed early in the AIP and CIP endoderm [38,49-51]. As the gut tube forms and undergoes morphogenesis, Shh expression expands and is maintained in the gut endoderm with the exception of the pancreatic bud endoderm [51-53]. The Shh null mouse shows hyperplastic epithelium in the stomach with intestinal transformation, indicating the remarkable promotion of the gastric phenotype by Shh [54]. Moreover, Shh is also expressed in the adult stomach in humans and the mouse. Mice treated with cyclopamine, a potent hedgehog signalling inhibitor [55-57], show reduced expression of HNF3β, Islet (Isl)-1 and BMP4, putative Shh target genes. Inhibition of Shh stimulated 5-bromo-2'-deoxyuridine (BrdU) incorporation in the precursor cells in the isthmus/neck and increased staining for proliferating cell nuclear antigen (PCNA) in gastric epithelial gland cells such as chief cells and parietal cells, where as PCNA expression in surface mucous cells remain unchanged, suggesting markedly enhanced gastric epithelial proliferation and reduced the cell cycle of gastric epithelial gland cells [58]. Moreover, Shh expression is observed in the fundic glands of the adult human stomach and also in fundic gland metaplasia and heterotopia. However, Shh expression is lost in intestinal metaplasia of the stomach, suggesting a role of Shh in maintenance of the gastric epithelial differentiation program [59]. The Shh null mouse also show abnormal and over-innervated villi which overgrow and clog the lumen of the duodenum, whereas growth of villi in the Ihh null mouse is strongly diminished and they often lack innervation [54]. In the adult small intestine, Shh mRNA is detected at the base of the crypts around the presumed location of the small intestinal stem cell [59].

The expression of Ihh is observed later in the gut endoderm in a partially overlapping pattern [54]. A very recent report [60] has revealed that Ihh is expressed by mature colonocytes in humans and rats (Figure 4 (a-f)). The Hh binding receptor Ptc is broadly expressed in the epithelial cells along the crypt axis and also on mesenchymal cells. Engrailed-1 (En-1) and BMP4, putative Wnt target genes, are both expressed in epithelial cells along the entire crypt axis, with highest expression at its base. However, in cyclopamine-treated rats, the gradient of expression of En-1 and BMP4 is lost and there was homogeneous expression throughout the crypt.

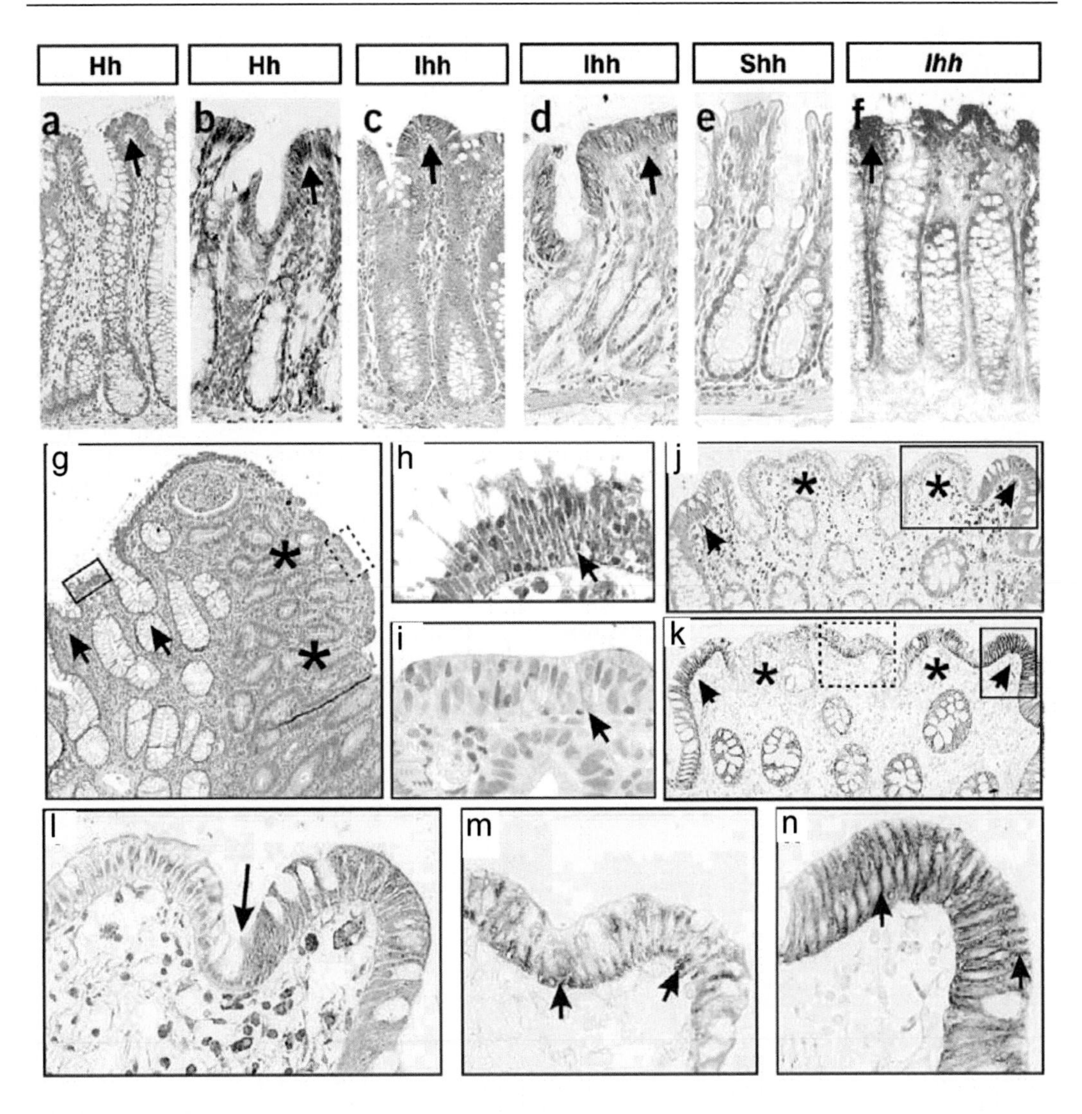

Figure 4. Hedgehog (Hh) expression in normal colon. (a, b) Immunohistochemistry of human (a) and rat (b) colon using an antibody for all Hh proteins. Hh protein is expressed by the absorptive columnar cells in both species (arrows). (c, d) The Indian Hedgehog (Ihh) specific antibody stained the same colonic cells in both humans (c) and rats (d). (e) No staining was observed with antibodies against Sonic Hedgehog (Shh). (f) *In situ* hybridisation with an Ihh probe on normal human colon. Terminally differentiated columnar cells (arrow) produces Ihh mRNA. (g-n) Loss of Ihh expression precedes the development of dysplasia in colon carcinogenesis. (g) An adenoma (asterisks) bordering histologically normal epithelium (arrows). (h) Histologically normal epithelial cells expresses Ihh. (i) Ihh is not expressed in the superficial dysplastic epithelial cells of the adenoma. (j-n) Immunohistochemistry of Ihh (j, l) and β-catenin (k, m, n) in serial sections of tissue from a familial adenomatous polyposis (FAP) patient. (j, l) Expression of Ihh was lost in a few morphologically normal crypts (asterisks), with a sharp transition to normal Ihh expression in the adjacent crypts on both sides (arrows). (k, m, n) Abnormal localization of β-catenin was observed in the same crypts (Reprinted from Nat Genet 2004 36:277-282 with permission from Nature Publishing Group).

Western blot analysis also revealed increased expression of En-1, BMP4 and Cyclin D1 in colonic homogenates from cyclopamine-treated rats, suggesting that Ihh signalling restricts expression of Wnt targets such as En-1, BMP4 and Cyclin D1 to the precursor cell compartment at the base of the colonic crypt. Transfection of Ihh into colon cancer cells leads to a downregulation of both components of the nuclear β-catenin/Tcf-4 complex and abrogates endogenous Wnt signalling *in vitro*. In turn, expression of Ihh is downregulated in polyps of Familial Adenomatous Polyposis (FAP) patients (Figure 4 (g-n)), suggesting that Ihh is an antagonist of Wnt signalling in colonic epithelial differentitation.

(iv) BMP Pathway

BMPs are members of the transforming growth factor-β (TGF-β) superfamily of signalling molecules. BMP ligands act via specific receptors in a complex which ultimately, by phosphorylation, activates target molecules, Smad1/5 and 8, that in turn moves to the nucleus to activate transcription of target genes [61]. The phosphorylation of these molecules elucidates the activation of BMP signalling pathway. Smad1/5/8 phosphorylation occurs in the ventral part of the foregut endoderm [62], suggesting early participation of BMP signalling in the development of endodermal AIP structures. Moreover, BMP-4 is shown to be expressed throughout the mesoderm of the chick gut, except for only in the avian muscular stomach, the gizzard [36, 63, 64]. Retroviral misexpression experiments suggest that the level of BMP activity may have a fundamental role in the control of gut muscular development, in pyloric sphincter development, and in stomach gland formation [36, 63-66]. Endogenous activation of this pathway is specifically found in the gut mesenchyme layer but also in the developing endoderm. In the normal adult colon, BMP-6 was uniformly expressed throughout crypts, but, expression was greatly reduced in carcinomas [67] in contrast to the Wnt pathway.

Recent reports have indicated that the BMP pathway plays an important role in intestinal epithelial homeostasis. Mutations in different members of the BMP signalling pathway were found to be associated with the human Juvenile Polyposis Syndrome (JPS). A specific Smad4 mutation is found in some JPS patients and results in the formation of a truncated protein [68, 69]. Smad4 is a common molecule in both TGF-β and BMP pathways. Germline mutations of the gene encoding bone morphogenic protein receptor 1A (BMPR1A), a serine-threonine kinase type I receptor, are also found in JPS patients. These mutations resulted in protein truncation due to a deletion of the intracellular serine-threonine kinase domain necessary for Smad protein phosphorylation and signal transduction.

Origin of the Intestinal Epithelium

The origin of the intestinal epithelium has been the subject of attention for some time. The 'Unitarian Theory' proposed by Cheng *et al* (1974) has been widely accepted [70-72] (Figure 5); here, all the differentiated cell lineages within the gastrointestinal structural units, intestinal crypts or gastric glands emanate from a common stem cell.

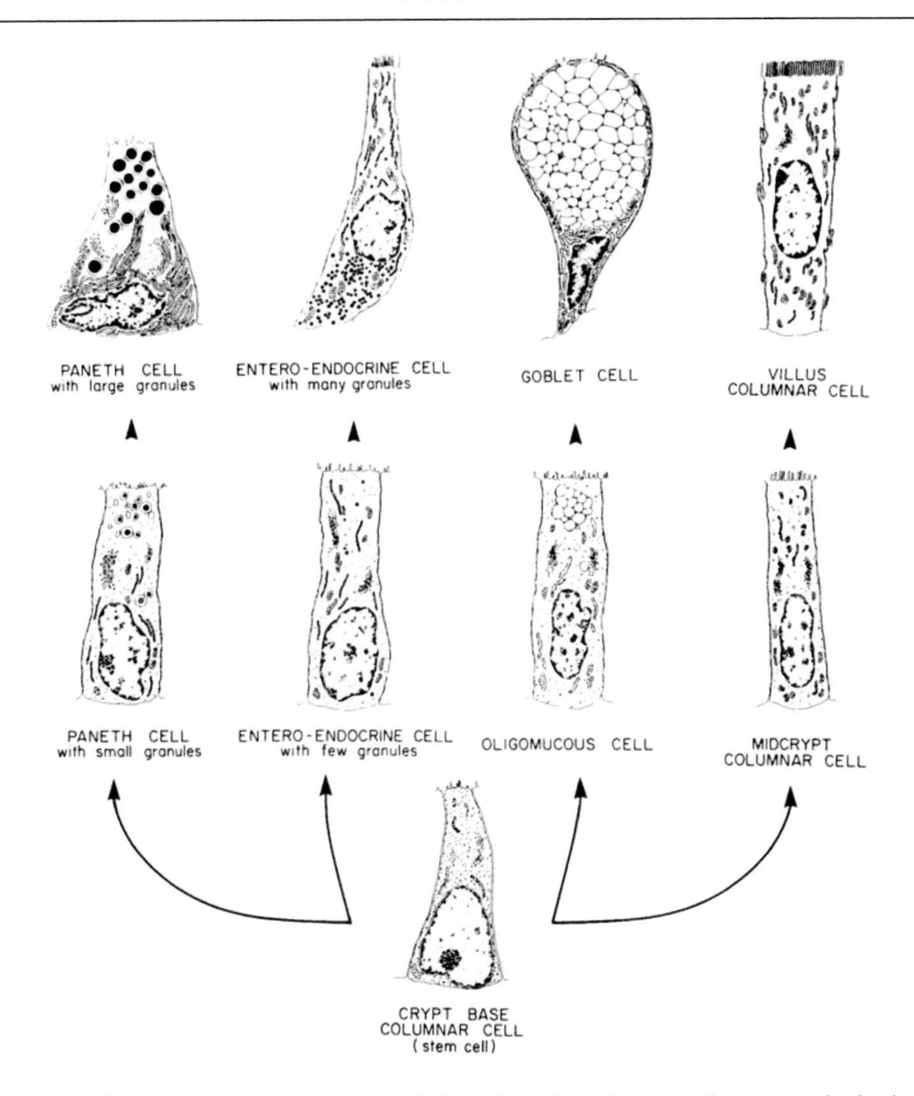

Figure 5. A diagrammatic representation of the 'Unitarian Theory' of cytogenesis in intestine (Reprinted from Origin, differentiation and renewal of the four main epithelial cell types in the mouse small intestine. V. Unitarian Theory of the origin of the four epithelial cell types, Cheng H and Leblond CP, Am J Anat, 141:537-561, (1974) by permission of Wiley-Liss, Inc., a subsidiary of John Wiley and Sons, Inc.).

While widely pronounced, until very recently little definitive evidence existed to underpin this hypothesis [73]. Moreover, Pearse *et al* (1976) proposed that gastrointestinal endocrine cells are derived from the neural crest, and these are ectodermal origin [74]. Although studies of quail neural crest cells transplanted into chick embryos [75], or experiments where the neural crest is eradicated [76], show gut endocrine cells to be of endodermal origin, Pearse subsequently suggested that the endoderm is colonised by 'neuroendocrine-programmed stem cells' from the primitive epiblast which give rise to gut endocrine cells [77]. This hypothesis was not ruled out by the chick/quail chimaera experiments and therefore other models were utilised to ascertain the gut endocrine cell origins.

(1) Evidence for the 'Unitarian Theory'

The use of mouse embryo aggregation chimaeras to investigate the clonal origin of epithelial cells in the murine gastrointestinal mucosa has provided many important insights. The lectin *Dolichos biflorus* agglutinin (DBA) binds to N-acetylgalactosamine residues present on the surface of intestinal epithelial cells derived from C57BL/6J (B6) mice. There is no binding site on these cells derived from SWR mice: consequently, in B6/SWR mice embryo aggregation chimaeras, DBA-binding can distinguish the two parental strains in the gut epithelium. The intestinal crypts studied were either completely positive or negative for DBA [78] (Figure 6 (a-c)).

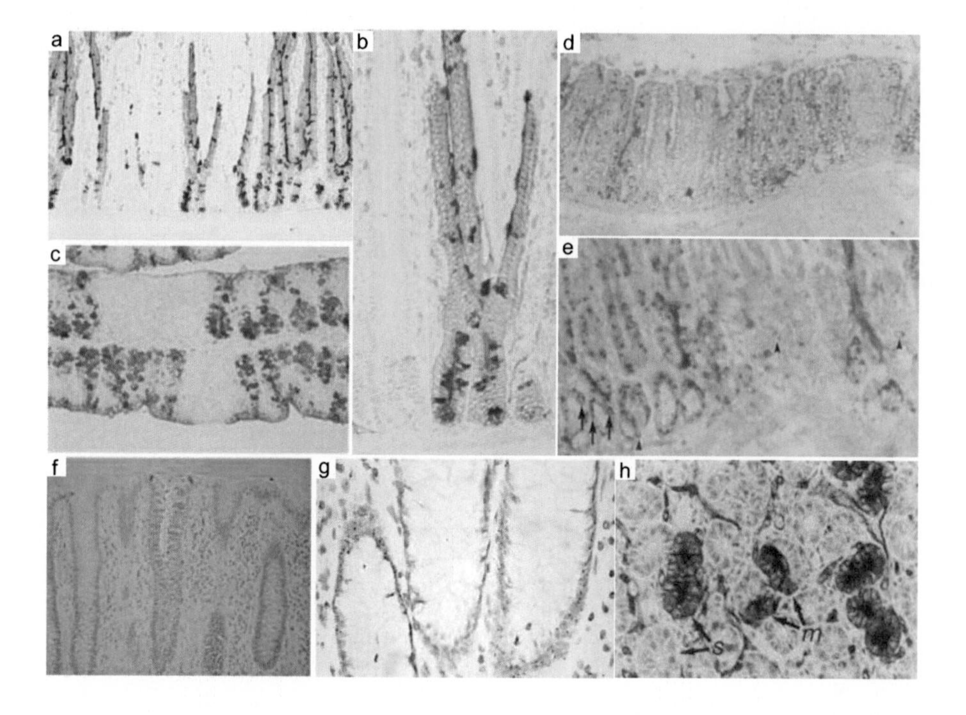

Figure 6. Crypt clonality. (a-c) Murine crypt clonality. Section from the jejunal mucosa (a, b) and colon (c) of a B6/SWR chimaeric mouse stained with the lectin *Dolichos biflorus* agglutinin, showing that crypts are either totally positive or negative (Reprinted from Int J Exp Pathol 2000 81:117-143 with permission from Blackwell Publishing). (d-e) Murine gastric gland clonality. Section from the stomach of an XX/XY chimaeric mouse. (d) Low power view showing *in situ* hybridisation for Y chromosome using digoxigenin. ~Again, glands are either positive or negative for the Y-chromosome. (e) High power view of antrum stained by immunohistochemistry for gastrin combined with *in situ* hybridisation for Y chromosome showing Y negative gastrin cells, and Y positive gastrin cells (large arrow) confined within the Y positive glands (Reprinted from Development 1990 110:477-481 with permission). (f-g) Human crypt clonality. (f) A human colonic crypt stained with the mPAS method, showing crypt-restricted loss of the ability to acylate O-sialomucin (Reprinted from Int J Exp Pathol 2000 81:117-143 with permission from Blackwell Publishing). (g) Section of colon from a patient with the XO/XY phenotype stained by non-isotopic *in situ* hybridisation for the Y chromosome showing an XO crypt (central) surrounded by two XY crypts (Reprinted from Gastrointestinal stem cells, Brittan M and Wright NA, J Pathol, 197:492-509 with permission from John Wiley and Sons Ltd on behalf of PathSoc). (h) Crypt clonality in chimaeric mice. Transverse section of neonatal (P6) B6/SWR chimaeric

mouse duodenum stained by DBA peroxidase showing the B6 component (black) and the SWR component (unstained). The contribution of both components to mixed crypts (*m*) and monoclonal crypts (*s*) (Reprinted from Development 1988 103:785-790 with permission).

This chimaeric mouse study revealed that each crypt forms a monoclonal population which included Paneth, mucous, and columnar cells. However, endocrine cells, because of their very small luminal surface area, could not be studied by this method. Winton *et al* (1988) confirmed that a single stem cell can give rise to all epithelial lineages within an intestinal crypt of the small intestine, using the *Dlb*-1 assay. *Dolichos biflorus* agglutinin (DBA) binding to intestinal epithelial cells of C57BL/6J-SWR F1 (first filial generation) mice is abolished when the *Dlb*-1 locus on chromosome 11 becomes mutated either spontaneously or by the chemical mutagen ethylnitrosourea (ENU). The results of this assay showed that all crypts are either entirely negative or positive for DBA staining and therefore mutation of the *Dlb*-1 locus in a stem cell within the small intestine crypt produces a clone of cells that cannot bind DBA and remain unstained [79].

To exclude the possibility that crypts from distinct strains segregate differentially during organogenesis, Ponder *et al* (1985) examined mosaic expression of the electrophoretic isoenzymes PGK-1A and PGK-1B, X-linked enzymes in colonic crypts of heterozygous mice [78]. In conformity with the results from the chimaeric studies, no mixed crypts were seen, thus eliminating the possibility that these crypts are monophenotypic and verifying that they are indeed derived from a single progenitor cell. Griffiths *et al* (1988) also reported that heterozygous mice which express a defective glucose-6-phosphate dehydrogenase (G6PD) gene, also carried on the X-chromosome, have a crypt-restricted pattern of G6PD expression [80]; thus, in these mice also, crypts derive from a single stem cell.

(2) Evidence for the 'Unitarian Theory' in the Stomach

Similar reports have been made also in the stomach. Thompson *et al* (1990) showed that epithelial cell lineages in the antral gastric mucosa of the mouse stomach, including the endocrine cells, derive from a common stem cell, using XX/XY chimaeric mice and combining immunohistochemistry for gastrin, an endocrine cell marker, with *in situ* hybridisation to detect the Y-chromosome [81] (Figure 6 (d-e)). Tatematsu *et al* (1994) also demonstrated that the gastric glands in the mouse are clonally derived by using C3H-BALB/c chimaeric mice using specific antibody against C3H strain specific antigen (CSA) [82]. Nomura *et al* (1998) used X-inactivation mosaic mice, expressing a lacZ reporter gene, showing the clonality of gastric glands in the fundic and pyloric regions of developing mouse stomach [83]. They proposed that most glands are initially polyclonal with three or four stem cells per gland, but that these become monoclonal during the first 6 weeks of life, 'monoclonal conversion' (see below under Clonal expansion and crypt fission).

Karam and Leblond (1993) approached the Unitarian Theory using classical continuous and flash-labelling methods with tritiated thymidine and concluded that the undifferentiated 'granule-free' cells located in the isthums of the gland are self-renewing and give rise to all lineages in the gastric tubule via three committeed precursors - the prepit cells, the preparietal cells, and the preneck cells. with the peptic or zymogenic cells as the end-stage differentiated progeny of the neck cells (Figure 7) [13]. These results also supported the Unitarian Theory in the stomach, although none of these experiments definitively demonstrates that a single stem cell gave rise to all epithelial cell lineages.

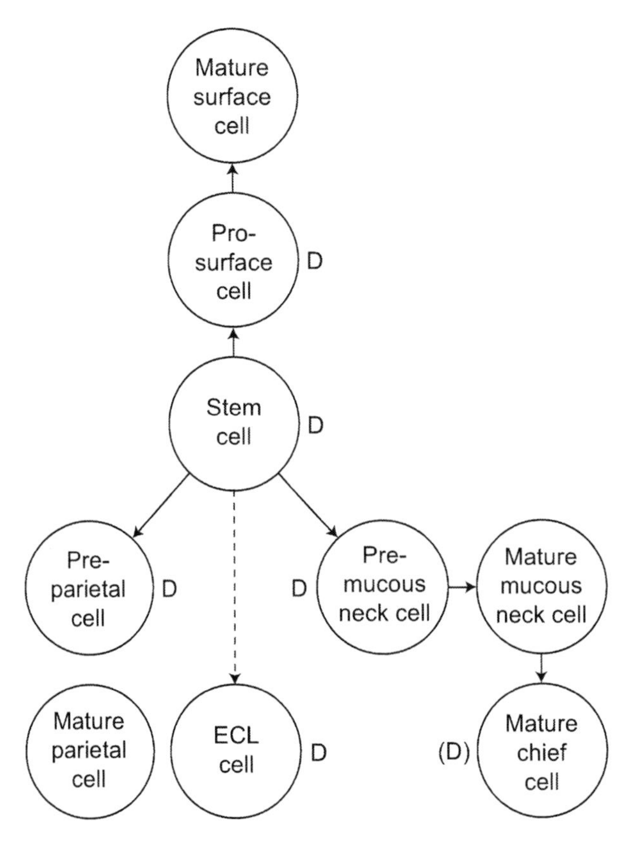

Figure 7. The Unitarian Theory in gastric fundic gland. The crypt-base columnar stem cells in the intestinal crypt give rise to the cell lineages shown while undifferentiated granule-free cells in the isthmus of the gastric gland are regarded as stem cells and give origin to the three committed precursors. In the gastric gland there is a bi-directional flux of cells, upwards to the surface (the presurface or prepit cells giving rise to the surface or pit cells), and downwards upwards the base of the gland (parietal, chief, ECL and other endocrine cells). The mucous neck cells remain in the neck (Redrawn from [13]).

Clonality in Human Gastrointestinal Epithelia

(1) Small Intestinal and Colonic Epithelia

Fuller *et al* (1990) studied crypt clonality using changes in the gene encoding the enzyme O-acetyl transferase (OAT), which produces O-acetylated sialic acid [84]. Approximately 9% of the human Caucasian population have a homozygous genetic mutation in OAT, which is expressed as negativity for O-acetylated sialic acid. The goblet cells, secreting non-O-acetylated sialic acid, are positive with the mild periodic acid-Schiff (mPAS) stain [85, 86] (Figure 6 (f)). Among heterozygotes ($OAT^{+/-}$), loss of the remaining active OAT gene converts the functional phenotype to $OAT^{-/-}$, resulting in occasional, random positively mPAS-stained crypts, with uniform staining of goblet cells from the base to the luminal surface, an effect that increases with age [84]. This could be due to a somatic mutation or non-dysjunction in a single crypt stem cell and subsequent colonization of the crypt by the mutated stem cell. The frequency of these events is ethnically-determined [87] and increases after irradiation [88]. Moreover, Campbell *et al* (1996) also analysed somatic mutations in O-acetyl transferase among heterozygotes ($OAT^{+/-}$) who had undergone colectomy up to 34 years after radiotherapy. This study revealed that crypts wholly containing the mutant phenotype are stable and persistent while partially-involved crypts are transient, and that this 'clonal stabilisation time' is approximately one year in the human colon [88], compared with four weeks in the mouse colon. The most likely reason for this is a difference in the number of stem cells in the crypt stem cell niche, although differences in stem cell cycle time and crypt fission may also contribute.

Novelli *et al* (1996) studied the colon of very rare XO/XY patients who had received a prophylactic colectomy for familial adenomatous polyposis (FAP) [89]. Non-isotopic *in situ* hybridisation (NISH) using Y-chromosome-specific probes showed the patient's normal intestinal crypts to be composed practically entirely of either Y-chromosome-positive or negative cells (Figure 6 (g)). Immunostaining for neuroendocrine-specific markers and Y-chromosome NISH used in combination showed that crypt neuroendocrine cells shared the same genotype as other crypt cells. The villus epithelium was a mixture of XO and XY cells, which was expected, as the villi are believed to derive from cells of more than one crypt. Of the 12614 crypts examined, only four crypts were composed of XO and XY cells, which could explained by non-disjunction with loss of the Y-chromosome in a crypt stem cell. Importantly, there were no mixed crypts at patch boundaries. These observations agree with previous results, using chimaeric mice that intestinal crypts, including neuroendocrine cells, are monoclonal and derive originally from a single multipotential stem cell. Consequently, the Unitarian Theory appears to apply to both mice and humans. These observations have recently been confirmed in Sardinian women heterozygous for a defective G6PD gene [90].

(2) Gastric Epithelium

In situ analyses of glandular clonality in the human stomach have been less straightforward. Nomura *et al* (1996) used X-chromosome-linked inactivation to study fundic and pyloric glands in human female stomachs [91]. Studies using polymophisms on X-linked genes such as the androgen receptor (HUMARA, human androgen receptor gene) to distinguish between the two X-chromosomes revealed that, while pyloric glands appear homotypic and thus monoclonal, about half of the fundic glands studied were heterotypic at the HUMARA locus and are consequently polyclonal. This finding suggests that a more complex situation occurs in humans than studies of gastric gland clonality in chimaeric mice indicate.

(3) Evidence for Multiple Stem Cells in a Single Colonic Crypt

In the normal human colon, methylation patterns are somatically-inherited endogenous sequences that randomly change and increase in frequency with age. Methylation patterns are useful markers for the investigation of crypt histories and allow fate mapping, as this method allows the detection of many methylation sites compared with histological methods that that show either the presence or absence of a marker. Examination of the methylation tags of three neutral loci in cells from normal human colon showed variations in sequence between crypts with a mosaic methylation patterns within single crypts. Multiple unique sites were present in morphologically-identical crypts; for example, one patient had no identical methylation sequences of one locus within any of the crypts studied, although all sequences were related [92]. This indicates that some apparently normal human colonic crypts have multiple stem cells, possibly up to 64 per crypt.

Very recently, Taylor *et al* (2003) examined the mutations in the mitochondrial DNA (mtDNA) of human colonic crypts [93]. Sporadic mtDNA mutations are fairly frequent events, a view supported by the high sequence divergence within the human population. They revealed that colonic crypt stem cells accumulate mtDNA mutations resulting in cytochrome c oxidase deficiency. In some crypts, in which in excess of 80% of the mitochondria were mutated, *partial* respiratory chain deficiency was observed (Figure 8). These findings indicate that such mutations in stem cells are transferred to their progeny and that a single colonic crypt contains several stem cells.

Clonal Expansion and Crypt Fission

In the small intestine of neonatal B6/SWR chimaeras mentioned above, there were mixed, and therefore polyclonal crypts for the first 2 weeks after birth, implicating multiple stem cells during development [94] (Figure 6 (h)). However, all crypts ultimately become monoclonal, thus deriving from a single stem cell between birth and postnatal day 14, so-called 'monoclonal conversion'. 'Monoclonal conversion' is a process of clonal expansion of one stem cell which eventually overrides all other stem cells by division, or by crypt and gland fission. Nomura *et al* (1998) used X-inactivation mosaic mice, and reported a similar phenomenon in the stomach [83], where most glands are initially polyclonal with three or four stem cells per gland, but become monoclonal during the first 6 weeks of life, displaying a longer 'monoclonal conversion' time than the small intestine, evidently due to organ specificity.

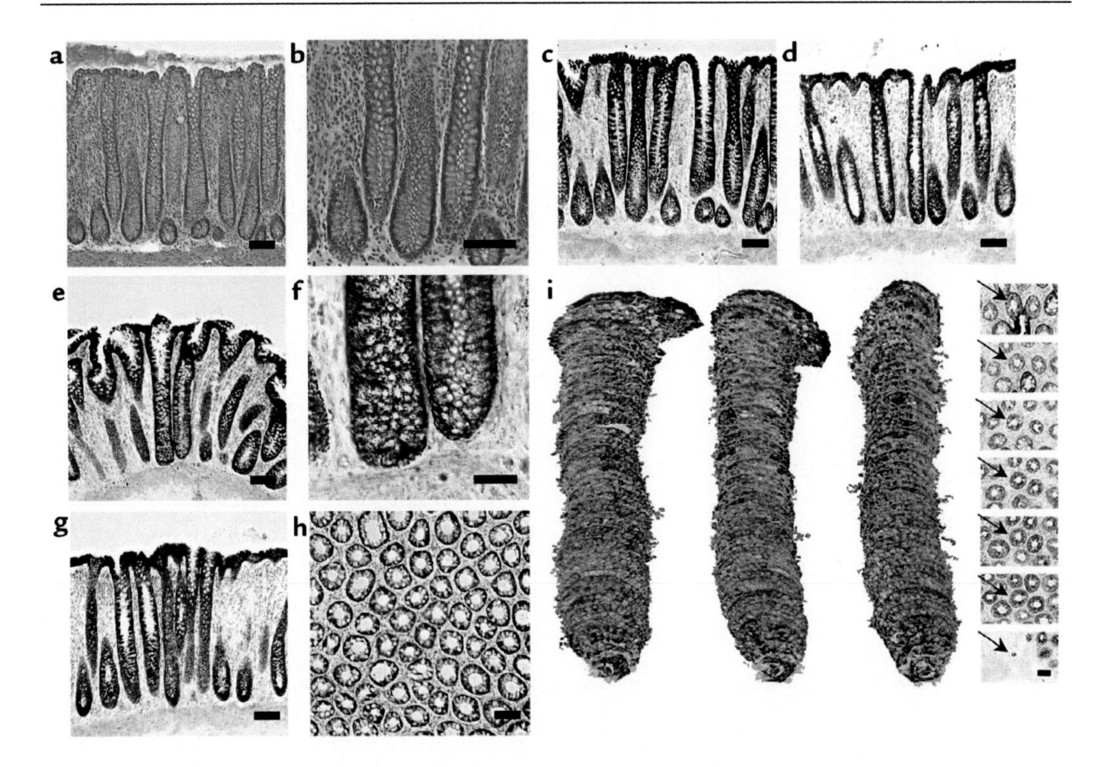

Figure 8. Respiratory chain deficiency in morphologically normal human colonic mucosa. (a) HandE staining showing normal mucosal structure. Scale bar: 100µm. (b) High power view of (a), showing normal mucosa. Scale bar: 50µm. (c) Normal cytochrome c oxidase activity (brown) in colonic crypts. Scale bar: 100µm. (d) Blue crypts showing the absence of immunohistochemically-detectable cytochrome c oxidase activity in colonic crypts. Scale bar: 100µm. (e) Single cytochrome c oxidase-deficient colonic crypt. Scale bar: 100µm. (f) High power view of (e). Scale bar: 20µm. (g) Multiple adjacent crypts showing the absence of cytochrome c oxidase. Scale bar: 100µm. (h) Transverse section showing a similar cluster of cytochrome c oxidase-deficient crypts. Scale bar: 100µm (i) 3D reconstruction produced from 50 transverse sections showing a partially cytochrome c oxidase-deficient crypt. The images of the reconstructed colonic crypt are each rotated clockwise through 45° to show a continuous ribbon of cytochrome c oxidase-deficient cells originating at the crypt base (from [93] with permission).

There are further examples of monoclonal conversion, for example in C3H/Heston mice treated with the colon carcinogen dimethylhydrazine (DMH) or the mutagen ethylnitrosourea (ENU). These mice display crypts that are partially or wholly negative for G6PD [95] (Figure 9). After ENU, the time taken for the decrease in partially mutated crypts and the emergence and rise in the number of entirely negative crypts to reach a plateau takes between 4.6 and 7 weeks in the colon and 12 weeks in the small intestine of the mouse. These differences could be explained by the stem cell niche hypothesis, wherein multiple stem cells occupy a crypt with random loss after stem cell division and thus 'niche succession' of a mutant stem cell. Alternatively the numbers of stem cells may differ between the small and large intestine – larger numbers in the small intestine could also explain the difference in time taken for phenotypic changes following mutagen treatment [96].

The crypt fission index, or the percentage of crypts in fission, is about four times higher in the colon than in the small intestine at the time of ENU injection. Thus crypt fission could cause the increase in the emergence of wholly-negative crypts and the decrease in partially-negative crypts, since the rate of which was dependent on the fission index at the time of mutagen administration. During crypt fission, crypts would divide longitudinally, thereby possibly 'cleansing' the partially-mutated crypts by segregating the mutated and non-mutated cells, and duplicating wholly negative crypts to create monoclonal crypts.

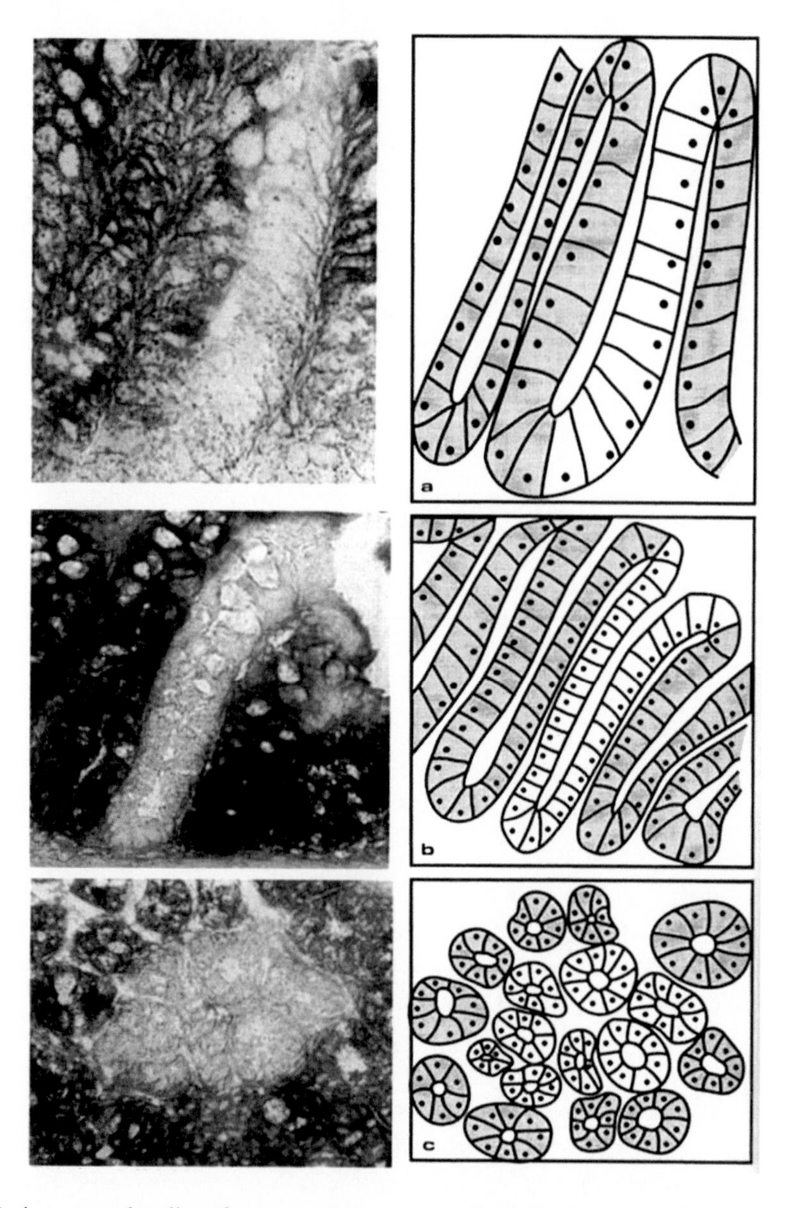

Figure 9. Murine crypt clonality after mutagen treatament. Sections from the colon of a mouse treated with a single injection of a mutagen (ethyl nitrosourea, ENU), and histochemically stained for glucose-6-phosphatase activity: (a) a partially negative crypt; (b) a wholly negative crypt; (c) a patch of negative-staining cells (Reprinted from Am J Pathol 1995, 147:1416-1427 with permission from the American Society for Investigative Pathology).

Yatabe *et al* (2001), studying the methylation profiles in the normal human colon [92] (described above), suggested that there is a 'bottleneck' effect, and all cells within a crypt are closely related to a single stem cell every 8.2 years in humans (95% confidence intervals of 2.7 to 19 years) [92], where all crypt descendants are derived from the same stem cell by gradual niche succession. These data suggest a crypt stabilisation time (described already under Clonality in human gastrointestinal epithelia) of around 220 days, not inconsistent with the year described above [88].

Crypt fission is a process that begins by basal bifurcation, is usually symmetrical (or, if asymmetrical, called budding), and is followed by longitudinal division of the crypt. The increase of crypt fission occurs in the postnatal period [97], during the recovery of the intestine from irradiation [98] and cytotoxic chemotherapy [99]. The dynamics of crypt fission have been described in a series of seminal papers by Bjerknes *et al* [100-102]: these studies led to the concept of *the crypt cycle*. The crypt cycle - crypts born by fission gradually increase in size until they undergo fission again - is about 108 days long in the mouse, in a fission process which takes about 12 hours. Loeffler *et al* suggested a simple concept; the threshold for triggering crypt fission is a doubling of stem cell number [103,104].

Little is known about the regulation of crypt fission, however Cell proliferation, and thus crypt size, and crypt fission maybe independently controlled [105]: in the rat colon, epidermal growth factor (EGF), which increases cell proliferation and also crypt size, does not increase the crypt fission index, whereas dimethylhydrazine (DMH), which does not significantly increase crypt size, does elevate the crypt fission index.

Recent human studies have shown significant increases in the rate of crypt fission in the colorectal mucosae in a range of pathological conditions, usually interpreted as a regenerative phenomenon in inflammatory conditions such as ulcerative colitis and Crohn's disease [102], and in as the flat (non-adenomatous) mucosa of FAP patients [106,107].

Gastrointestinal Neoplasms and Clonal Succession

Multiple alterations are thought to accumulate in a stepwise manner during tumour progression [108]. A well-established model of this is the adenoma-carcinoma sequence [109] in the colon recognized as by morphological changes and progressing from *aberrant crypt foci* (ACF), through adenoma, and finally to carcinoma. Prior to loss of the wild-type allele, the crypts of humans and mice with germline mutations in tumour suppressor genes can be morphologically normal, although these mutations, such as $APC^{+/-}$(human) [110], $TP53^{+/-}$(human) [111], $Trp53^{-/-}$ [112], $Apc^{+/-}$; $Trp53^{-/-}$ [113], $MLH1^{-/-}$(human) [114], $Apc^{+/-}$;$MLH1^{-/-}$ [115], and $Tgfbr2^{+/-}$ [116], are thought to have critical roles in cancers and may progress to carcinomas. However, in normal individuals, such somatic alterations that accumulate before loss of heterozygosity would be difficult to directly detect because it would be nearly impossible to identify or isolate the rare phenotypically normal cells containing the alterations.

The understanding of the molecular pathology of the adenoma, its origin, and its mode of growth or expansion thus becomes important. The initial genetic change in the development of most colorectal adenomas is thought to be at the *Apc* locus, and the molecular events associated with these stages are clear: a second hit in the *Apc* gene is sufficient to give microadenoma development, at least in familial adenomatous polyposis (FAP) [117].

In the following section we would like to describe firstly the role of *Apc* and, subsequently two basic models for adenoma morphogenesis, both of which closely involve basic concepts of stem cell biology in the colon: the 'top-down' and 'bottom-up' theories.

(1) The Role of the *Apc* Gene in Tumorigenesis

The *Apc* gene codes for a 312 kDa protein comprising 2843 amino acids. *Apc* inhibits members of the Wnt signalling pathway, which promote the expression of β-catenin, an enhancer of cell division within crypts [118]. *Apc* specifically promotes the phosphorylation of β-catenin. The importance of this pathway is illustrated by the observation that a mutant of β-catenin lacking this phosphorylation site produced a 4-fold increase in cell proliferation within murine crypts [119,120]. An increase in E-cadherin, a component of intercellular adherens junctions and therefore a regulator of cell contacts, was also noted [121]. This process may indicate how *Apc*-mutated cells are able to manipulate their cellular contacts to remain within the crypt, thereby avoiding migration to the luminal surface and consequent death/exfoliation. Mutated *Apc* allows β-catenin to translocate to the nucleus, where it associates with members of the DNA-binding high-mobility group (HMG) box family of transcription factors, T-cell factor (Tcf) and lymphoid enhancer factor (Lef) [118,121]. These complexes regulate the transcription of target genes in the nucleus. The presence of mutated *Apc* has been shown to increase the transcriptional activity of targets containing a DNA-binding site recognized by Tcf family members [118,122].

β-catenin is an important factor in determining cell adhesion and proliferative signalling. Increased, unregulated expression may prove to be a vital early factor in the development of colorectal cancer. Adding weight to this proposal, amongst the minority of sporadic colorectal tumours with wild-type *Apc* gene function, 50% have a dominant mutation of the β-catenin gene rendering it resistant to degradation [122-124].

van de Wetering *et al* [125] have revealed that the nuclear expression of β-catenin, which is confined to the bottom of the crypts in the normal colon, is strongly observed in the nuclei of early adenoma cells and is strongly correlated with the expression of the EphB2 tyrosine kinase receptor. Moreover, the disruption of β-catenin/Tcf-4 activity by a dominant-negative Tcf (dnTcf) in carcinoma cells induces a rapid G_1 arrest and blocks Tcf target genes such as CD44, endodermal-neural cortex 1 (ENC1), BMP4 and claudin 1 that are physiologically active in the proliferative compartment of colonic crypt. Contemporaneously, an intestinal differentiation program is induced. The Tcf-4 target gene c-MYC directly represses the $p21^{waf/cip1}$ promoter but following disruption of β-catenin/Tcf-4 activity, the decreased expression of c-MYC releases $p21^{waf/cip1}$ transcription, which in turn mediates G_1 arrest and differentiation. Cyclin D1 is also a Tcf-4 target gene [126], to be activated through a Lef-1 binding site resulting in cell cycle entry, although Cyclin D1 was not affected by the expression of dnTcf [125].

Eph receptors and their ephrin ligands (Eph/ephrin) have been shown to be essential for migration of many cell types [127] and maintenance of pattern boundaries [128] during embryogenesis. Eph receptors constitute a large family of transmembrane tyrosine kinase receptors [129]. Binding and activation of Eph receptors to ephrin ligands require cell-cell interaction [130] and Eph/ephrin signalling converges to regulate the cytoskeleton [131]. Members of the Eph/ephrin signalling pathway were found expressed in the small intestine epithelium [132]. EphB2 and EphB3 receptors are expressed in the proliferative compartment, whereas their ligand ephrin-B1 displays the complementary expression pattern. Ephrin-B1 is expressed in all epithelial cells excluding those localized at the bottom of intervillus pockets. This suggests that the Eph/ephrin system may regulate epithelial cell migration.

In colorectal cancer, the expression of EphB2 is upregulated. β-catenin and Tcf inversely control the expression of the EphB2/EphB3 receptors and their ligand ephrin-B1 in colorectal cancer and along the crypt-villus axis [125]. Disruption of EphB2 and EphB3 genes reveals that their gene products restrict cell intermingling and re-allocate cell populations within the intestinal epithelium. In EphB3/EphB3 null mice, the proliferative and differentiated populations intermingle and in adult EphB3$^{-/-}$ mice, Paneth cells do not follow their downward migratory path, but scatter along the crypt and villus [132]. These findings suggest that in the intestinal epithelium β-catenin and Tcf couple proliferation and differentiation to the sorting of cell populations through the EphB/ephrin-B system.

(2) 'Top-Down' Hypothesis

In the 'top-down' theory, mutant cells appear in the intracryptal zone between crypt orifices, and the clone expands laterally and downwards to displace the normal epithelium of adjacent crypts [4]. A slight modification of this proposal is that a mutant cell in the crypt base, the site of the stem cell compartment [133], migrates to the intercryptal zone, where it expands as before. This theory originates from the observations in some early non-FAP adenomas, where dysplastic cells were seen only at the orifices and luminal surface of colonic crypts [4]. Assessment of loss of heterozytosity (LOH) for *Apc* and nucleotide sequence analysis of the mutation cluster region of the *Apc* gene was applied to microdissected, well-oriented histological sections of these adenomas. Half the sample showed LOH in the upper, dysplastic, portion of the crypts, and most of these had a truncating *Apc* mutation. Shih *et al* showed that dysplastic cells at the crypt apex contained genetic alterations of the Apc gene at a high rate of 90%, and that, in contrast, cells located at the bases of these same crypts appeared normal morphologically, and did not have such genetic alterations. However, they cannot exclude the possibility that these dysplastic cells initially originated from the dysplastic progenitors at the base of the adjacent crypts. Moreover, only these superficial cells showed intense proliferative activity, with nuclear localization of β-catenin, supporting the presence of an *Apc* mutation in these apical, dysplastic cells. Several previous morphological studies have demonstrated the same appearances [134], including those in FAP. Such a 'top-down morphogenesis' has profound implications for concepts of stem cell biology in the gut.

Most evidence indicates that crypt stem cells are found at the origin of the cell flux, near the crypt base [135]. These proposals by Shih *et al* either reestablish the stem cell compartment in the intracryptal zone or make the intracryptal zone a locus where stem cells, having acquired a second mutating hit, clonally expand.

Observations on relatively older adenomas have suggested that the distribution of proliferating, apoptotic, and TGF-β immunoreactive cells was strikingly reversed [136-138]. In these adenomas, the increased number of proliferating cells was mainly located at the luminal surface and TGF-β immunoreactive and apoptotic cells were located principally at the crypt base, suggesting that migration kinetics along the crypt-villus axis had been reversed inwards toward the polyp base. These observations could support a 'top-down' mechanism. However, there is evidence from examining the methylation histories of cells in adenomas, for a discrete stem cell architecture [139].

(3) 'Bottom-up' Theory

An alternative hypothesis, the 'bottom-up' theory, is a proposal that transformation takes place among the stem cell population in the crypt base, the transformed stem cell expands stochastically, and monoclonal conversion gives the monoclonal, *monocryptal* adenoma, which itself expands by crypt fission [140].

The recognition of this earliest lesion, the monocryptal adenoma, where the dysplastic epithelium occupies an entire single crypt, is of pivotal importance in this proposal. These lesions are very common in FAP, and although they are rare in non-FAP patients, they have certainly been described [141], and are thought to arise from a second hit of a stem cell in the crypt base, which expands stochastically or more likely because of their acquired properties such as resistance to apoptotic signals. Thus monocryptal adenomas should indeed be clonal [89]. We have seen crypt-restricted expansion of mutated stem in mice after ethylnitrosourea (ENU) treatment [95] and in humans heterozygous for the O-acetyl transferase (OAT) gene, where, after LOH, initially half, and then the whole, crypt is colonized by the progeny of the mutant stem cell [88]. This also occurs in individuals with Cox1 mutations (Figure 8). Interestingly, $OAT^{+/-}$ individuals with FAP show increased rates of stem cell mutation with clustering of mutated crypts. In this scenario, in sharp contrast to the 'top-down' theory, the mutated clone further expands, not by lateral migration but by *crypt fission*, where the crypt divides, usually symmetrically at the base, or by budding. In several studies, fission of adenomatous crypts is regarded as the main mode of adenoma progression, certainly in FAP, where such events are readily evaluated [106,142], but also in sporadic adenomas [143]. In fact, the non-adenomatous mucosa in FAP, with only one *Apc* mutation, shows a large increase in the incidence of crypts in fission [106]. Aberrant crypt foci, lesions that are putative precursors of adenomas, which can show k-RAS and *Apc* mutations, grow by crypt fission [144,145], as do hyperplastic polyps [146]. However, this concept does not exclude the possibility that the clone later expands in addition by lateral migration and downwards spread into adjacent crypts, with the initial lesion being the monocryptal adenoma. However, this model of morphogenesis is conceptually very different from that proposed by Shih *et al.*

Close examination of early sporadic adenomas supported their 'bottom-up' spread. In the normal colon, the nuclear expression of β-catenin is confined to the bottom of the crypts [125] (Figure 10). However, in early adenomas loss of the function in one of the genes in the

Wnt signalling pathway, most likely *Apc*, occurrs, with subsequent translocation of β-catenin to the nucleus. Fig. 11 (a, b) shows that the nuclear β-catenin expression extends to the apex of the adenomatous crypts and is especially present in crypts in the process of crypt fission (Figure 11 (c, d)); nuclear β-catenin expression is particularly marked in the nuclei of buds. Moreover, at the surface, there is a sharp cutoff between the adenomatous cells filling the crypt and showing nuclear β-catenin, and surface cells that do not (Figure 11 (e, f)).

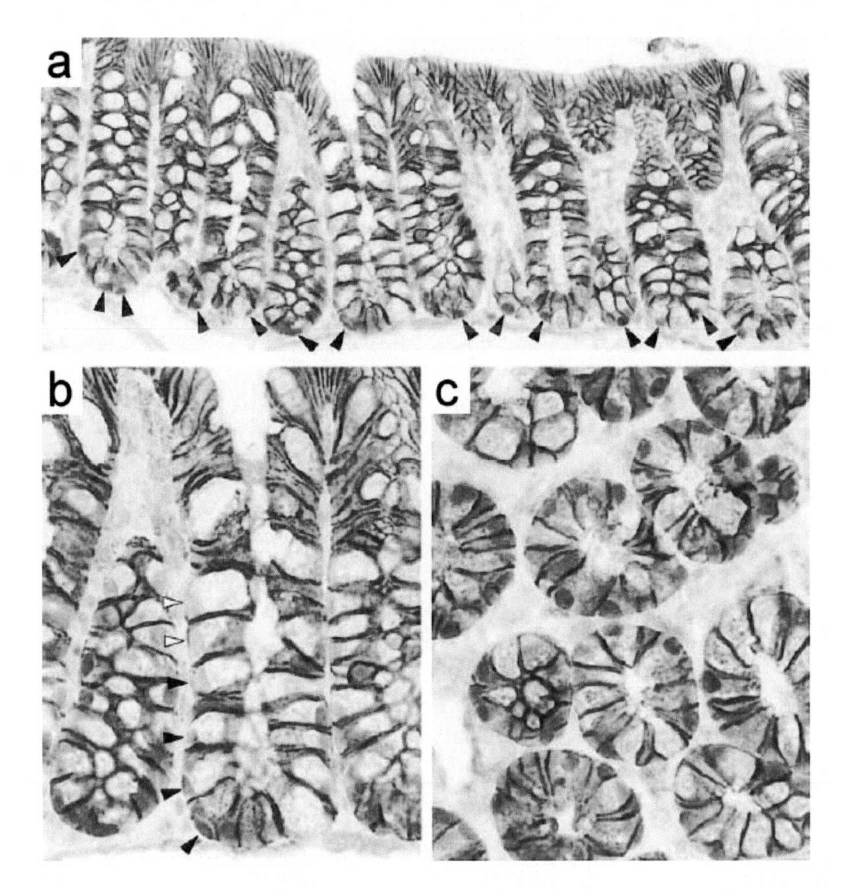

Figure 10. Nuclear β-catenin staining in the crypt base of normal murine colon epithelium. (a) Strong nuclear β-catenin expression is confined to the bottom of the crypts. (b) Nuclear β-catenin staining decreases in a gradient as cells migrate upwards (black arrowheads). Nuclei from cells at the midcrypts region are negative (white arrowheads). (c) Transverse section at the base of colon crypts showing nuclei positive for β-catenin staining (Reprinted from Cell, 111, van de Wetering M et al, The beta-catenin/TCF-4 complex imposes a crypt progenitor phenotype on colorectal cancer cells, 241-250 (2002) with permission from Elsevier).

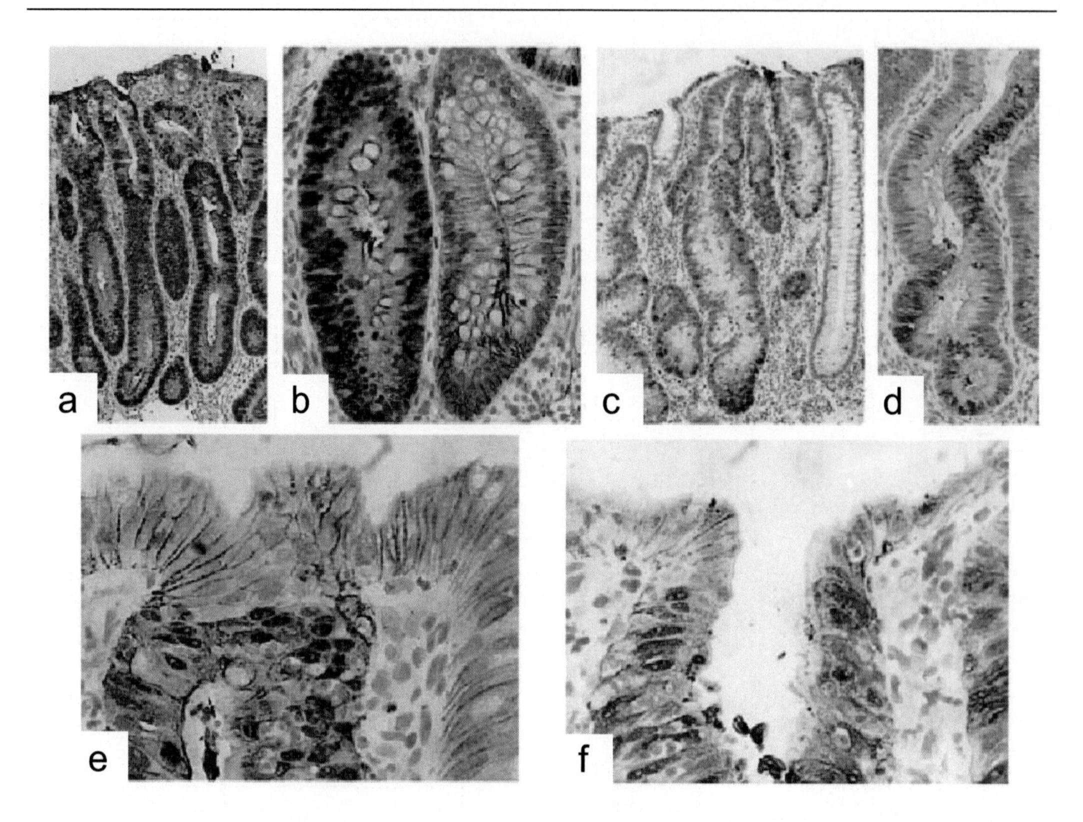

Figure 11. β-catenin in the nuclei of adenomatous crypts in human colon. (a) Low power view of immunohistochemistry for β-catenin in human adenoma. (b) High power view showing the strong staining for β-catenin in the left adenoma crypt. (c, d) β-catenin in nuclei of budding crypts. (e, f) Serial section of the junction of an adenomatous crypt with the surface epithelium, showing a clear boundary for nuclear expression of β-catenin (Reprinted from Cancer Res 2003 63:3819-3825 with permission from the American Association for Cancer Research, Inc.).

Crypt fission was rare in normal and noninvolved mucosa and began with basal bifurcation at the base of the gland (Figure 12 A). In adenomas, fission was mostly asymmetrical (Figure 12 B). Budding from the superficial and mid-crypt (Figure 12 B) was commonly seen, and multiple fission events were regularly observed in adenomas, in contrast with normal controls, in which all budding was basal (Figure 12 C).

The crypt fission index (proportion of crypts in fission) for adenomas was significantly greater than in normal mucosa [146]. These findings suggest that (1) the earliest lesion in the development of both sporadic and FAP adenomas is likely to be the monocryptal adenoma, (2) early clonal expansion is through crypt fission, and (3) spread via the surface is likely to be a phenomenon confined the later stages of evolution of colorectal adenomas. These findings suggest that the developmental process of the earliest adenomas is very similar to the clonal expansion processes we have described above. 'Top-down' spread of the adenomas may occur at more later stages in the progression.

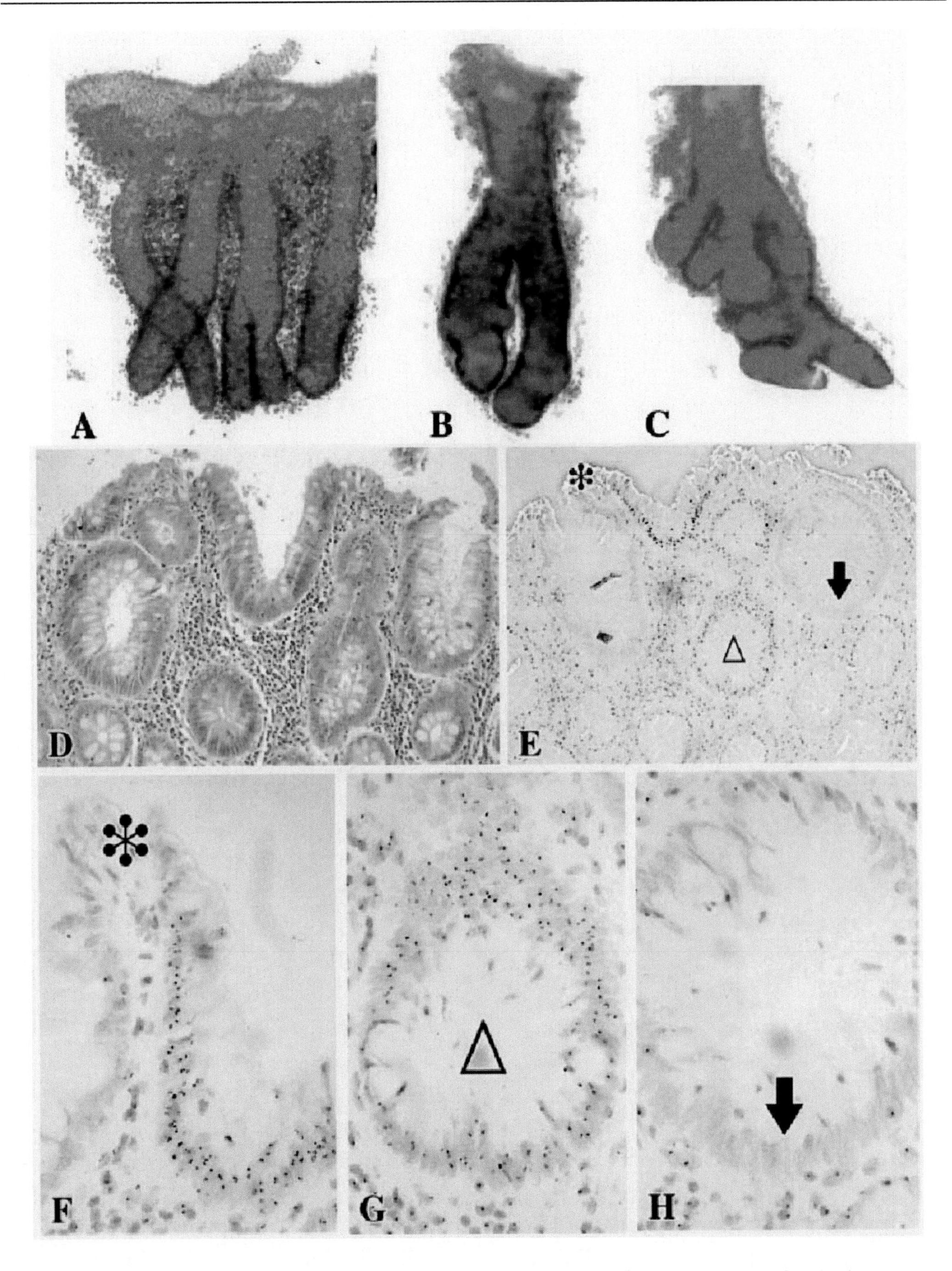

Figure 12. Clonal expansion in adenomas. A-C, Microdissected crypts from normal colonic mucosae and adenomas. A, symmetrical fission of normal colonic crypts. B, a crypt isolated from an adenoma, showing asymmetrical branching. C, another crypt from an adenoma, showing asymmetrical branching and multiple budding. D-H, Colonic sections of a XO/XY individual with FAP. D, HandE staining of a microadenoma. E, adjacent section with non-isotopic *in situ* hybridisation (NISH) staining for the Y chromosome showing adenomatous crypts. F-H, high power views of the labelled adenomatous crypts from E; note the sharp demarcation of Y-positive epithelium showing no overgrowth either way (Reprinted from Cancer Res 2003 63:3819-3825 with permission from the American Association for Cancer Research, Inc.).

Moreover, examination of the adenomas of an XO/XY individual with FAP revealed that monocryptal adenomas all showed either the XO or XY genotype, with no mixed monocryptal adenomas being seen and thus, as expected, monocryptal adenomas are clonal. However 76% of microadenomas were polyclonal [89] (Figure 12 D-H). These findings are supported by observations in Rosa26/Min chimaeric mice [147] which showed that 18 out of 260 adenomas are mixed tumours. But, how can the polyclonality of early adenomas be explained? A possible mechanism could be that adenomatous transformation of non-involved crypts occurs under the influence of transformed crypts; in fact, this was the explanation chosen for the observation that normal crypts were apparently seen in direct continuity with transformed crypts [106].

Crypt Stem Cell Hierarchy

Adult tissue stem cells, like intestinal stem cells, are undifferentiated compared to the other cells in the tissue, but more differentiated than embryonic stem (ES) cells. An adult stem cell undergoes an asymmetrical division to produce an identical daughter cell, and thus replicate itself, and a committed progenitor cell which further differentiates into a more mature cell type. On the other hand, in developing tissues and damaged adult tissues, stem cells temporarily suspend their asymmetrical division and start symmetrical division to augment stem cells [148] (Figure 13).

In the intestinal epithelium several reports indicate that the intestinal stem cells are found in specific zones, just above the crypt base in the small intestine and towards the crypt base in the colon. Crypts are thus composed of a hierarchy of more proliferating cells near the base, to less proliferating cells near the crypt-villus junction. The crypt of the mouse forms a flask-like shape, which is about 16 cells in cross-section, and approximately 250 cells form one crypt. The tissue has a distinct polarity, with new cells created by mitosis in the crypt, migrating predominantly upwards onto the villus or intercrypt table in the colon.

Therefore, any given position along the crypt-villus axis is therefore a reflection of the age of that cell i.e., its position within the lineage or hierarchy. Intestinal stem cells produce transient amplifying cells, but the distinction between these two cell types is difficult to detect before 1 or 2 mitoses have occurred. Developing this concept, Potten *et al* proposed that there are two different stem cell types: *actual* stem cells, which maintain tissue homeostasis; and *potential* stem cells, which are mobilized in case of actual stem cell depletion, and which undergo symmetrical divisions to restore stem cell numbers [7,149,150]. They also proposed a three-tiered hierarchial system of stem cell organization. In the first tier, actual stem cells reside in the crypt base, which are extremely sensitive to DNA damage, cannot repair the damage from1 Gy gamma irradiation [151], and subsequently undergo apoptosis mediated by p53 [152]. In the second and third tiers there are potential stem cells with more effective repair for irradiation. The clonogenic cells in the second tier can survive 1 Gy but undergo apoptosis at higher doses of irradiation. These cells retain some stem-cell function, and can become actual stem cells if required. Additionally, these cells also appear to have acquired the ability to repair their DNA, involving p53-mediated cell cycle arrest. Ultimately, these cells may regenerate the first-tier compartment and a normal crypt.

At higher doses of irradiation, the additional radioresistant cells in the third tier survive. These cells are the final clonogenic resource.

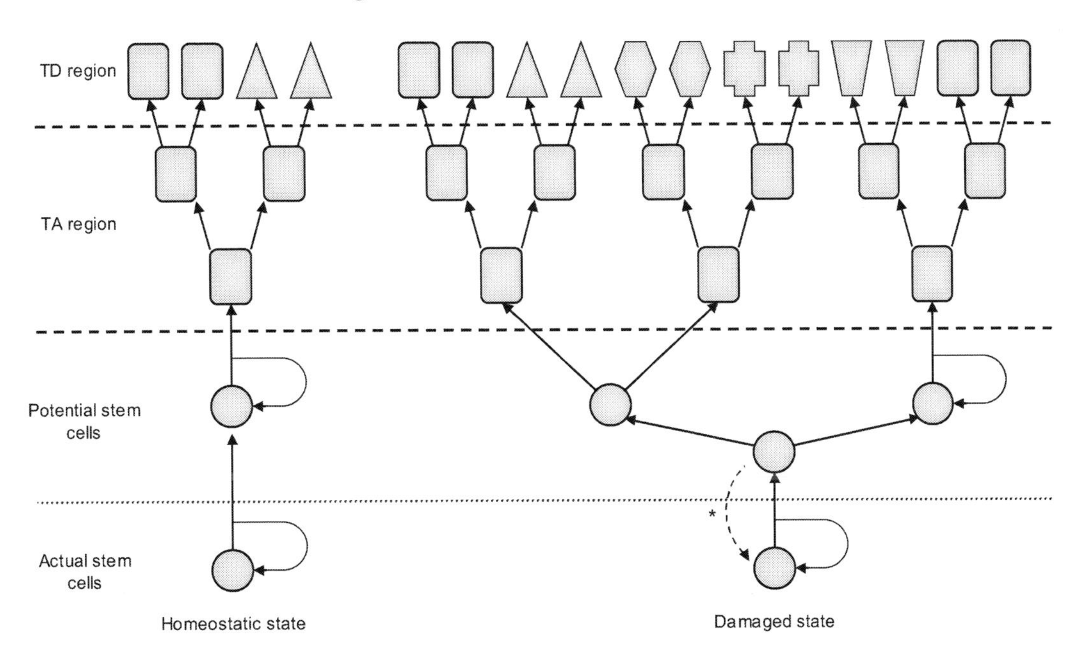

Figure 13. A model for stem cell hierarchy. In a homeostatic state, actual stem cells undergo cell division in an asymmetrical manner. However, in a damaged state, actual stem cells undergo a symmetrical division to maintain the integrity of the epithelium. Actual stem cells maybe especially sensitive to DNA damage. In the case of stem cell depletion, potential stem cells are thought to become actual stem cells (*) to reproduce their progeny (Redrawn and modified from [148]).

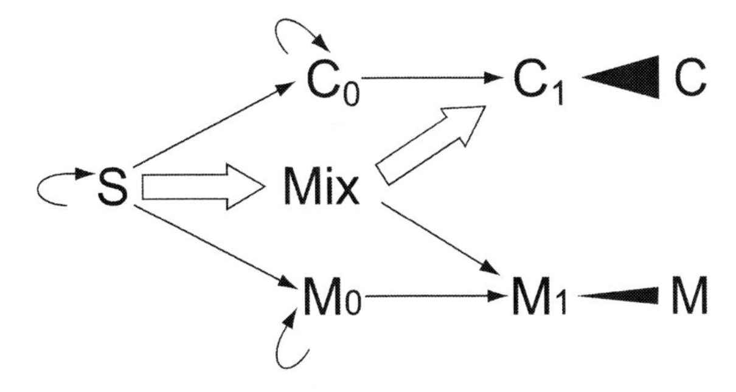

Figure 14. Cell lineages in the colon. Lineage diagram showing columnar (C) and mucous (M) cell production, generated by the stem cell (S). Large arrows indicate dominant pathways. Mix can produce both columnar and mucous short-lived progenitors (C_1 and M_1 respectively). M_0 and C_0 are long-lived (months) mucous and columnar progenitors with mucous and columnar phenotypes, respectively. The arrowheads leading from C_1 to C and M_1 to M indicate potential amplification of the population by several divisions of the respective committed progenitor (Redrawn from [135]).

However, little is known about the distinction between actual stem cells and potential stem cells; although the presence of actual stem cells can be inferred [153], the possibility cannot be excluded that actual stem cells can be induced from potential stem cells by extrinsic factors such as the specific stem cell microenvironment, the 'niche'. The number of stem cells located within the crypts of the small intestine and colon is not known precisely.

However, estimates have been made based upon cell cycle times, tissue regeneration studies and the pattern of expression of cells of differing genotypes within a crypt. In the murine small intestine, there are thought to be 4-6 actual stem cells in the first tier and about 30 potential stem cells (approximately 6 in the second tier, with 16-24 in the third tier) [103,154]. Bjerknes *et al* (1999) also proposed about 4-5 long-lived progenitors in a single crypt of the murine intestine using the *Dlb*-1 assay described above [135]. In this model, there are 4 to 5 long-lived progenitors, which consist of mucous cell progenitors (M0), columnar cell progenitors (C0), and multipotential stem cells (S) capable of giving rise to all epithelial cell types. These long-lived progenitors can differentiate into short-lived progenitors which remain for a few days (C1, M1, and Mix), before yielding one or two cell types (Figure 14).

Maintenance of Stem Cell Integrity

Stem cells undertake cell replacement and restoration of injury throughout an individual's life and thus intestinal stem cells probablydivide1000 and 5000~6000 times throughout mouse and human lifetimes, respectively. Intestinal stem cells need to keep dividing without any mutations which can lead to carcinogenesis, which is especially true in the small intestine [149], although not for the colon.

(1) Selective Segregation of Template DNA Strands (Cairns' Hypothesis)

Replicating of DNA is at its most susceptible state for mutations to occur. In 1975, Cairns proposed that, when undergoing asymmetrical division, stem cells retain the parental (template) strand within the new stem cell, and the newly synthesised strand is passed onto the daughter cell destined for differentiation [155]. If the DNA strands containing errors are incorporated into the non-stem daughter cells, and those cells continue dividing and differentiating, eventually becoming apoptotic or are shed from the villus and lost. Potten *et al* (1978) demonstrated that stem cells have the ability to retain tritiated thymidine (^{3}HTdR), suggesting that such label-retaining stem cells pass selective DNA onto their progeny [156]. This proposal was supported by experiments using ^{3}HTdR and bromodeoxyuridine (BrdU) to double-label DNA [153]. To produce the new actual stem cells labelled with ^{3}HTdR, they gave ^{3}HTdR after irradiation with 8 Gy, which is enough to kill the actual stem cells. Subsequently, after an estimated stem-cell cycle time of 8 days, BrdU was administrated. After one more stem-cell cycle of 8 days after BrdU administration, none of the ^{3}HTdR labelled cells contained BrdU; however, ^{3}HTdR-labelled stem cells were still present at the crypt base. This observation can be explained by the ability of the stem cells to selectively segregate and retain the ^{3}HTdR-labelled template strands of DNA and pass the new strand onto their progeny – in accord with the Cairns' hypothesis.

(2) The Role of p53 and bcl-2 in Maintaining Stem Cell Integrity

As mentioned above, stem cells are thought to retain the parental strand, and are long-lived populations, which means that the possibility of mutations occurring in the parental strand is increased with age, and that multiple mutations are likely to accumulate in stem cells. To prevent these mutations from being passed on to their progeny, stem cells carrying DNA errors introduced into the template strands by genotoxic agents and irradiation could undergo altruistic cell suicide, or apoptosis. Several studies indicate that the ultimate, actual, stem cells do not attempt to repair such damage but rather initiate a p53-dependent apoptosis [151,152,157]. p53 also regulates 'cell cycle arrest' (an event which could allow the cell time to repair) among potential stem cells. The genes involved in this processes still remain to be fully identified but certainly include p53 and related genes and the anti-apoptotic gene bcl-2 family.

Merritt *et al* (1994) demonstrated that a very small dosage of irradiation (0.01-0.05 Gy) induced marked p53-dependent apoptosis amongst small intestinal basal crypt cells within the first 24 hours [152]. The number of apoptotic cells increased as the dose was raised to 1 Gy. At this point approximately 6 cells per crypt were killed and above this dose few additional cells could be seen to die via apoptosis over the first few hours. Within colonic crypts a more diffuse pattern of apoptosis was seen, not specifically localised to the putative stem cell compartment. These data suggest that small intestinal stem cells have a lower threshold for initiating apoptosis in response to DNA damage and do not attempt to effect DNA repair.

In both the small intestine and colon, p53 protein is expressed 2-4 hours after irradiation exposure , and in the small intestine its expression, in terms of time and cell position, is coincident with that observed for apoptosis [152]. However, it is not expressed in many of the apoptotic cells but can be found in other cells in stem cell positions. The p53-related gene, $p21^{waf/cip1}$ is also expressed at this time, broadly over the same cell positions within the crypt. $p21^{waf/cip1}$ protein mediates p53-dependent cell cycle arrest [158-161] and inhibits cell cycle progression by binding to and inhibiting the function of cyclin-dependent kinase and proliferating cell nuclear antigen [162,163]. In this context, p53 plays a role in regulating cell fate, either positively or negatively.

The anti-apoptotic gene bcl-2 is expressed at the base of murine and human colonic crypts, whereas expression is not seen in small intestine [164,165]. In the colon, bcl-2 prevents apoptosis and it is thought that the colonic stem cell relies on cell cycle check-point genes such as p53 and p21 to arrest cycle progression; this may increase the carcinogenic risk in the colon.

IDENTIFICATION OF INTESTINAL STEM CELLS

Many studies using mouse models have suggested that intestinal stem cells exist in the crypt base, just superior to the Paneth cells (approximately the 4th or 5th cell position in mice), in the small intestine, and in the crypt base in the colon [134]. This was first suggested by Cairnie *et al* (1965) [166,167] in the studies of the variation in cell cycle time with cell position in the crypt using tritiated thymidine (^{3}HTdR) labelling. Those basal crypt cells that retain label for longer than the average (*label-retaining cells, LRC*) have divided less frequently and are slowly cycling: other cell kinetic approaches and modelling studies have suggested that stem cells are more slowly cycling, with a prolonged cell cycle time inlcuding studies on the label-retaining cells in these basal locations (Figure 15 (a, b)).

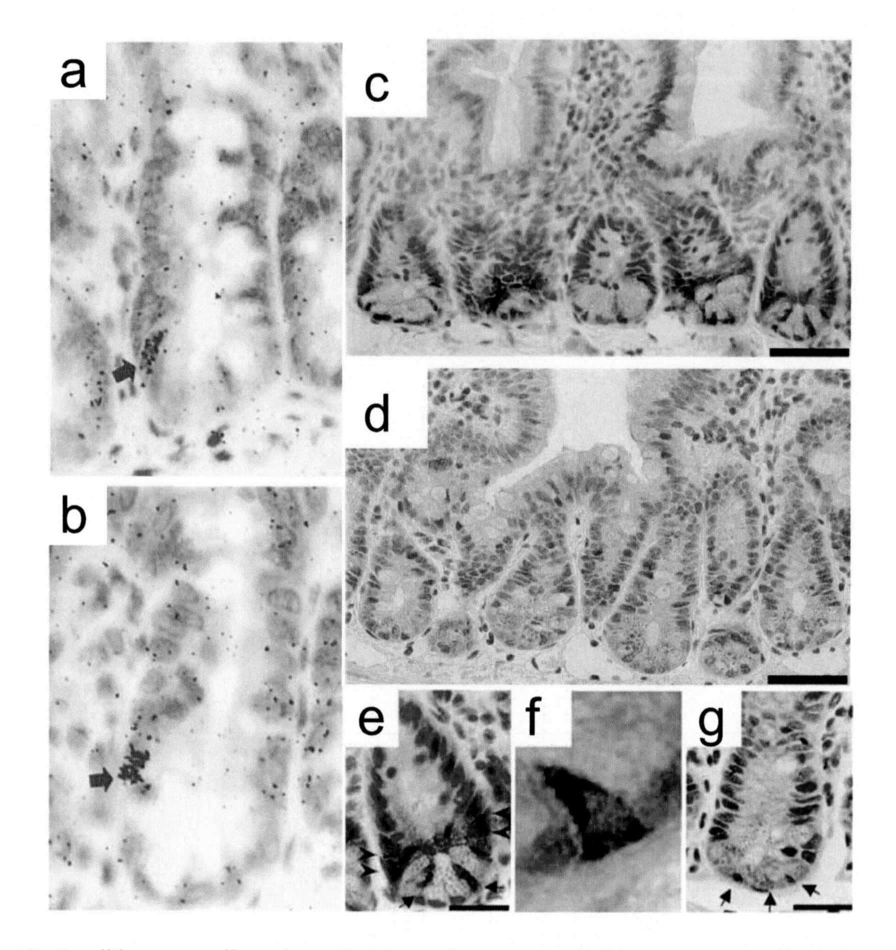

Figure 15. Possible stem cell markers. Sections of mouse small intestinal crypts displaying putative stem cell markers. (a, b) A crypt which had been irradiated (8Gy), followed by tritiated thymidine application every 6 hours for 2 days and then examined 8 days later. Differential DNA strand segregation can cause the parental (template) strand to be retained with the stem cell, and the newly synthesised strand to be passed onto the daughter cell, which is destined for differentiation. At cell positions 4th to 5th, label-retaining cells are clearly labelled (Reprinted from The intestinal epithelial stem cell, Marshman E, Booth C and Potten CS, Bioessays, 24:91-98, (2002) by permission of Wiley-Liss, Inc., a subsidiary of John Wiley and Sons, Inc.). (c-f) Immunohistochemical analysis of Musashi (Msi)-1 (c-e) and Hes-1 (e, f) in adult mouse small intestine. (c, d). A few crypt cells just above the Paneth cells (arrowhead) and the crypt base columnar cells between the Paneth cells (arrow) were stained with Msi-1 antibody. Msi-1 is absent in the Paneth cells and the villous epithelium. (e) High power view of an Msi-1 positive cell showing the Msi-1 reactivity is present in the cytoplasm, not in the nucleus. (f, g) Hes-1 was expressed dominantly in mid to lower crypt cell nuclei. Goblet and Paneth cells are negative for Hes-1 antibody. Crypt base columnar cells are also stained with Hes-1 antibody (arrow) (Reprinted by permission of Federation of the European Biochemical Societies from Candidate markers for stem and early progenitor cells, Musashi-1 and Hes1, are expressed in crypt base columnar cells of mouse small intestine, by Kayahara T et al, FEBS Letters 535:131-135 (2003)).

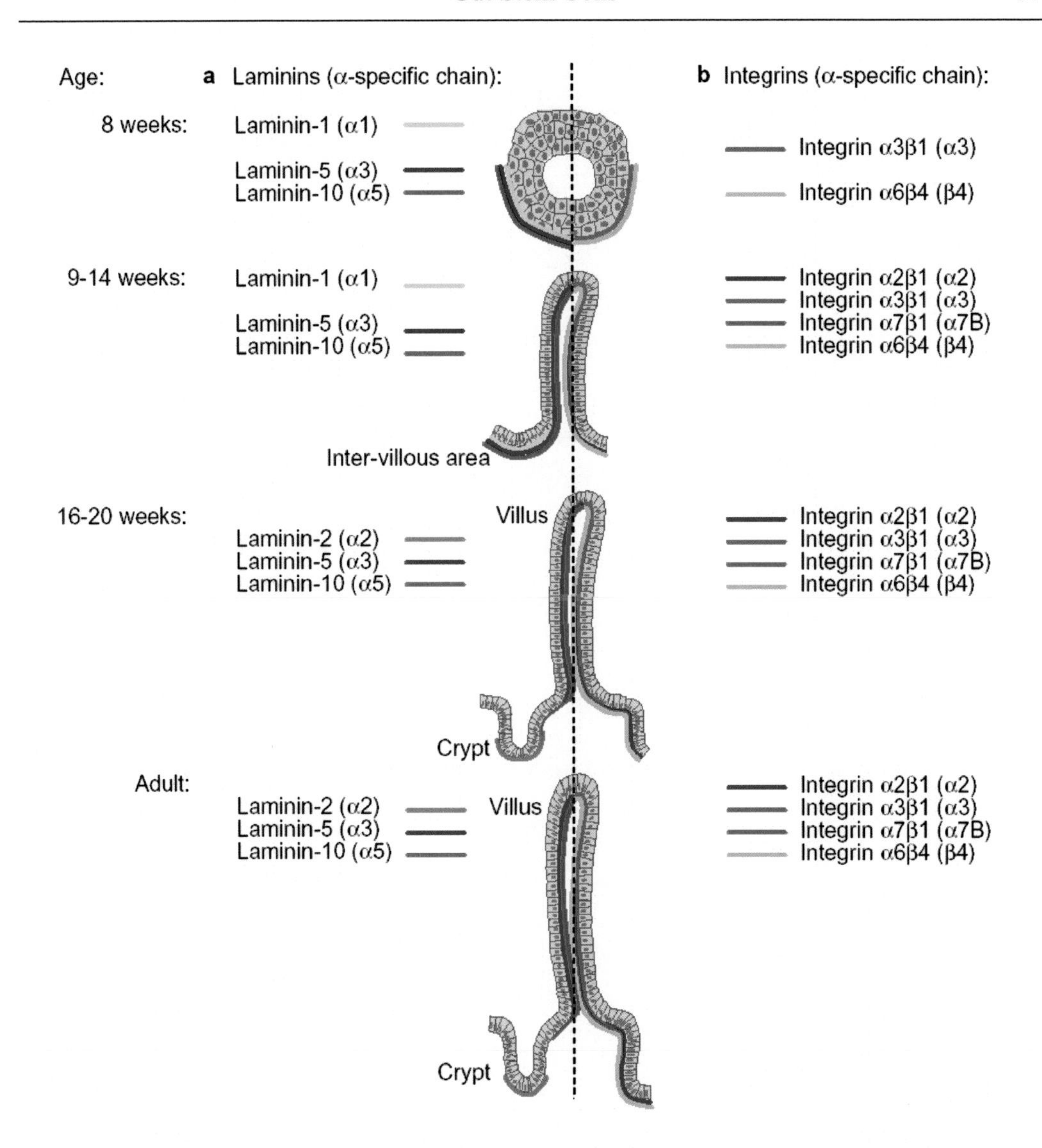

Figure 16. The expression of laminins and integrins along the crypt-villus axis in the developing and adult small intestine. (a) Laminins (left). Laminin-1, -5, and -10 are present early during embryogenesis. However, from the time that the crypts start to form (16 weeks), laminin-1 is gradually replaced by laminin-2, while laminin-5 and -10 become more restricted to the villus. (b) Integrins (right). Integrin $\alpha_2\beta_1$ is not detected in the gut anlage. From10 weeks, it is present in the inter-villous axis, and gradually confined to the crypts around 16 weeks onwards. The integrin $\alpha_6\beta_4$ is found ubiquitously throughout the crypt-villus axis (Reprinted from Interactions between laminin and epithelial cells in intestinal health and disease, Teller IC and Beaulieu JF, Expert Reviews in Molecular Medicine, 28:1-18, (2001) with permission from Cambridge University Press).

Recently, several molecules have been reported to be expressed specifically in the predicted stem cell region within the crypt [125,168,169]. β-catenin is a ubiquitously expressed molecule within the intestinal epithelium, but β-catenin located within the nucleus is reported to be restricted to the crypt base region [125]. This suggests that the intestinal stem cell requires constitutive Wnt signalling also supported by the finding that Tcf-4 knockout mice fail to develop an intestinal stem cell region [170]. Musashi (Msi)-1 is an

RNA binding protein reported to be enriched in neural stem cells [171]. This molecule appears to have a close relation to stem cell properties, especially in terms of self renewal [172] and recent two studies have revealed that this molecule is expressed predominantly in the basal stem cell region of the mouse intestinal crypt [168,169] (Figure 15 (c, e, f)). One of the molecular functions of Msi-1 as an RNA-binding protein is to modulate Notch signalling by repressing the expression of a Notch signal repressor molecule, m-Numb, in a post-transcriptional manner [173]. Kayahara *et al* (2003) examined this in the mouse small intestine, and reported that Hes-1, a major transcription factor downstream of Notch signalling, was also expressed in the intestinal crypt stem cell region [169] (Figure 15 (d, g)). This suggests that Notch signalling, as well as Wnt signalling, is important in the intestinal crypts of mice, and that this may be regulated or modified by Msi-1. These molecules might be used as stem cell markers for identifying and isolating intestinal stem or progenitor cells.

Other putative markers for gastrointestinal stem cells may include the integrin family, which are important mediators of cell-laminin interactions. Integrins are differentially expressed along the crypt-villus axis, except for the integrin $\alpha_6\beta_4$, which is thought to be ubiquitously expressed by intestinal cells. In the lower region of the crypts, the $\alpha_2\beta_1$ integrin is expressed [25] (Figure 16).

ISEMFs Maintain a Gastrointestinal Stem Cell 'Niche'

Stem cells within many different tissues are thought to reside within a 'niche' formed by a group of surrounding cells and their extracellular matrix (ECM), which provide an optimal microenvironment for stem cells to function. Within a niche, a lost stem cell can be replenished when a remaining stem cell divides symmetrically [174]. Moreover, niches can be colonized after the transplantation of isolated stem cells [175] or the immigration of cells from other niches [176,177]. The ability of niches to dynamically modulate cellular behaviour suggests that they may play important roles in cellular transdifferentiation [178-180] and in proliferative disorders [181]. Kai *et al* (2003) [182] studied *Drosophila* ovarioles, which maintain two to three germline stem cells in a niche requiring adhesive stromal cap cells. After experimentally emptying the germline stem cell niche, cap cell activity persists for several weeks. Subsequently follicle cell progenitors, including somatic stem cells, enter the niche and cap cells provide support to the incorporated cells. In this context, it becomes a matter of great significance that ISEMFs influence epithelial cell proliferation and regeneration through epithelial-mesenchymal cross-talk, and ultimately they may determine epithelial cell fate. ISEMFs express several growth factors including Wnt, while the receptors are found on the epithelial cells, underlining their pivotal role in maintaining the stem cell niche.

Role of the Epithelial Stem Cells in the Restoration of Damaged Epithelium

Irradiation is one of the classical experimental models for producing epithelial injury. At higher radiation dose levels exceeding 9 Gy only single cells survive in each crypt, resulting regeneration of all epithelial cell populations of that crypt from *microcolonies* which regenerate from these single cells [183]. Furthermore, humoral factors such as Fibroblast Growth Factor (FGF)-7 (KGF; Keratinocyte Growth Factor) [184], FGF-2 (b-FGF; basic-FGF), prostaglandins [185,186], and Interleukin (IL)-11 [187,188] are reported to attenuate epithelial injury induced by chemical agents or irradiation, although, it is not clear whether they are acting directly on epithelial stem cells. Other models indicate that several growth factor families such as Transforming Growth Factor (TGF)-α/Epidermal Growth Factor (EGF), and TGF-β play an important role in epithelial restoration. In particular, TGF-β is though to be a powerful modulator of epithelial cell migration and regulates the epithelial proliferation [189]. On the other hand, Trefoil Factor family (TFF) peptides, secreted from goblet cells distributed in the intestine, regulate restitution via TGF-β-dependent pathways. These two humoral factors, TGF-β and TFF, may cooperate with each other to restore mucosal integrity.

Moreover, Cheng *et al* (2001) reported that Glucagon-Like Peptide-2 (GLP-2) attenuates indomethacin-induced colitis and also promotes proliferation and differentiation of the long lived progenitor cells described above. The receptor for GLP-2 was located on enteric neurons [190] and GLP-2 activation of these neurons produces a rapid induction in c-Fos expression which signals growth of columnar epithelial cell progenitors and stem cells, since both cell types give rise to adult columnar cell types; there was no effect on the mucous cell lineage, which is stimulated by KGF [190]. Thus, committed progenitor cells are involved in regeneration of damaged epithelium, possibly via a neural regulatory pathway.

Thus most of our knowledge emanates from *in vivo* experiments: the difficulty of long-term culture of purified and isolated normal intestinal epithelium does not yet allow the investigation of these cells *in vitro*.

The Plasticity of Adult Bone Marrow Stem Cells: Gastrointestinal Aspects

The plasticity of the adult tissue stem cells is a topic of heated controversy. Tissue stem cells were previously regarded as organ-specific stem cells, but recent studies have demonstrated unequivocally that the bone marrow stem cells can transform into different cell lineages, including hepatocytes [180,191-193], biliary epithelial cells [194], skeletal muscle cells [195], cardiomyocytes [196], central nervous system cells [197], renal tubular epithelial cells [198], and cells which make up the gastrointestinal mucosa [199,200]. Reverse transformation is also reported: muscle and neuronal stem cells can form bone marrow [201,202], and persuasively Shen *et al* (2000) demonstrated that pancreatic cells can be converted into hepatocytes *in vitro* [203]. Moreover, Clarke *et al* (2000) showed that neural stem cells from adult mouse brain can contribute to the formation of chimaeric chick and mouse embryos and give rise to cells of all germ layers [204].

However, there are problems with reproducing some of these phenomena bone marrow-derived mesenchymal stem cells, so-called multipotent adult progenitor cells (MAPCs), can also contribute to most somatic cell types [205] and transformation of haematopoietic stem cells has been claimed to be an extremely rare event [206]. Moreover, cell fusion, first shown in ES cells using an *in vitro* model [207,208] has now been demonstrated by *in vivo* studies in mice with a fatal metabolic liver disease (fumarylaceroacetate hydrolase deficiency). Bone marrow cells from normal donors generate healthy hepatocytes by forming hybrid cells that contain both donor and host genes [209,210], with most of the fused cells possessing a tetraploid or hexaploid DNA content. In contrast, cytogenetic analysis of bone marrow-derived cells specific to solid organs such as kidney and muscle in allogeneic transplant studies reveal a diploid karyotype [211,212]. Donor-derived, solid-organ-specific cells in patients who had received allogeneic bone marrow or peripheral-blood stem-cell transplants have also been identified as diploid [193,194,213-215], except in liver tissue, where polyploidy in not uncommon. Furthermore, numerous cytogenetic analyses of bone marrow-biopsy specimens from patients undergoing allogeneic stem-cell transplantation have been characterized by euploidy, except in diseased tissue. However, it is possible that hybrid cells undergo a reduction division, thus converting the hyperploid cell into a diploid karyotype and thereby concealing the fusion event [209]. Further studies of experimental liver damage in mice, showed that about 50% of donor engraftment events are non-fusions [216].

Thus, cell fusion appears to account for a proportion of cells in the tissues of solid-organ that show donor characteristics, but not completely. Fusion could even be seen as a physiological ongoing repair mechanism by which cells deliver healthy and new genes to highly specialized cells to prevent them from dying and to correct genetically-defective cells.

(1) Epithelial Cells

The possible contribution of the bone marrow to the intestinal epithelium has been examined for many years. This possibility denied by Cheng and Leblond (1974) using a tritiated thymidine-labelled bone marrow transfusion model [71] and similar experiments by Cairnie (1976) using a genetic marker [217]. Thus the 'Unitarian Theory' proposed by Cheng *et al* (1974) has been widely accepted; all of the intestinal epithelial cells are provided by intestinal stem cells.

However, there are several reports that bone marrow stem cells (BMSCs) can repopulate both epithelial and mesenchymal lineages in the gut [200,213,218]. Okamoto *et al* (2002) claimed that BMSCs contribute to the human gastric and intestinal epithelium [200] (Figure 17) while Korbling *et al* (2002) also reported that donor-derived cells exist in the gastric epithelium after peripheral-blood stem cell transplantation [213]. Moreover, Krause *et al* (2001) demonstrated that a single bone marrow-derived stem cell can engraft into the intestinal epithelium in the mouse [218].

In perhaps the most convincing report [200], the same patient with graft versus host disease (GVHD) was followed at different time points and a 10 to 15-fold increase in engrafted bone marrow-derived cells in the small intestine was observed in the recovery phase. There was also a 40 to 50-fold increase of engrafted bone marrow derived-epithelial cells in the regenerative epithelium of gastric ulcers among bone marrow transplanted patients. However it is not clear whether the bone marrow derived-cells were engrafted as short-lived epithelial cells or as long-lived epithelial cells which were still undifferentiated and multipotent, although the absence of clonally-derived units would suggest that crypt stem cells are not so formed.

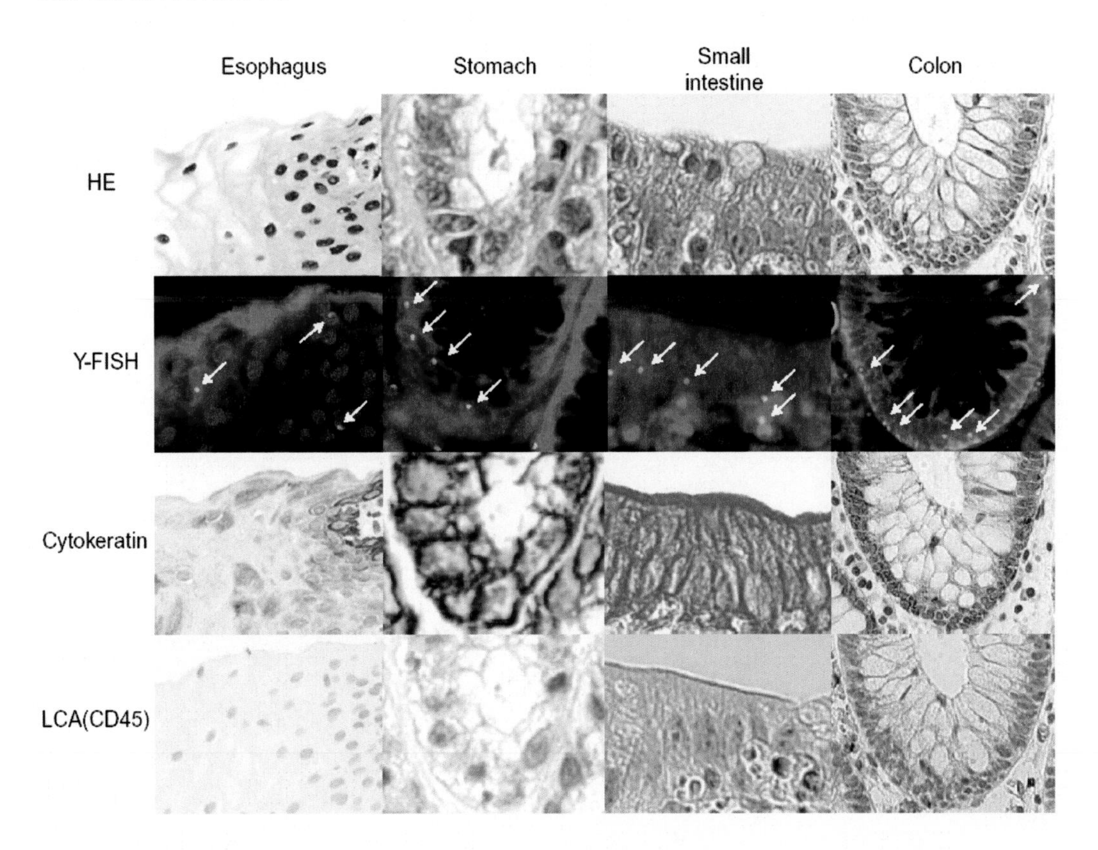

Figure 17. Contribution of bone marrow-derived cells into human gastrointestinal epithelia. Serial-section analysis by HandE staining, *in situ* hybridisation for Y-chromosome (Y-FISH), and immunoperoxidase staining in endoscopic biopsy specimens of gastrointestinal tract tissues from female recipients 26-381 days after BMT from male donors. Y-FISH positive cells (white arrows) are shown in the tissue from the oesophagus (magnification ×800), stomach (magnification ×1600), small intestine (magnification ×1200) and colon (magnification ×800) (Reprinted from Nat Med 2002 8:1011-1017 with permission from Nature Publishing Group).

(2) Intestinal Subepithelial Myofibroblasts (ISEMF)

Brittan *et al* (2002) analysed the colons and small intestines of female mice that had received a bone marrow transplantation from male donors, as well as gastrointestinal biopsies from female patients with graft-versus-host disease following bone marrow transplantation from male donors [199]. BMSCs frequently engrafted into the mouse small intestine and colon, and differentiated to form ISEMFs within the lamina propria. *In situ* hybridisation confirmed the presence of Y-chromosomes in these cells, and their positive immunostaining for α-smooth muscle actin (α-SMA), with negativity for desmin, the mouse macrophage marker F4/80, and the haematopoietic precursor marker CD34, determined their phenotype as pericryptal myofibroblasts in the lamina propria and were derived from the transplanted bone marrow (Figure 18). This engraftment occurred as early as 1 week after bone marrow transplantation (BMT), and almost 60% of ISEMFs were bone marrow-derived at 6 weeks after BMT, thereby indicating that transplanted bone marrow cells are apparently capable of a sustained turnover of the ISEMFs in the lamina propria.

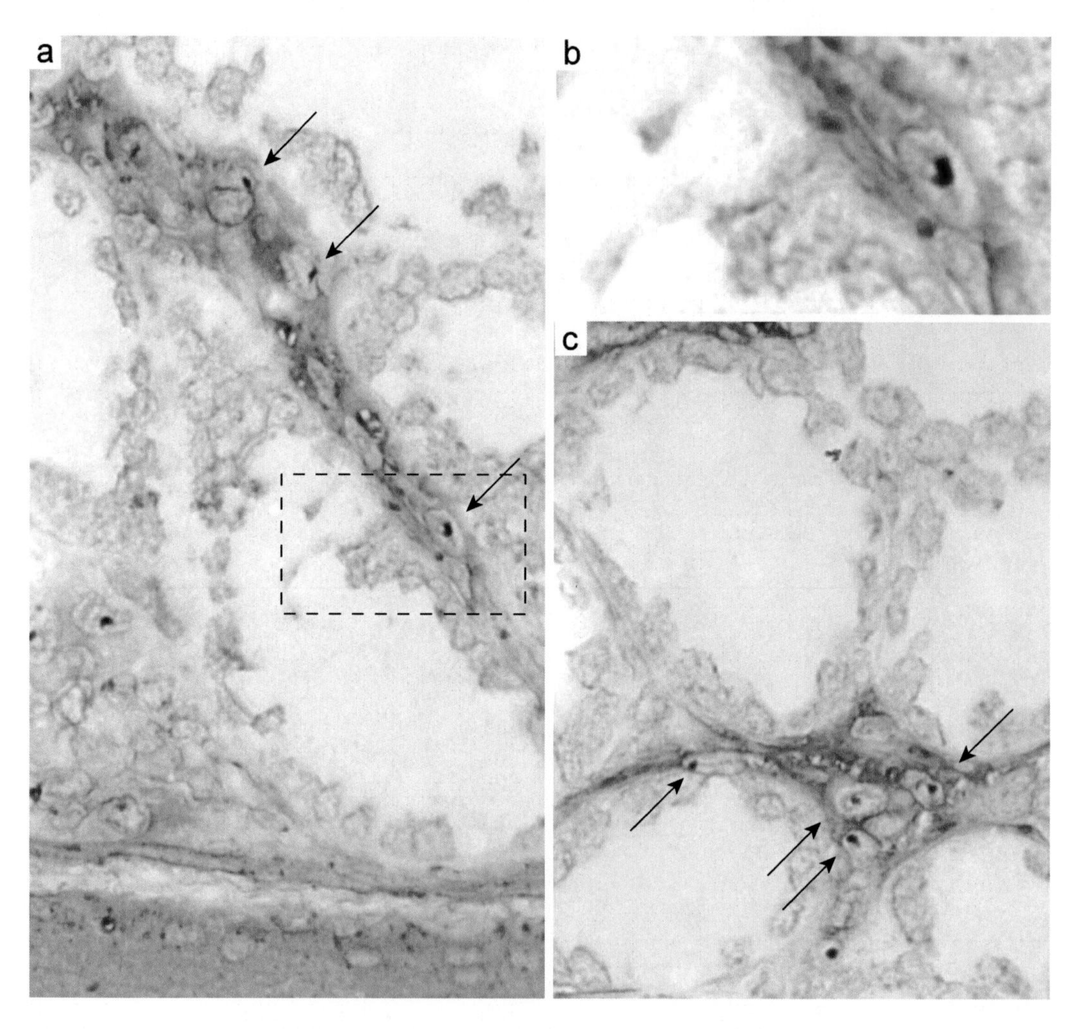

Figure 18. Contribution of bone marrow-derived cells into murine colonic subepithelial myofibroblasts. (a) Low power view picture showing immunohistochemistry for α-smooth muscle actin combined with *in situ* hybridisation for Y-chromosome. Double positive cells (arrows) displaying engraftment of donor-derived cells into ISEMFs. The boxed area is magnified in (b). (c) A large number of ISEMFs are donor-derived (Courtesy of. M. Brittan).

In the human gut, Y-chromosome positive ISEMFs were also found in the small intestinal biopsy specimens for suspected graft-versus-host disease from female patients who had received unprocessed bone marrow transplants from male donors. In mice donor-derived myofibroblasts were also found in stomach: approximately 65% of myofibroblasts were donor derived [219]. Relatively high levels of engraftment into murine and human myofibroblasts suggest that myofibroblasts may generally derive from bone marrow stem cells, although whether from haematopoietic stem cells or mesenchymal stem cells is yet to be determined.

Molecular Mechanisms Regulating Epithelial Proliferation and Differentiation

Recent studies using transgenic mice have revealed considerable information on the molecular aspects of the control of epithelial proliferation and differentiation. Broadly, there are three main signalling pathways: (1) Delta/Notch signalling, (2) Wnt signalling and (3) other pathways which include GTP binding proteins and the other transcription factors. These pathways interact with each other, mutually regulating proliferation and differentiation. Here we review how Notch signalling leads multipotent stem cells towards differentiation, and how Wnt signalling keeps stem cells in an undifferentiated state. Interestingly suggest these two signalling antagonize each other to regulate proliferation and differentiation in both *Drosophila* [220] and mammalian vertebrae [221]. Furthermore, Notch and Wnt signalling hold presenilin-1 in common, which is required for the proteolytic processing of both β-catenin and the Notch receptor, which translocate into the nucleus to activate transcription of downstream genes [222-226]. These two signalling pathways may thus modulate each other in the cytoplasm.

(1) Delta/Notch Signalling

As mentioned above, the genome of stem cells is extremely well protected against DNA-replication-induced errors. Delta/Notch signalling is thought to play an important role in stem cell division because Musashi (Msi)-1, regulator of asymmetrical division, is expressed specifically in putative intestinal stem cells and up-regulates Delta/Notch signalling by down-regulating the transcription of m-Numb [227-230] (Figure 19 (a)). Importantly, m-Numb is often segregated between daughter cells during mitosis: m-Numb localizes asymmetrically in mitotic precursor cells, and is subsequently distributed to one of the daughter cells [231,232]. As a result, m-Numb creates a difference in sensitivity to Notch signal activation in the two daughter cells, and also brings about two different cell fates [2] one an undifferentiated cell fate and other designed to differentiate into mature cells.

Moreover, knockout animals show that several transcription factors downstream of Notch signalling, when lost, have been revealed to lead to depletion of intestinal epithelial cells in a lineage-specific manner. The Notch pathway works by using lateral inhibition in cell-cell interactions with its cell membrane-anchored ligand Delta [233]. Feedback amplification of relative differences in Notch and Delta results in subsets of cells with high levels of Notch and others with high Delta levels. Elevated cellular Notch levels induce expression of transcription factors, such as Hes-1 [234] (Figure 19 (b)), and Hes-1 positive cells have been shown to remain in the precursor population [169,235]. Downstream targets of Hes-1 include Math-1, a basic-Helix-Loop-Helix (bHLH) transcription factor. Math-1 null mice have a depleted secretory cell complement (Paneth, goblet and enteroendocrine cells) in the epithelium [236], and thus Hes-1 and Math-1 are major factors involved in intestinal epithelial cell development [236,237]. Hes-1 seems to drive progenitor cells to an absorptive columnar cell fate, whereas Math-1 drives them towards a secretory fate.

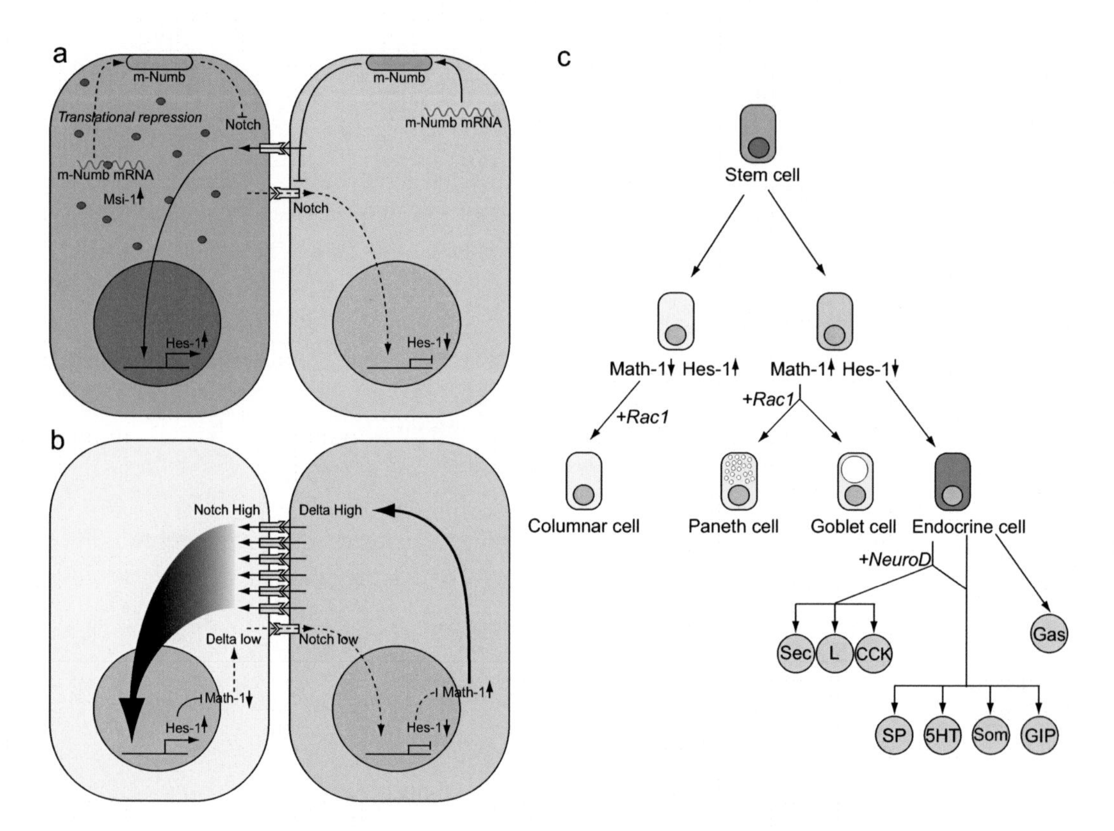

Figure 19. Delta/Notch signalling. (a) A model of Musashi (Msi)-1 function in a precursor population. The Notch signalling pathway is triggered by Delta expressed in a neighbouring cell. m-Numb blocks activation of the Notch signal (right). In Msi-1-expressing immature cells (left), Msi-1 (blue dots in the cytoplasm) binds to m-Numb mRNA and blocks its translation. This potential of Delta/Notch signalling by Msi-1 should maintain the immature proliferation status of cells expressing Msi-1. (b) Lateral inhibition mediated by the Delta/Notch signalling pathway. Two adjacent cells in a precursor compartment are shown. One cell (left), the winner in the lateral-inhibition competition, expresses high levels of Notch and the Hes-1 transcription factor is switched on and the expression of Math-1 and of other 'prosecretory' genes is blocked. The result is that the precursor cells become columnar cells. In cells expressing low amounts of Notch (right), levels of Delta are high, production of Hes-1 is blocked, and Math-1 expression is induced. (c) Math-1 is essential for secretory cells. Whether Math-1-expressing cells descend directly from stem cells or an intermediate progenitor remains unknown. Abbreviations: Sec, secretin; L, glucagon/peptide YY; CCK, cholecystokinin; SP, substance P; 5HT, serotonin; Som, somatostatin; GIP, gastric inhibitory peptide; Gas, gastrin (Redrawn and modified from [236]).

In crypt progenitor stem cells expressing high levels of Notch, Hes-1 transcription is switched on and the expression of Math-1 and of other 'prosecretory' genes is blocked. The result is that the precursor cells become enterocytes. In cells expressing low amounts of Notch, levels of Delta are high, production of Hes-1 is blocked, and Math-1 expression is induced, bringing about the secretory cell fate in small intestine [238].

These results suggest that downstream transcription factors of Delta/Notch signalling act to regulate stepwise the differentiation from multipotent stem cells to all the mature epithelial cell types (Figure 19 (c)).

The other bHLH type transcription factors reported to be associated with cell phenotype in the intestine are Neurogenin-3 and BATA2/neuroD. Neurogenin-3$^{-/-}$ mice are reported to have decreased numbers of neuroendocrine cells in the small intestine [239], and while BETA2/neuroD$^{-/-}$ mice show reduced serotonin- or cholecystokinin-producing neuroendocrine cells [240].

These bHLH transcription factors seem to regulate binary cell fate decisions by suppressing or enhancing the expression of cell lineage-specific genes. In contrast with Wnt signalling, so far Notch signalling seems to have little involvement in the maintenance of stem-cell properties, but, rather, plays a critical role in the cell fate decision during the maturation of epithelial cells.

(2) Wnt Signalling

The Wnt family of signalling protein is known to play an important role in embryogenesis and carcinogenesis. There are now 19 different Wnt genes known in humans and in the mouse. The interaction of extracellular Wnt ligands with the Frizzled (Fzd)/LDL-receptor-related protein (LRP) receptor complex results in increased intracellular levels of β-catenin in the target cell [241-243].

(i) The Canonical Wnt Pathway and Apc

In unstimulated cells, free cytoplasmic β-catenin is destabilized by a multiprotein complex containing axin (or its homologue conductin), glycogen synthase kinase 3 (GSK3) and the *Apc* tumour suppressor [118,244-248] (Figure 20). Axin/Conductin has a scaffold function in this complex, binding to *Apc*, GSK3, and β-catenin. Interaction between axin and GSK3 in the complex facilitates efficient phosphorylation of β-catenin by GSK3 [249], most likely at critical serine and threonine residues in its N terminus [250]. This phosporylation event earmarks β-catenin for ubiquitination by the SCF complex (containing the F box protein βTrCP/Slimb) [251,252], and for subsequent rapid degradation by the proteasome pathway [250], resulting in the low cytoplasmic level of β-catenin. Thus Tcf is repressed.

Once cells are stimulated by Wnt ligands, the cytoplasmic protein Dishevelled is recruited to the membrane [253,254] and, by so far unknown mechanisms, inhibits the Axin complex by direct binding to Axin [248,255]. The unphosphorylated stable β-catenin molecules accumulate in the cytoplasm and translocate into the nucleus. Nuclear β-catenin forms heterocomplexes with Tcf/Lef proteins [118,121,256], and the Tcf/β-catenin heterodimers act as bipartite transcription factors and to activate the expression of specific Wnt responsive genes [122,257] such as c-myc [258], Cyclin D1 [126,259], BMP4 [260], and claudin-1 [261]. A more detailed list can be found at 'The Wnt gene Homepage' hosted by R. Nusse *(http://www.stanford.edu/~rnusse/wntwindow.html)*.

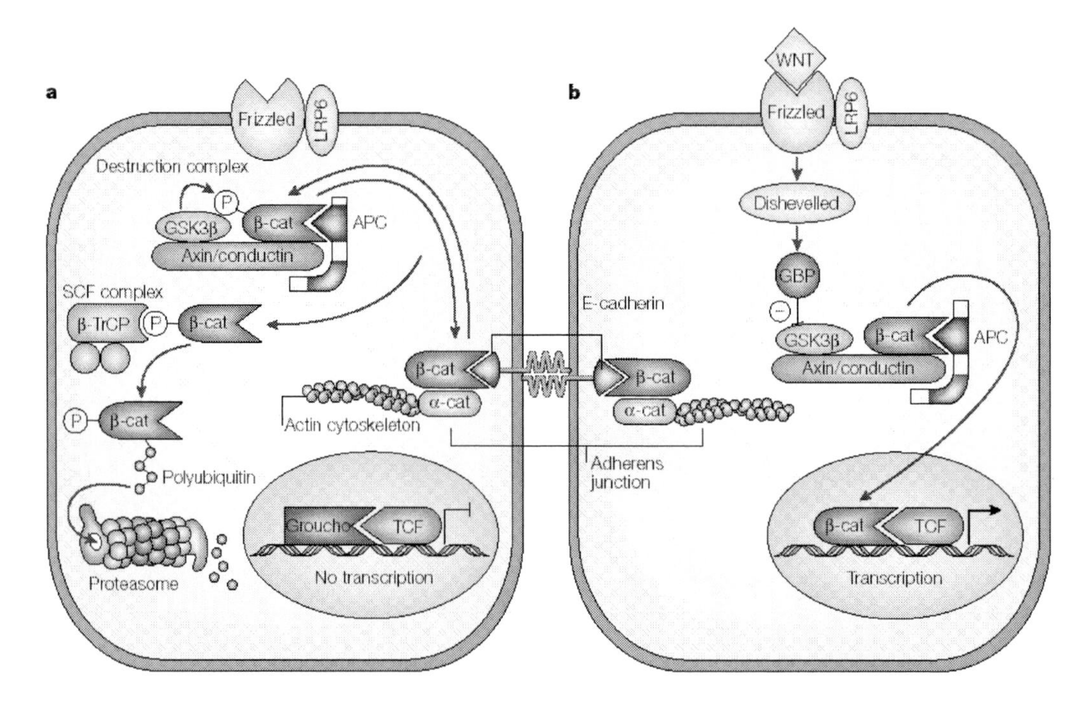

Figure 20. The canonical Wnt signalling pathway. (a) In the absence of a Wnt signal, the level of cytoplasmic β-catenin is minimized by sending it for degradation in the proteasome. Free cytoplasmic β-catenin, which is in equilibrium with β-catenin at adherens junctions, is recruited to a 'destruction complex' containing *Apc*, axin/conductin and glycogen synthase kinase 3 (GSK3β). GSK3β phosphorylates β-catenin, allowing it to be recognized by an SCF complex contaning the F-box protein βTrCP. Other proteins in the SCP complex in addition catalyse a polyubiquitin chain to β-catenin, allowing it to be recognized and degraded by the proteasome. Consequently, β-catenin cannot reach the nucleus, and cannot co-activate Tcf-responsive genes. Groucho also prevents the activation of Tcf-responsive genes in the absence of β-catenin. (b) In the presence of Wnt, its receptor, Frizzled, in complex with LRP6, is activated. This leads to a poorly understood signalling cascade in which Dishevelled activates GBP - an inhibitor of GSK3β. Consequently, β-catenin cannot be targeted for destruction and is free to diffuse into the nucleus, where it acts as a co-activator for Tcf-responsive genes (Reprinted from Nat Rev Cancer 2001 1:55-67 with permission from Nature Publishing Group).

In addition to factors regulating the activation of Wnt pathways, there are some innate molecules inhibiting this pathway, such as the secreted Frizzled-related proteins (sFRP's) which interact directly with Wnt ligands to block their activity. There is some specificity to the Wnt signalling pathway: for example, the antagonists Frzb1 and sFRP2 can both inhibit Wnt1 [262], and while Frzb1 can additionally inhibit Wnt8, sFRP2 cannot, although sFRP2 can inhibit Wnt4. This specificity exists despite the fact that Wnt 1, 4, and 8 are all ligands known to activate signalling through the canonical Wnt pathway [263].

Wnt signalling has long been studied in the development of colon cancer, as it is well known that mutations in the gene *Apc*, have a critical role of Wnt signalling in tumorigenesis [244, 264]. Mutations of the *Apc* tumour suppressor gene are present in up to 80% of human sporadic colorectal tumours [265,266]. This mutation prevents normal β-catenin turnover by the GSK3/Axin/*Apc* complex, resulting in increased nuclear β-catenin/Tcf/Lef gene transcription, and a subsequent increase in β-catenin-induced Tcf/Lef transcription [267,268].

(ii) Role of Wnt Signalling Related Molecules in the Gut

Recent studies have revealed that Wnt signalling has distinct functions within the intestinal crypt, especially in the stem cell region.

Pinto *et al* (2003) investigated the contribution of Wnt signalling to gastrointestinal epithelial proliferation using an adenovirus which mediated Dickkopf1 expression (Ad Dkk1), the soluble canonical Wnt inhibitor [269]. Ad Dkk1 inhibited epithelial proliferation in small intestine and colon, accompanied by progressive architectural degeneration with reduced number and size of crypts and villi [270].

Transgenic and knock-out mice studies of Tcf-4 and Fkh-6, transcription factors downstream of Wnt signalling, and of the Cdx-1 and Cdx-2, homeobox (Hox) genes have been revealing. The Tcf/Lef family of transcription factors has four members; Tcf-1, Lef1, Tcf-3, and Tcf-4. Tcf-4 is expressed in high levels in the epithelium of small intestine, colon, and colon cancers. Tcf-4$^{-/-}$ mice displayed no proliferating cells within their small intestinal crypts and a depletion in the stem-cell compartment in the small intestine [170]. This suggests that Tcf-4 is responsible for establishing stem cell populations within intestinal crypts. Although Lef1 is a transcriptional mediator of canonical Wnt signalling, Lef1 is not normally expressed in the colon, and indeed mice lacking the Lef1 gene do not display any gut abnormalities [271], although Lef1 has been found ectopically expressed in human colon carcinomas [272]. In contrast, Min/+Tcf-1$^{-/-}$ mice demonstrated a 10-fold increase in the formation of adenomatous intestinal polyps compared with Min/+ mice. [273], suggesting that Tcf-1 acts as a feedback repressor of β-catenin/Tcf-4 target genes and thus cooperates with *Apc* to suppress malignant transformation of epithelial cells.

The winged helix/forkhead family of transcription factors are essential for proper development of the ectodermal and endodermal regions of the gut. Foxl1 (forkhead box protein) is expressed in the mesenchyme of the gastrointestinal tract. Foxl1 (previously Fkh6 [274]) null mice showed postnatal growth retardation secondary to severe structural abnormalities of the stomach, duodenum, and jejunum and the dysregulation of epithelial cell proliferation in these organs resulted in an approximately four-fold increase in the number of dividing intestinal epithelial cells and marked expansion of the proliferative zone, suggesting that Fkh6 directs a signalling cascade that mediates communication between the mesenchyme and endoderm of the gut to regulate cell proliferation [275]. Moreover, Foxl1 activates the Wnt pathway by increasing extracellular proteoglycans such as heparan sulphate proteoglycans (HSPG), which act as co-receptors for Wnt [276]. HSPGs are abundant on the cell surface and in the extracellular matrix surrounding gastrointestinal cells and several kinds of HSPGs have been shown to regulate the Wnt signalling in *Drosophila* [277,278] and mammalian vertebrae [279].

The homeobox genes Cdx-1 and Cdx-2 display specific regional expression in developing and mature colon and small intestine. During embryogenesis, Cdx-1 localises to the proliferating cells of the crypts and maintains this expression during adulthood. The Tcf-4 knockout mouse does not express Cdx-1 in the small intestinal epithelium, and thus the Wnt/β-catenin complex appears to induce Cdx-1 transcription in association with Tcf-4 during the development of intestinal crypts [280]. Cdx-2$^{+/-}$ mice develop multiple colonic hamartomatous polyps which do not express Cdx-2, particularly in the proximal colon. These polyps occasionally contain keratinizing stratified squamous epithelium, similar to that occurring in the mouse esophagus and forestomach [281] with heterotopic stomach and small intestinal mucosa in juxta-position, suggesting that Cdx-2 directs endodermal differentiation towards a caudal phenotype and that haploinsufficient levels of expression in the developing distal intestine lead to homeotic transformation of a more rostral endodermal phenotype, such as gastric mucosa that does not express Cdx-2 during normal development [282]. This could indicate a possible homeotic shift in stem cell phenotype. Region-specific genes such as Cdx-1, Cdx-2, and Tcf-4 appear to define the morphological features of differential regions of the intestinal epithelium and regulate the proliferation and differentiation of the stem cells.

Based upon this knowledge, Wnt signalling seems to be closely related to the stem cell properties and cell proliferation of immature cells, at a more early stage than Delta/Notch signalling.

(iii) Wnt Expression and Regulation in the Gut

During embryogenesis, mRNA expression of several Wnt and Frizzled (Fzd) genes has been reported using whole mount *in situ* hybridisation in the chick and the mouse [283,284] (Figure 21). In the human normal colon, differential expression of Wnt5a was noted, with increased expression at the base of the crypts compared with the luminal surface and slightly increased expression in colon cancer. In the normal and malignant colon, the mRNA expression of Wnt 1, 4, 5b, 6, 7b, and 10b were equally detected using *in situ* antisense RNA hybridisation.

On the other hand, Wnt2 mRNA expression was detected only in colon cancer [67]. The upregulation of Wnt2 mRNA in gastric cancer, colorectal polyps and cancers has been also reported [285].

So far, the source of Wnt in the colon is not well documented. However, Wnt proteins are considered to be very important in the maintenance of the stem cell niche, and intestinal subepithelial myofibroblasts (ISEMFs) are believed to produce the Wnt proteins [125]. A recent report has demonstrated that in situ hybridisation revealed the expression of Wnt 2b, 4, 5a and 5b and Fzd 4 and 6 mRNAs in the mesenchyme of the adult murine small intestine and large intestine. In contrast, Wnt 3, 6 and 9b and Fzd 4, 6 and 7, LRP-5 and sFRP-5 mRNAs were observed in intestinal epithelial cells [286].

We have demonstrated expression of Wnt 2, 3, 4, 5a and 5b mRNAs in cultured ISEMFs from C57/BL6 mice, and except for Wnt5a, all Wnt expression was observed only in ISEMFs and not in the isolated crypt epithelium. On the other hand, the expression of Fzd 1, 2, 3, 4, 5, 6 and 7 mRNA was observed both ISEMFs and crypt epithelium, implying that the Wnt proteins secreted from ISEMFs can act not only in a paracrine manner but also in an autocrine manner (Bamba, unpublished).

A very recent report has demonstrated that the inhibition of Wnt1 signalling using a neutralising monoclonal antibody induced apoptosis in human cancer cells, including non small cell lung cancer, breast cancer, mesothelioma, and sarcoma [286], implying a survival role may be important. However, the regulation, roles and interplay of these individual Wnts and Fzds still remains obscure.

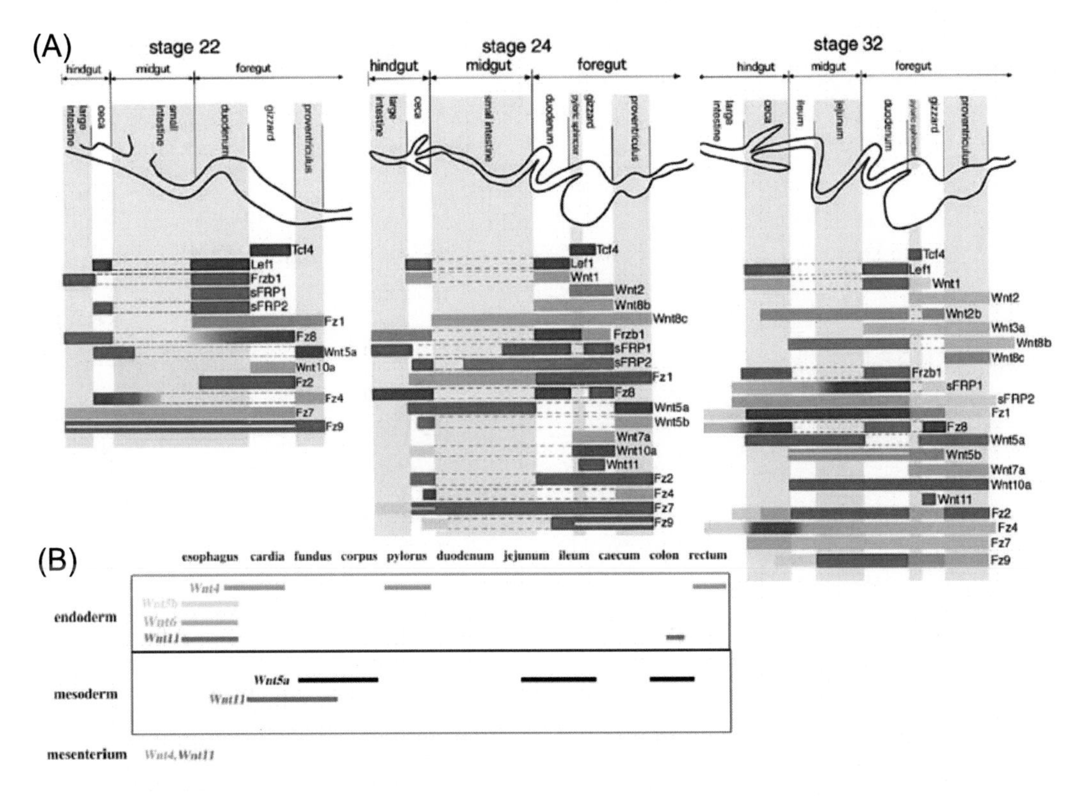

Figure 21. Wnt mRNA expression during embryogenesis. (A) Schematic of the temporospatial expression patterns of Wnt expression along the anterior-posterior axis of E12.5-16.5 mouse embryos (Reprinted from Mechanisms of Development, 105, Lickert H et al, Expression patterns of Wnt genes in mouse gut development, 181-184 (2001) with permission from Elsevier). (B) Expression profile of genes from the Wnt signalling pathway during gut development in chick. Green bars identify expression in the gut mesoderm, while purple bars refer to expression in the gut endoderm. The shades of colour are corresponding to relative intensities of signal from light (low signal) to dark (strong signal) (Reprinted from Developmental Biology, 259, Theodosiou CA et al, Wnt signaling during development of the gastrointestinal tract, 258-271 (2003) with permission from Elsevier).

(3) Others (GTP Binding Proteins and the other Transcription Factors)

Paneth and columnar cell fate choices involve the Rho GTPase family members in addition to Delta/Notch signalling molecules. These factors play a central role in all eukaryotic cells by controlling the organization of the actin cytoskeleton [287]. Rac1 is a member of the Rho family of GTP-binding proteins which can activate the Jun N-terminal kinase (JNK) and p38 mitogen-activated protein (MAP) kinase pathways [288]. Expression of either constitutively active and/or dominant-negative Rac1 forms in mice results in perturbation of cell differentiation in the intestinal epithelium [289]. Sustained Rac1 activation brings about early differentiation of Paneth and columnar cells within the small intestine intervillous epithelium in late fetal mice, but no affect was observed on goblet and endocrine cells. In the adult mouse, forced Rac1 activation increases cell proliferation in intestinal crypts and leads to unusually wide villi [290]. Activated Rac1 specifically increases phosphorylation of JNK in both intervillous and villous epithelial cells and alters the actin cytoskeleton [290]. The cytoskeleton is also involved in cell migration and the position of cells along the crypt-villus axis is one of the important factors regulating cellular differentiation [291].

In addition, Klf4 and E2F4, transcription factors, are reported to affect epithelial development of the gut. Klf4 is a zinc-finger transcription factor expressed in the epithelia of the skin, lung and gastrointestinal tract and several other organs [292-294]. *In vitro* studies have suggested a role for Klf4 in cell proliferation and/or differentiation [295]. $Klf4^{-/-}$ mice demonstrate a 90% decrease in the number of goblet cells in the colon [296].

On the other hand, the E2F family of transcription factors are regulators of cell proliferation, allowing transit from the G_1 to S phase. $E2F4^{-/-}$ mice showed a substantial reduction in the height of the small intestinal villi as well as a reduction in their density, and an increase in the number of goblet cells [297].

Recently, mutations in the LKB1, a gene encoding a serine/threonine kinase, have been implicated in Peutz-Jehger syndrome [298,299], a rare autosomal-dominant disorder characterized by melanocytic macules of the lips, multiple gastrointestinal hamartomatous polyps and a increased risk for several neoplasms, and disturbances in apoptosis of the gut epithelium [300]. The cytoplasmic expression of LKB1 shows a gradient along the villus in the small intestine since LKB1 expression is higher in older epithelial cells compared to newly-differentiated epithelial cells. LKB1 has been shown to regulate the specific p53-dependent cell death pathway in the intestinal epithelium [300], suggesting that LKB1 may control apoptosis in the gut.

Conclusions

As we have discussed, a large number of molecules have been incriminated over the years, which collectively provides an increasingly detailed characterization of epithelial stem cells and a more comprehensive knowledge of gastrointestinal tumorigenesis. However, what we can explain so far may be only the tip of the iceberg in what is actually happening in the gut. For example, we do not even have a definitive stem cell marker in the gastrointestinal epithelium.

In colon tumorigenesis, the accumulation of genetic alterations triggers the adenoma-carcinoma sequence, although several mechanisms exist in crypt stem cells to maintain epithelial homeostasis and integrity. We have argued that the early clonal expansion of adenoma occurs in a 'bottom-up' manner, and 'top-down' spread of the adenomas occurs at later stages in the progression: however, this cannot explain the polycolonality of early adenoma formation. Mesenchymal cells, such as subepithelial myofibroblasts, are involved in aberrant epithelial-mesenchymal interaction which may drive tumour development. In epithelial cancer tissues, stromal cells are known to change their own morphology and genetic expression patterns to influence tumorigenesis [301-303]. However, further studies are requied to confirm the contribution of stromal cells in gut tumorigenesis.

From the viewpoint of clinical medicine, bone marrow transplantation is a good model to examine the contribution of bone marrow-derived cells to tissue homeostasis and repair. In Crohn's disease and ulcerative colitis, several case reports have suggested that allogeneic or autologous bone marrow transplantation occasionally has some ameliorating effect [304-308]. These results could be explained by the elimination of pathogenic inflammatory cells, but the contribution of bone marrow-derived cells into gastrointestinal epithelial or stromal cells may have a role in this process. The number of cells engrafting into gut epithelium is relatively fewer compared to subepithelial myofibroblasts [200]. To increase the contribution of bone marrow cells, the assistance of humoral factors, such as granulocyte-macrophage colony-stimulating factor (GM-CSF) may be required. GM-CSF has been effective in refractory Crohn's disease [309,310]. The mechanism of this therapy is yet to be understood, but this factor may stimulate bone marrow stem cell efflux and encourage their engraftment into damaged epithelium.

ACKNOWLEDGEMENTS

We are grateful to Cancer Research UK for funding this work. We thank Mairi Brittan for her helpful advice during the preparation of this manuscript. One author (SB) is supported by Daiwa Foundation Small Grant, Shiga International Foundation of Medicine, and Mitsubishi Pharma Research Foundation.

REFERENCES

[1] Ahrens E. H., Jr., Blankenhorn D. H., Hirsch J. Measurement of the human intestinal length in vivo and some causes of variation. *Gastroenterology.* Sep 1956;31(3):274-284.

[2] Shen Q., Zhong W., Jan Y. N., Temple S. Asymmetric Numb distribution is critical for asymmetric cell division of mouse cerebral cortical stem cells and neuroblasts. *Development.* Oct 2002;129(20):4843-4853.

[3] Yamashita Y. M., Jones D. L., Fuller M. T. Orientation of asymmetric stem cell division by the APC tumor suppressor and centrosome. *Science.* Sep 12 2003;301(5639):1547-1550.

[4] Shih I. M., Wang T. L., Traverso G., et al. Top-down morphogenesis of colorectal tumors. *Proc. Natl. Acad. Sci. USA.* Feb 27 2001;98(5):2640-2645.

[5] Bannister L. H. Gray's anatomy. In: Williams PL, ed. *The anatomical basis of medicine and surgery.* 38th ed. London: Churchill Livingstone; 1995:1748.

[6] Cheng H., Bjerknes M., Amar J. Methods for the determination of epithelial cell kinetic parameters of human colonic epithelium isolated from surgical and biopsy specimens. *Gastroenterology.* Jan 1984;86(1):78-85.

[7] Potten C. S., Loeffler M. Stem cells: attributes, cycles, spirals, pitfalls and uncertainties. Lessons for and from the crypt. *Development.* Dec 1990;110(4):1001-1020.

[8] Potten C. S., Kellett M., Roberts S. A., Rew D. A., Wilson G. D. Measurement of in vivo proliferation in human colorectal mucosa using bromodeoxyuridine. *Gut.* Jan 1992;33(1):71-78.

[9] Booth C., Potten C. S. Gut instincts: thoughts on intestinal epithelial stem cells. *J. Clin. Invest.* Jun 2000;105(11):1493-1499.

[10] Potten C. S. *function and proliferative organisation of the mammalian gut*. Amsterdam: Elsevier Science; 1995.

[11] Lala S., Ogura Y., Osborne C., et al. Crohn's disease and the NOD2 gene: a role for paneth cells. *Gastroenterology.* Jul 2003;125(1):47-57.

[12] Ayabe T., Satchell D. P., Wilson C. L., Parks W. C., Selsted M. E., Ouellette A. J.. Secretion of microbicidal alpha-defensins by intestinal Paneth cells in response to bacteria. *Nat. Immunol.* Aug 2000;1(2):113-118.

[13] Karam S. M., Leblond C. P. Dynamics of epithelial cells in the corpus of the mouse stomach. I. Identification of proliferative cell types and pinpointing of the stem cell. *Anat. Rec.* Jun 1993;236(2):259-279.

[14] Karam S. M. Lineage commitment and maturation of epithelial cells in the gut. *Front Biosci.* Mar 15 1999;4:D286-298.

[15] Mahida Y. R., Beltinger J., Makh S., et al. Adult human colonic subepithelial myofibroblasts express extracellular matrix proteins and cyclooxygenase-1 and -2. *Am J Physiol.* Dec 1997;273(6 Pt 1):G1341-1348.

[16] Okuno T., Andoh A., Bamba S., et al. Interleukin-1beta and tumor necrosis factor-alpha induce chemokine and matrix metalloproteinase gene expression in human colonic subepithelial myofibroblasts. *Scand. J. Gastroenterol.* Mar 2002;37(3):317-324.

[17] Rogler G., Gelbmann C. M., Vogl D., et al. Differential activation of cytokine secretion in primary human colonic fibroblast/myofibroblast cultures. *Scand J. Gastroenterol.* Apr 2001;36(4):389-398.

[18] Andoh A., Fujino S., Bamba S., et al. IL-17 selectively down-regulates TNF-alpha-induced RANTES gene expression in human colonic subepithelial myofibroblasts. *J. Immunol.* Aug 15 2002;169(4):1683-1687.

[19] Karlsson L., Lindahl P., Heath J. K., Betsholtz C. Abnormal gastrointestinal development in PDGF-A and PDGFR-(alpha) deficient mice implicates a novel mesenchymal structure with putative instructive properties in villus morphogenesis. *Development.* Aug 2000;127(16):3457-3466.

[20] Goke M., Kanai M., Podolsky D. K. Intestinal fibroblasts regulate intestinal epithelial cell proliferation via hepatocyte growth factor. *Am. J. Physiol.* May 1998;274(5 Pt 1):G809-818.

[21] Plateroti M., Rubin D. C., Duluc I., et al. Subepithelial fibroblast cell lines from different levels of gut axis display regional characteristics. *Am. J. Physiol.* May 1998;274(5 Pt 1):G945-954.

[22] Powell D. W., Mifflin R. C., Valentich J. D., Crowe S. E., Saada J. I., West A. B. Myofibroblasts. II. Intestinal subepithelial myofibroblasts. *Am. J. Physiol.* Aug 1999;277(2 Pt 1):C183-201.

[23] Yasui H., Andoh A., Bamba S., Inatomi O., Ishida H., Fujiyama Y. Role of Fibroblast Growth Factor-2 in the Expression of Matrix Metalloproteinases and Tissue Inhibitors of Metalloproteinases in Human Intestinal Myofibroblasts. *Digestion.* 2004;69(1):34-44.

[24] Bamba S., Andoh A., Yasui H., Makino J., Kim S., Fujiyama Y. Regulation of IL-11 expression in intestinal myofibroblasts: role of c-Jun AP-1- and MAPK-dependent pathways. *Am. J. Physiol. Gastrointest Liver Physiol.* Sep 2003;285(3):G529-538.

[25] Teller I. C., Beaulieu J. F. Interactions between laminin and epithelial cells in intestinal health and disease. *Expert Rev. Mol. Med.* Sep 28 2001;2001:1-18.

[26] Kanai Y., Kanai-Azuma M., Noce T., et al. Identification of two Sox17 messenger RNA isoforms, with and without the high mobility group box region, and their differential expression in mouse spermatogenesis. *J. Cell Biol.* May 1996;133(3):667-681.

[27] Hudson C., Clements D., Friday R. V., Stott D., Woodland H. R. Xsox17alpha and -beta mediate endoderm formation in Xenopus. *Cell.* Oct 31 1997;91(3):397-405.

[28] Engleka M. J., Craig E. J., Kessler D. S. VegT activation of Sox17 at the midblastula transition alters the response to nodal signals in the vegetal endoderm domain. *Dev. Biol.* Sep 1 2001;237(1):159-172.

[29] Kanai-Azuma M., Kanai Y., Gad J. M., et al. Depletion of definitive gut endoderm in Sox17-null mutant mice. *Development.* May 2002;129(10):2367-2379.

[30] Ishii Y., Rex M., Scotting P. J., Yasugi S. Region-specific expression of chicken Sox2 in the developing gut and lung epithelium: regulation by epithelial-mesenchymal interactions. *Dev. Dyn.* Dec 1998;213(4):464-475.

[31] Takash W., Canizares J., Bonneaud N., et al. SOX7 transcription factor: sequence, chromosomal localisation, expression, transactivation and interference with Wnt signalling. *Nucleic. Acids Res.* Nov 1 2001;29(21):4274-4283.

[32] Katoh M. Expression of human SOX7 in normal tissues and tumors. *Int. J. Mol. Med.* Apr 2002;9(4):363-368.

[33] de Santa Barbara P., van den Brink G. R., Roberts D. J. Development and differentiation of the intestinal epithelium. *Cell Mol. Life Sci.* Jul 2003;60(7):1322-1332.

[34] Katoh M. Molecular cloning and characterization of human SOX17. *Int. J. Mol. Med.* Feb 2002;9(2):153-157.

[35] Saitoh T., Katoh M. Expression of human SOX18 in normal tissues and tumors. *Int. J. Mol. Med.* Sep 2002;10(3):339-344.

[36] Roberts D. J., Smith D. M., Goff D. J., Tabin C. J. Epithelial-mesenchymal signaling during the regionalization of the chick gut. *Development.* Aug 1998;125(15):2791-2801.

[37] de Santa Barbara P., Roberts D. J. Tail gut endoderm and gut/genitourinary/tail development: a new tissue-specific role for Hoxa13. *Development.* Feb 2002;129(3):551-561.

[38] Roberts D. J., Johnson R. L., Burke A. C., Nelson C. E., Morgan B. A., Tabin C. Sonic hedgehog is an endodermal signal inducing Bmp-4 and Hox genes during induction and regionalization of the chick hindgut. *Development.* Oct 1995;121(10):3163-3174.

[39] Yokouchi Y., Sakiyama J., Kuroiwa A. Coordinated expression of Abd-B subfamily genes of the HoxA cluster in the developing digestive tract of chick embryo. *Dev Biol.* May 1995;169(1):76-89.

[40] Kondo T., Dolle P., Zakany J., Duboule D. Function of posterior HoxD genes in the morphogenesis of the anal sphincter. *Development.* Sep 1996;122(9):2651-2659.

[41] Warot X., Fromental-Ramain C., Fraulob V., Chambon P., Dolle P. Gene dosage-dependent effects of the Hoxa-13 and Hoxd-13 mutations on morphogenesis of the terminal parts of the digestive and urogenital tracts. *Development.* Dec 1997;124(23):4781-4791.

[42] Beck F., Tata F., Chawengsaksophak K. Homeobox genes and gut development. *Bioessays.* May 2000;22(5):431-441.

[43] Sekimoto T., Yoshinobu K., Yoshida M., et al. Region-specific expression of murine Hox genes implies the Hox code-mediated patterning of the digestive tract. *Genes Cells.* Jan 1998;3(1):51-64.

[44] Bitgood M. J., McMahon A. P. Hedgehog and Bmp genes are coexpressed at many diverse sites of cell-cell interaction in the mouse embryo. *Dev. Biol.* Nov 1995;172(1):126-138.

[45] Bilder D., Scott M. P. Hedgehog and wingless induce metameric pattern in the Drosophila visceral mesoderm. *Dev. Biol.* Sep 1 1998;201(1):43-56.

[46] Murone M., Rosenthal A., de Sauvage F. J. Hedgehog signal transduction: from flies to vertebrates. *Exp. Cell Res.* Nov 25 1999;253(1):25-33.

[47] Taipale J., Cooper M. K., Maiti T., Beachy P. A. Patched acts catalytically to suppress the activity of Smoothened. *Nature.* Aug 22 2002;418(6900):892-897.

[48] Litingtung Y., Lei L., Westphal H., Chiang C. Sonic hedgehog is essential to foregut development. *Nat. Genet.* Sep 1998;20(1):58-61.

[49] Levin M., Johnson R. L., Stern C. D., Kuehn M., Tabin C. A molecular pathway determining left-right asymmetry in chick embryogenesis. *Cell.* Sep 8 1995;82(5):803-814.

[50] Roberts D. J. Molecular mechanisms of development of the gastrointestinal tract. *Dev. Dyn.* Oct 2000;219(2):109-120.

[51] Grapin-Botton A., Melton D. A. Endoderm development: from patterning to organogenesis. *Trends Genet.* Mar 2000;16(3):124-130.

[52] Sukegawa A., Narita T., Kameda T., et al. The concentric structure of the developing gut is regulated by Sonic hedgehog derived from endodermal epithelium. *Development.* May 2000;127(9):1971-1980.

[53] Apelqvist A., Ahlgren U., Edlund H. Sonic hedgehog directs specialised mesoderm differentiation in the intestine and pancreas. *Curr. Biol.* Oct 1 1997;7(10):801-804.

[54] Ramalho-Santos M., Melton D. A., McMahon A. P. Hedgehog signals regulate multiple aspects of gastrointestinal development. *Development.* Jun 2000;127(12):2763-2772.

[55] Incardona J. P., Gaffield W., Kapur R. P., Roelink H. The teratogenic Veratrum alkaloid cyclopamine inhibits sonic hedgehog signal transduction. *Development.* Sep 1998;125(18):3553-3562.

[56] Cooper M. K., Porter J. A., Young K. E., Beachy P. A. Teratogen-mediated inhibition of target tissue response to Shh signaling. *Science.* Jun 5 1998;280(5369):1603-1607.

[57] Taipale J., Chen J. K., Cooper M. K., et al. Effects of oncogenic mutations in Smoothened and Patched can be reversed by cyclopamine. *Nature.* Aug 31 2000;406(6799):1005-1009.

[58] van den Brink G. R., Hardwick J. C., Tytgat G. N., et al. Sonic hedgehog regulates gastric gland morphogenesis in man and mouse. *Gastroenterology.* Aug 2001;121(2):317-328.

[59] van den Brink G. R., Hardwick J. C., Nielsen C., et al. Sonic hedgehog expression correlates with fundic gland differentiation in the adult gastrointestinal tract. *Gut.* Nov 2002;51(5):628-633.

[60] van den Brink G. R., Bleuming S. A., Hardwick J. C., et al. Indian Hedgehog is an antagonist of Wnt signaling in colonic epithelial cell differentiation. *Nat Genet.* Mar 2004;36(3):277-282.

[61] Whitman M. Smads and early developmental signaling by the TGFbeta superfamily. *Genes Dev*. Aug 15 1998;12(16):2445-2462.

[62] Faure S., Lee M. A., Keller T., ten Dijke P., Whitman M. Endogenous patterns of TGFbeta superfamily signaling during early Xenopus development. *Development.* Jul 2000;127(13):2917-2931.

[63] Smith D. M., Nielsen C., Tabin C. J., Roberts D. J. Roles of BMP signaling and Nkx2.5 in patterning at the chick midgut-foregut boundary. *Development.* Sep 2000;127(17):3671-3681.

[64] Nielsen C., Murtaugh L. C., Chyung J. C., Lassar A., Roberts D. J. Gizzard formation and the role of Bapx1. *Dev Biol.* Mar 1 2001;231(1):164-174.

[65] Smith D. M., Tabin C. J. BMP signalling specifies the pyloric sphincter. *Nature.* Dec 16 1999;402(6763):748-749.

[66] Narita T., Saitoh K., Kameda T., et al. BMPs are necessary for stomach gland formation in the chicken embryo: a study using virally induced BMP-2 and Noggin expression. *Development.* Mar 2000;127(5):981-988.

[67] Holcombe R. F., Marsh J. L., Waterman M. L., Lin F., Milovanovic T., Truong T. Expression of Wnt ligands and Frizzled receptors in colonic mucosa and in colon carcinoma. *Mol. Pathol.* Aug 2002;55(4):220-226.

[68] Houlston R, Bevan S, Williams A, et al. Mutations in DPC4 (SMAD4) cause juvenile polyposis syndrome, but only account for a minority of cases. *Hum. Mol. Genet*. Nov 1998;7(12):1907-1912.

[69] Howe J. R., Roth S., Ringold J. C., et al. Mutations in the SMAD4/DPC4 gene in juvenile polyposis. *Science.* May 15 1998;280(5366):1086-1088.

[70] Cheng H., Leblond C. P. Origin, differentiation and renewal of the four main epithelial cell types in the mouse small intestine. I. Columnar cell. *Am. J. Anat.* Dec 1974;141(4):461-479.

[71] Cheng H., Leblond C. P. Origin, differentiation and renewal of the four main epithelial cell types in the mouse small intestine. V. Unitarian Theory of the origin of the four epithelial cell types. *Am J Anat.* Dec 1974;141(4):537-561.

[72] Cheng H., Leblond C. P. Origin, differentiation and renewal of the four main epithelial cell types in the mouse small intestine. III. Entero-endocrine cells. *Am. J. Anat*. Dec 1974;141(4):503-519.

[73] Kirkland S. C. Clonal origin of columnar, mucous, and endocrine cell lineages in human colorectal epithelium. *Cancer.* Apr 1 1988;61(7):1359-1363.

[74] Pearse A. G., Takor T. T. Neuroendocrine embryology and the APUD concept. *Clin. Endocrinol (Oxf).* 1976;5 Suppl:229S-244S.

[75] LeDouarin N. M., Jotereau F. V. Origin and renewal of lymphocytes in avian embryo thymuses studied in interspecific combinations. *Nat. New Biol.* Nov 7 1973;246(149):25-27.

[76] Andrew A., Kramer B., Rawdon B. B. The origin of gut and pancreatic neuroendocrine (APUD) cells--the last word? *J. Pathol.* Oct 1998;186(2):117-118.

[77] Pearse A. G. The common peptides and the cytochemistry of their cells of origin. *Basic Appl. Histochem.* 1980;24(2):63-73.

[78] Ponder B. A., Schmidt G. H., Wilkinson M. M., Wood M. J., Monk M., Reid A. Derivation of mouse intestinal crypts from single progenitor cells. *Nature.* Feb 21-27 1985;313(6004):689-691.

[79] Winton D. J., Blount M. A., Ponder B. A. A clonal marker induced by mutation in mouse intestinal epithelium. *Nature*. Jun 2 1988;333(6172):463-466.

[80] Griffiths D. F., Davies S. J., Williams D., Williams G. T., Williams E. D. Demonstration of somatic mutation and colonic crypt clonality by X-linked enzyme histochemistry. *Nature.* Jun 2 1988;333(6172):461-463.

[81] Thompson M., Fleming K. A., Evans D. J., Fundele R., Surani M. A., Wright N. A. Gastric endocrine cells share a clonal origin with other gut cell lineages. *Development.* Oct 1990;110(2):477-481.

[82] Tatematsu M., Fukami H., Yamamoto M., et al. Clonal analysis of glandular stomach carcinogenesis in C3H/HeN<==>BALB/c chimeric mice treated with N-methyl-N-nitrosourea. *Cancer Lett.* Aug 15 1994;83(1-2):37-42.

[83] Nomura S., Esumi H., Job C., Tan S. S. Lineage and clonal development of gastric glands. *Dev. Biol.* Dec 1 1998;204(1):124-135.

[84] Fuller C. E., Davies R. P., Williams G. T., Williams E. D. Crypt restricted heterogeneity of goblet cell mucus glycoprotein in histologically normal human colonic mucosa: a potential marker of somatic mutation. *Br. J. Cancer*. Mar. 1990;61(3):382-384.

[85] Jass J. R., Roberton A. M. Colorectal mucin histochemistry in health and disease: a critical review. *Pathol. Int.* Jul 1994;44(7):487-504.

[86] Sugihara K., Jass J. R. Colorectal goblet cell sialomucin heterogeneity: its relation to malignant disease. *J. Clin. Pathol.* Oct 1986;39(10):1088-1095.

[87] Campbell F., Appleton M. A., Fuller C. E., et al. Racial variation in the O-acetylation phenotype of human colonic mucosa. *J. Pathol.* Nov 1994;174(3):169-174.

[88] Campbell F., Williams G. T., Appleton M. A., Dixon M. F., Harris M., Williams E. D. Post-irradiation somatic mutation and clonal stabilisation time in the human colon. *Gut.* Oct 1996;39(4):569-573.

[89] Novelli M. R., Williamson J. A., Tomlinson I. P., et al. Polyclonal origin of colonic adenomas in an XO/XY patient with FAP. *Science.* May 24 1996;272(5265):1187-1190.

[90] Novelli M., Cossu A., Oukrif D., et al. X-inactivation patch size in human female tissue confounds the assessment of tumor clonality. *Proc. Natl. Acad. Sci. USA.* Mar 18 2003;100(6):3311-3314.

[91] Nomura S., Kaminishi M., Sugiyama K., Oohara T., Esumi H. Clonal analysis of isolated single fundic and pyloric gland of stomach using X-linked polymorphism. *Biochem. Biophys. Res. Commun.* Sep 13 1996;226(2):385-390.

[92] Yatabe Y., Tavare S., Shibata D. Investigating stem cells in human colon by using methylation patterns. *Proc Natl Acad Sci U S A.* Sep 11 2001;98(19):10839-10844.

[93] Taylor R. W., Barron M. J., Borthwick G. M., et al. Mitochondrial DNA mutations in human colonic crypt stem cells. *J. Clin. Invest.* Nov 2003;112(9):1351-1360.

[94] Schmidt G. H., Winton D. J., Ponder B. A. Development of the pattern of cell renewal in the crypt-villus unit of chimaeric mouse small intestine. *Development.* Aug 1988;103(4):785-790.

[95] Park H. S., Goodlad R. A., Wright N. A. Crypt fission in the small intestine and colon. A mechanism for the emergence of G6PD locus-mutated crypts after treatment with mutagens. *Am. J. Pathol.* Nov 1995;147(5):1416-1427.

[96] Williams E. D., Lowes A. P., Williams D., Williams G. T. A stem cell niche theory of intestinal crypt maintenance based on a study of somatic mutation in colonic mucosa. *Am. J. Pathol.* Oct 1992;141(4):773-776.

[97] Maskens A. P., Dujardin-Loits R. M. Kinetics of tissue proliferation in colorectal mucosa during post-natal growth. *Cell Tissue Kinet.* Sep 1981;14(5):467-477.

[98] Cairnie A. B., Millen B. H. Fission of crypts in the small intestine of the irradiated mouse. *Cell Tissue Kinet.* Mar 1975;8(2):189-196.

[99] Wright N. A., Al-Nafussi A. The kinetics of villus cell populations in the mouse small intestine. II. Studies on growth control after death of proliferative cells induced by cytosine arabinoside, with special reference to negative feedback mechanisms. *Cell Tissue Kinet.* Nov 1982;15(6):611-621.

[100] Bjerknes M. A test of the stochastic theory of stem cell differentiation. *Biophys J.* Jun 1986;49(6):1223-1227.

[101] Totafurno J., Bjerknes M., Cheng H. The crypt cycle. Crypt and villus production in the adult intestinal epithelium. *Biophys J.* Aug 1987;52(2):279-294.

[102] Cheng H., McCulloch C., Bjerknes M. Effects of 30% intestinal resection on whole population cell kinetics of mouse intestinal epithelium. *Anat. Rec.* May 1986;215(1):35-41.

[103] Loeffler M., Birke A., Winton D., Potten C. Somatic mutation, monoclonality and stochastic models of stem cell organization in the intestinal crypt. *J. Theor. Biol.* Feb 21 1993;160(4):471-491.

[104] Loeffler M., Bratke T., Paulus U., Li Y. Q., Potten C. S. Clonality and life cycles of intestinal crypts explained by a state dependent stochastic model of epithelial stem cell organization. *J. Theor. Biol.* May 7 1997;186(1):41-54.

[105] Park H. S., Goodlad R. A., Ahnen D. J., et al. Effects of epidermal growth factor and dimethylhydrazine on crypt size, cell proliferation, and crypt fission in the rat colon. Cell proliferation and crypt fission are controlled independently. *Am. J. Pathol.* Sep 1997;151(3):843-852.

[106] Wasan H. S., Park H. S., Liu K. C., et al. APC in the regulation of intestinal crypt fission. *J. Pathol.* Jul 1998;185(3):246-255.
[107] Bjerknes M., Cheng H., Hay K., Gallinger S. APC mutation and the crypt cycle in murine and human intestine. *Am. J. Pathol.* Mar 1997;150(3):833-839.
[108] Nowell P. C. The clonal evolution of tumor cell populations. *Science.* Oct 1 1976;194(4260):23-28.
[109] Morson B. C. Evolution of cancer of the colon and rectum. *Cancer.* Sep 1974;34(3):suppl:845-849.
[110] Groden J., Thliveris A., Samowitz W., et al. Identification and characterization of the familial adenomatous polyposis coli gene. *Cell.* Aug 9 1991;66(3):589-600.
[111] Malkin D., Li F. P., Strong L. C., et al. Germ line p53 mutations in a familial syndrome of breast cancer, sarcomas, and other neoplasms. *Science.* Nov 30 1990;250(4985):1233-1238.
[112] Donehower L. A., Harvey M., Slagle B. L., et al. Mice deficient for p53 are developmentally normal but susceptible to spontaneous tumours. *Nature.* Mar 19 1992;356(6366):215-221.
[113] Clarke A. R., Cummings M. C., Harrison D. J. Interaction between murine germline mutations in p53 and APC predisposes to pancreatic neoplasia but not to increased intestinal malignancy. *Oncogene.* Nov 2 1995;11(9):1913-1920.
[114] Parsons R., Li G. M., Longley M., et al. Mismatch repair deficiency in phenotypically normal human cells. *Science.* May 5 1995;268(5211):738-740.
[115] Shoemaker A. R., Haigis K. M., Baker S. M., Dudley S., Liskay R. M., Dove W. F. Mlh1 deficiency enhances several phenotypes of Apc(Min)/+ mice. *Oncogene.* May 25 2000;19(23):2774-2779.
[116] Oshima M., Oshima H., Taketo M. M. TGF-beta receptor type II deficiency results in defects of yolk sac hematopoiesis and vasculogenesis. *Dev. Biol.* Oct 10 1996;179(1):297-302.
[117] Lamlum H., Papadopoulou A., Ilyas M., et al. APC mutations are sufficient for the growth of early colorectal adenomas. *Proc. Natl. Acad. Sci. USA.* Feb 29 2000;97(5):2225-2228.
[118] Behrens J., Jerchow B. A., Wurtele M., et al. Functional interaction of an axin homolog, conductin, with beta-catenin, APC, and GSK3beta. *Science.* Apr 24 1998;280(5363):596-599.
[119] Korinek V., Barker N., Willert K., et al. Two members of the Tcf family implicated in Wnt/beta-catenin signaling during embryogenesis in the mouse. *Mol. Cell Biol.* Mar 1998;18(3):1248-1256.
[120] Wong M. H., Rubinfeld B., Gordon J. I. Effects of forced expression of an NH2-terminal truncated beta-Catenin on mouse intestinal epithelial homeostasis. *J. Cell Biol.* May 4 1998;141(3):765-777.
[121] Huber O., Korn R., McLaughlin J., Ohsugi M., Herrmann B. G., Kemler R. Nuclear localization of beta-catenin by interaction with transcription factor LEF-1. *Mech. Dev.* Sep 1996;59(1):3-10.

[122] Morin P. J., Sparks A. B., Korinek V., et al. Activation of beta-catenin-Tcf signaling in colon cancer by mutations in beta-catenin or APC. *Science.* Mar 21 1997;275(5307):1787-1790.

[123] Ilyas M., Tomlinson I. P., Rowan A., Pignatelli M., Bodmer W. F. Beta-catenin mutations in cell lines established from human colorectal cancers. *Proc Natl Acad Sci U S A.* Sep 16 1997;94(19):10330-10334.

[124] Sparks A. B., Morin P. J., Vogelstein B., Kinzler K. W. Mutational analysis of the APC/beta-catenin/Tcf pathway in colorectal cancer. *Cancer Res.* Mar 15 1998;58(6):1130-1134.

[125] van de Wetering M., Sancho E., Verweij C., et al. The beta-catenin/TCF-4 complex imposes a crypt progenitor phenotype on colorectal cancer cells. *Cell.* Oct 18 2002;111(2):241-250.

[126] Shtutman M., Zhurinsky J., Simcha I., et al. The cyclin D1 gene is a target of the beta-catenin/LEF-1 pathway. *Proc. Natl. Acad. Sci. USA.* May 11 1999;96(10):5522-5527.

[127] Santiago A., Erickson C. A. Ephrin-B ligands play a dual role in the control of neural crest cell migration. *Development.* Aug 2002;129(15):3621-3632.

[128] Adams R. H., Diella F., Hennig S., Helmbacher F., Deutsch U., Klein R. The cytoplasmic domain of the ligand ephrinB2 is required for vascular morphogenesis but not cranial neural crest migration. *Cell.* Jan 12 2001;104(1):57-69.

[129] Bruckner K., Pasquale E. B., Klein R. Tyrosine phosphorylation of transmembrane ligands for Eph receptors. *Science.* Mar 14 1997;275(5306):1640-1643.

[130] Kullander K., Klein R. Mechanisms and functions of Eph and ephrin signalling. *Nat. Rev. Mol. Cell Biol.* Jul 2002;3(7):475-486.

[131] Shamah S. M., Lin M. Z., Goldberg J. L., et al. EphA receptors regulate growth cone dynamics through the novel guanine nucleotide exchange factor ephexin. *Cell.* Apr 20 2001;105(2):233-244.

[132] Batlle E., Henderson J. T., Beghtel H., et al. Beta-catenin and TCF mediate cell positioning in the intestinal epithelium by controlling the expression of EphB/ephrinB. *Cell.* Oct 18 2002;111(2):251-263.

[133] Wright N. A. Epithelial stem cell repertoire in the gut: clues to the origin of cell lineages, proliferative units and cancer. *Int. J. Exp. Pathol.* Apr 2000;81(2):117-143.

[134] Nakamura S., Kino I. Morphogenesis of minute adenomas in familial polyposis coli. *J. Natl. Cancer Inst.* Jul 1984;73(1):41-49.

[135] Bjerknes M., Cheng H. Clonal analysis of mouse intestinal epithelial progenitors. *Gastroenterology.* Jan 1999;116(1):7-14.

[136] Moss S. F., Liu T. C., Petrotos A., Hsu T. M., Gold L. I., Holt P. R. Inward growth of colonic adenomatous polyps. *Gastroenterology.* Dec 1996;111(6):1425-1432.

[137] Shiff S. J., Rigas B. Colon adenomatous polyps--do they grow inward? *Lancet.* Jun 28 1997;349(9069):1853-1854.

[138] Sinicrope F. A., Roddey G., McDonnell T. J., Shen Y., Cleary K. R., Stephens L. C. Increased apoptosis accompanies neoplastic development in the human colorectum. *Clin. Cancer Res.* Dec 1996;2(12):1999-2006.

[139] Tsao J. L., Zhang J., Salovaara R., et al. Tracing cell fates in human colorectal tumors from somatic microsatellite mutations: evidence of adenomas with stem cell architecture. *Am. J. Pathol.* Oct 1998;153(4):1189-1200.

[140] Preston S. L., Wong W. M., Chan A. O., et al. Bottom-up histogenesis of colorectal adenomas: origin in the monocryptal adenoma and initial expansion by crypt fission. *Cancer Res.* Jul 1 2003;63(13):3819-3825.

[141] Woda B. A., Forde K., Lane N. A unicryptal colonic adenoma, the smallest colonic neoplasm yet observed in a non-polyposis individual. *Am. J. Clin. Pathol.* Nov 1977;68(5):631-632.

[142] Chang W. W., Whitener C. J. Histogenesis of tubular adenomas in hereditary colonic adenomatous polyposis. *Arch. Pathol. Lab. Med.* Sep 1989;113(9):1042-1049.

[143] Wong W. M., Garcia S. B., Wright N. A. Origins and morphogenesis of colorectal neoplasms. *Apmis.* Jun 1999;107(6):535-544.

[144] Siu I. M., Robinson D. R., Schwartz S., et al. The identification of monoclonality in human aberrant crypt foci. *Cancer Res.* Jan 1 1999;59(1):63-66.

[145] Fujimitsu Y., Nakanishi H., Inada K., et al. Development of aberrant crypt foci involves a fission mechanism as revealed by isolation of aberrant crypts. *Jpn. J. Cancer Res.* Dec 1996;87(12):1199-1203.

[146] Araki K., Ogata T., Kobayashi M., Yatani R. A morphological study on the histogenesis of human colorectal hyperplastic polyps. *Gastroenterology.* Nov 1995;109(5):1468-1474.

[147] Merritt A. J., Gould K. A., Dove W. F. Polyclonal structure of intestinal adenomas in ApcMin/+ mice with concomitant loss of Apc+ from all tumor lineages. *Proc. Natl. Acad. Sci. USA.* Dec 9 1997;94(25):13927-13931.

[148] Potten C. S., Booth C., Hargreaves D. The small intestine as a model for evaluating adult tissue stem cell drug targets. *Cell Prolif.* Jun 2003;36(3):115-129.

[149] Potten C. S., Booth C., Pritchard D. M. The intestinal epithelial stem cell: the mucosal governor. *Int. J. Exp. Pathol.* Aug 1997;78(4):219-243.

[150] Marshman E., Booth C., Potten C. S. The intestinal epithelial stem cell. *Bioessays.* Jan 2002;24(1):91-98.

[151] Potten C. S. Extreme sensitivity of some intestinal crypt cells to X and gamma irradiation. *Nature.* Oct 6 1977;269(5628):518-521.

[152] Merritt A. J., Potten C. S., Kemp C. J., et al. The role of p53 in spontaneous and radiation-induced apoptosis in the gastrointestinal tract of normal and p53-deficient mice. *Cancer Res.* Feb 1 1994;54(3):614-617.

[153] Potten C. S., Owen G., Booth D. Intestinal stem cells protect their genome by selective segregation of template DNA strands. *J. Cell Sci.* Jun 1 2002;115(Pt 11):2381-2388.

[154] Potten C. S., Loeffler M. A comprehensive model of the crypts of the small intestine of the mouse provides insight into the mechanisms of cell migration and the proliferation hierarchy. *J. Theor. Biol.* Aug 21 1987;127(4):381-391.

[155] Cairns J. Mutation selection and the natural history of cancer. *Nature.* May 15 1975;255(5505):197-200.

[156] Potten C. S., Hume W. J., Reid P., Cairns J. The segregation of DNA in epithelial stem cells. *Cell.* Nov 1978;15(3):899-906.

[157] Potten C. S., Grant H. K. The relationship between ionizing radiation-induced apoptosis and stem cells in the small and large intestine. *Br J Cancer.* Oct 1998;78(8):993-1003.

[158] Dulic V., Kaufmann W. K., Wilson S. J., et al. p53-dependent inhibition of cyclin-dependent kinase activities in human fibroblasts during radiation-induced G1 arrest. *Cell.* Mar 25 1994;76(6):1013-1023.

[159] Brugarolas J., Chandrasekaran C., Gordon J. I., Beach D., Jacks T., Hannon G. J. Radiation-induced cell cycle arrest compromised by p21 deficiency. *Nature.* Oct 12 1995;377(6549):552-557.

[160] Waldman T., Kinzler K. W., Vogelstein B. p21 is necessary for the p53-mediated G1 arrest in human cancer cells. *Cancer Res.* Nov 15 1995;55(22):5187-5190.

[161] Del Sal G., Murphy M., Ruaro E., Lazarevic D., Levine A. J., Schneider C. Cyclin D1 and p21/waf1 are both involved in p53 growth suppression. *Oncogene.* Jan 4 1996;12(1):177-185.

[162] Xiong Y., Hannon G. J., Zhang H., Casso D., Kobayashi R., Beach D. p21 is a universal inhibitor of cyclin kinases. *Nature.* Dec 16 1993;366(6456):701-704.

[163] Waga S., Hannon G. J., Beach D., Stillman B. The p21 inhibitor of cyclin-dependent kinases controls DNA replication by interaction with PCNA. *Nature.* Jun 16 1994;369(6481):574-578.

[164] Merritt A. J., Potten C. S., Watson A. J., Loh D. Y., Nakayama K., Hickman J. A. Differential expression of bcl-2 in intestinal epithelia. Correlation with attenuation of apoptosis in colonic crypts and the incidence of colonic neoplasia. *J. Cell Sci.* Jun 1995;108 (Pt 6):2261-2271.

[165] Potten C. S. Stem cells in gastrointestinal epithelium: numbers, characteristics and death. *Philo.s Trans R. Soc. Lond B. Biol. Sci.* Jun 29 1998;353(1370):821-830.

[166] Cairnie A. B., Lamerton L. F., Steel G. G. Cell proliferation studies in the intestinal epithelium of the rat. I. Determination of the kinetic parameters. *Exp. Cell Res.* Sep 1965;39(2):528-538.

[167] Cairnie A. B., Lamerton L. F., Steel G. G. Cell proliferation studies in the intestinal epithelium of the rat. II. Theoretical aspects. *Exp. Cell Res.* Sep 1965;39(2):539-553.

[168] Potten C. S., Booth C., Tudor G. L., et al. Identification of a putative intestinal stem cell and early lineage marker; musashi-1. *Differentiation.* Jan 2003;71(1):28-41.

[169] Kayahara T., Sawada M., Takaishi S., et al. Candidate markers for stem and early progenitor cells, Musashi-1 and Hes1, are expressed in crypt base columnar cells of mouse small intestine. *FEBS Lett.* Jan 30 2003;535(1-3):131-135.

[170] Korinek V., Barker N., Moerer P., et al. Depletion of epithelial stem-cell compartments in the small intestine of mice lacking Tcf-4. *Nat Genet.* Aug 1998;19(4):379-383.

[171] Good P., Yoda A., Sakakibara S., et al. The human Musashi homolog 1 (MSI1) gene encoding the homologue of Musashi/Nrp-1, a neural RNA-binding protein putatively expressed in CNS stem cells and neural progenitor cells. *Genomics.* Sep 15 1998;52(3):382-384.

[172] Sakakibara S., Nakamura Y., Yoshida T., et al. RNA-binding protein Musashi family: roles for CNS stem cells and a subpopulation of ependymal cells revealed by targeted

disruption and antisense ablation. *Proc. Natl. Acad. Sci. USA.* Nov 12 2002;99(23):15194-15199.

[173] Imai T., Tokunaga A., Yoshida T., et al. The neural RNA-binding protein Musashi1 translationally regulates mammalian numb gene expression by interacting with its mRNA. *Mol. Cell Biol.* Jun 2001;21(12):3888-3900.

[174] Xie T., Spradling A. C. A niche maintaining germ line stem cells in the Drosophila ovary. *Science.* Oct 13 2000;290(5490):328-330.

[175] Brinster R. L., Zimmermann J. W. Spermatogenesis following male germ-cell transplantation. *Proc. Natl. Acad. Sci. USA.* Nov 22 1994;91(24):11298-11302.

[176] Nishimura E. K., Jordan S. A., Oshima H., et al. Dominant role of the niche in melanocyte stem-cell fate determination. *Nature.* Apr 25 2002;416(6883):854-860.

[177] Oshima H., Rochat A., Kedzia C., Kobayashi K., Barrandon Y. Morphogenesis and renewal of hair follicles from adult multipotent stem cells. *Cell.* Jan 26 2001;104(2):233-245.

[178] Gussoni E., Soneoka Y., Strickland C. D., et al. Dystrophin expression in the mdx mouse restored by stem cell transplantation. *Nature.* Sep 23 1999;401(6751):390-394.

[179] Brazelton T. R., Rossi F. M., Keshet G. I., Blau H. M. From marrow to brain: expression of neuronal phenotypes in adult mice. *Science.* Dec 1 2000;290(5497):1775-1779.

[180] Lagasse E., Connors H., Al-Dhalimy M., et al. Purified hematopoietic stem cells can differentiate into hepatocytes in vivo. *Nat. Med.* Nov 2000;6(11):1229-1234.

[181] Kim K. M., Shibata D. Methylation reveals a niche: stem cell succession in human colon crypts. *Oncogene.* Aug 12 2002;21(35):5441-5449.

[182] Kai T., Spradling A. An empty Drosophila stem cell niche reactivates the proliferation of ectopic cells. *Proc. Natl. Acad. Sci. USA.* Apr 15 2003;100(8):4633-4638.

[183] Hendry J. H., Roberts S. A., Potten C. S. The clonogen content of murine intestinal crypts: dependence on radiation dose used in its determination. *Radiat Res.* Oct 1992;132(1):115-119.

[184] Zeeh J. M., Procaccino F., Hoffmann P., et al. Keratinocyte growth factor ameliorates mucosal injury in an experimental model of colitis in rats. *Gastroenterology.* Apr 1996;110(4):1077-1083.

[185] Cominelli F., Nast C. C., Llerena R., Dinarello C. A., Zipser R. D. Interleukin 1 suppresses inflammation in rabbit colitis. Mediation by endogenous prostaglandins. *J. Clin. Invest.* Feb 1990;85(2):582-586.

[186] Tessner T. G., Cohn S. M., Schloemann S., Stenson W. F. Prostaglandins prevent decreased epithelial cell proliferation associated with dextran sodium sulfate injury in mice. *Gastroenterology.* Oct 1998;115(4):874-882.

[187] Qiu B. S., Pfeiffer C. J., Keith J. C., Jr. Protection by recombinant human interleukin-11 against experimental TNB-induced colitis in rats. *Dig. Dis. Sci.* Aug 1996;41(8):1625-1630.

[188] Peterson R. L., Wang L., Albert L., Keith J. C., Jr., Dorner A. J. Molecular effects of recombinant human interleukin-11 in the HLA-B27 rat model of inflammatory bowel disease. *Lab. Invest.* Dec 1998;78(12):1503-1512.

[189] Dignass A. U. Mechanisms and modulation of intestinal epithelial repair. *Inflamm Bowel. Dis.* Feb 2001;7(1):68-77.

[190] Bjerknes M., Cheng H. Modulation of specific intestinal epithelial progenitors by enteric neurons. *Proc. Natl. Acad. Sci. USA.* Oct 23 2001;98(22):12497-12502.

[191] Petersen B. E., Bowen W. C., Patrene K. D., et al. Bone marrow as a potential source of hepatic oval cells. *Science.* May 14 1999;284(5417):1168-1170.

[192] Theise N. D., Badve S., Saxena R., et al. Derivation of hepatocytes from bone marrow cells in mice after radiation-induced myeloablation. *Hepatology.* Jan 2000;31(1):235-240.

[193] Alison M. R., Poulsom R., Jeffery R., et al. Hepatocytes from non-hepatic adult stem cells. *Nature.* Jul 20 2000;406(6793):257.

[194] Theise N. D., Nimmakayalu M., Gardner R., et al. Liver from bone marrow in humans. *Hepatology.* Jul 2000;32(1):11-16.

[195] Ferrari G., Cusella-De Angelis G., Coletta M., et al. Muscle regeneration by bone marrow-derived myogenic progenitors. *Science.* Mar 6 1998;279(5356):1528-1530.

[196] Orlic D., Kajstura J., Chimenti S., et al. Bone marrow cells regenerate infarcted myocardium. *Nature.* Apr 5 2001;410(6829):701-705.

[197] Eglitis M. A., Mezey E. Hematopoietic cells differentiate into both microglia and macroglia in the brains of adult mice. *Proc. Natl. Acad. Sci. USA.* Apr 15 1997;94(8):4080-4085.

[198] Poulsom R., Forbes S. J., Hodivala-Dilke K., et al. Bone marrow contributes to renal parenchymal turnover and regeneration. *J. Pathol.* Sep 2001;195(2):229-235.

[199] Brittan M., Hunt T., Jeffery R., et al. Bone marrow derivation of pericryptal myofibroblasts in the mouse and human small intestine and colon. *Gut.* Jun 2002;50(6):752-757.

[200] Okamoto R., Yajima T., Yamazaki M., et al. Damaged epithelia regenerated by bone marrow-derived cells in the human gastrointestinal tract. *Nat. Med.* Sep 2002;8(9):1011-1017.

[201] Jackson K. A., Mi T., Goodell M. A. Hematopoietic potential of stem cells isolated from murine skeletal muscle. *Proc. Natl. Acad. Sci. USA.* Dec 7 1999;96(25):14482-14486.

[202] Bjornson C. R., Rietze R. L., Reynolds B. A., Magli M. C., Vescovi A. L. Turning brain into blood: a hematopoietic fate adopted by adult neural stem cells in vivo. *Science.* Jan 22 1999;283(5401):534-537.

[203] Shen C. N., Slack J. M., Tosh D. Molecular basis of transdifferentiation of pancreas to liver. *Nat. Cell Biol.* Dec 2000;2(12):879-887.

[204] Clarke D. L., Johansson C. B., Wilbertz J., et al. Generalized potential of adult neural stem cells. *Science.* Jun 2 2000;288(5471):1660-1663.

[205] Jiang Y., Jahagirdar B. N., Reinhardt R. L., et al. Pluripotency of mesenchymal stem cells derived from adult marrow. *Nature.* Jul 4 2002;418(6893):41-49.

[206] Wagers A. J., Sherwood R. I., Christensen J. L., Weissman I. L. Little evidence for developmental plasticity of adult hematopoietic stem cells. *Science.* Sep 27 2002;297(5590):2256-2259.

[207] Terada N., Hamazaki T., Oka M., et al. Bone marrow cells adopt the phenotype of other cells by spontaneous cell fusion. *Nature.* Apr 4 2002;416(6880):542-545.

[208] Ying Q. L., Nichols J., Evans E. P., Smith A. G. Changing potency by spontaneous fusion. *Nature.* Apr 4 2002;416(6880):545-548.

[209] Wang X., Willenbring H., Akkari Y., et al. Cell fusion is the principal source of bone-marrow-derived hepatocytes. *Nature.* Apr 24 2003;422(6934):897-901.

[210] Vassilopoulos G., Wang P. R., Russell D. W. Transplanted bone marrow regenerates liver by cell fusion. *Nature.* Apr 24 2003;422(6934):901-904.

[211] Masuya M., Drake C. J., Fleming P. A., et al. Hematopoietic origin of glomerular mesangial cells. *Blood.* Mar 15 2003;101(6):2215-2218.

[212] LaBarge M. A., Blau H. M. Biological progression from adult bone marrow to mononucleate muscle stem cell to multinucleate muscle fiber in response to injury. *Cell.* Nov 15 2002;111(4):589-601.

[213] Korbling M., Katz R. L., Khanna A., et al. Hepatocytes and epithelial cells of donor origin in recipients of peripheral-blood stem cells. *N. Engl. J. Med.* Mar 7 2002;346(10):738-746.

[214] Hematti P., Sloand E. M., Carvallo C. A., et al. Absence of donor-derived keratinocyte stem cells in skin tissues cultured from patients after mobilized peripheral blood hematopoietic stem cell transplantation. *Exp. Hematol.* Aug 2002;30(8):943-949.

[215] Srivatsa B., Srivatsa S., Johnson K. L., Samura O., Lee S. L., Bianchi D. W. Microchimerism of presumed fetal origin in thyroid specimens from women: a case-control study. *Lancet.* Dec 15 2001;358(9298):2034-2038.

[216] Vig P, Russo FP, Edwards RJ, et al. The sources of parenchymal regeneration after chronic hepatocellular liver injury in mice. *Hepatology.* Feb 2006;43(2):316-324.

[217] Cairnie AB. *Homeostasis in the small intestine.* New York: Academic Press; 1976.

[218] Krause DS, Theise ND, Collector MI, et al. Multi-organ, multi-lineage engraftment by a single bone marrow-derived stem cell. *Cell.* May 4 2001;105(3):369-377.

[219] Direkze NC, Forbes SJ, Brittan M, et al. Multiple organ engraftment by bone-marrow-derived myofibroblasts and fibroblasts in bone-marrow-transplanted mice. *Stem Cells.* 2003;21(5):514-520.

[220] Arias AM. New alleles of Notch draw a blueprint for multifunctionality. *Trends Genet.* Apr 2002;18(4):168-170.

[221] Pourquie O. The segmentation clock: converting embryonic time into spatial pattern. *Science.* Jul 18 2003;301(5631):328-330.

[222] Soriano S, Kang DE, Fu M, et al. Presenilin 1 negatively regulates beta-catenin/T cell factor/lymphoid enhancer factor-1 signaling independently of beta-amyloid precursor protein and notch processing. *J Cell Biol.* Feb 19 2001;152(4):785-794.

[223] De Strooper B, Annaert W, Cupers P, et al. A presenilin-1-dependent gamma-secretase-like protease mediates release of Notch intracellular domain. *Nature.* Apr 8 1999;398(6727):518-522.

[224] De Strooper B, Saftig P, Craessaerts K, et al. Deficiency of presenilin-1 inhibits the normal cleavage of amyloid precursor protein. *Nature.* Jan 22 1998;391(6665):387-390.

[225] Brown MS, Ye J, Rawson RB, Goldstein JL. Regulated intramembrane proteolysis: a control mechanism conserved from bacteria to humans. *Cell.* Feb 18 2000;100(4):391-398.

[226] Xia X, Qian S, Soriano S, et al. Loss of presenilin 1 is associated with enhanced beta-catenin signaling and skin tumorigenesis. *Proc Natl Acad Sci U S A.* Sep 11 2001;98(19):10863-10868.

[227] Zhong W, Feder JN, Jiang MM, Jan LY, Jan YN. Asymmetric localization of a mammalian numb homolog during mouse cortical neurogenesis. *Neuron.* Jul 1996;17(1):43-53.

[228] Zhong W, Jiang MM, Weinmaster G, Jan LY, Jan YN. Differential expression of mammalian Numb, Numblike and Notch1 suggests distinct roles during mouse cortical neurogenesis. *Development.* May 1997;124(10):1887-1897.

[229] Uemura T, Shepherd S, Ackerman L, Jan LY, Jan YN. numb, a gene required in determination of cell fate during sensory organ formation in Drosophila embryos. *Cell.* Jul 28 1989;58(2):349-360.

[230] Guo M, Jan LY, Jan YN. Control of daughter cell fates during asymmetric division: interaction of Numb and Notch. *Neuron.* Jul 1996;17(1):27-41.

[231] Rhyu MS, Jan LY, Jan YN. Asymmetric distribution of numb protein during division of the sensory organ precursor cell confers distinct fates to daughter cells. *Cell.* Feb 11 1994;76(3):477-491.

[232] Spana EP, Kopczynski C, Goodman CS, Doe CQ. Asymmetric localization of numb autonomously determines sibling neuron identity in the Drosophila CNS. *Development.* Nov 1995;121(11):3489-3494.

[233] Lewis J. Notch signalling and the control of cell fate choices in vertebrates. *Semin Cell Dev Biol.* Dec 1998;9(6):583-589.

[234] Kageyama R, Ohtsuka T, Tomita K. The bHLH gene Hes1 regulates differentiation of multiple cell types. *Mol Cells.* Feb 29 2000;10(1):1-7.

[235] Skipper M, Lewis J. Getting to the guts of enteroendocrine differentiation. *Nat Genet.* Jan 2000;24(1):3-4.

[236] Yang Q, Bermingham NA, Finegold MJ, Zoghbi HY. Requirement of Math1 for secretory cell lineage commitment in the mouse intestine. *Science.* Dec 7 2001;294(5549):2155-2158.

[237] Jensen J, Pedersen EE, Galante P, et al. Control of endodermal endocrine development by Hes-1. *Nat Genet.* Jan 2000;24(1):36-44.

[238] van Den Brink GR, de Santa Barbara P, Roberts DJ. Development. Epithelial cell differentiation--a Mather of choice. *Science.* Dec 7 2001;294(5549):2115-2116.

[239] Lee CS, Perreault N, Brestelli JE, Kaestner KH. Neurogenin 3 is essential for the proper specification of gastric enteroendocrine cells and the maintenance of gastric epithelial cell identity. *Genes Dev.* Jun 15 2002;16(12):1488-1497.

[240] Naya FJ, Huang HP, Qiu Y, et al. Diabetes, defective pancreatic morphogenesis, and abnormal enteroendocrine differentiation in BETA2/neuroD-deficient mice. *Genes Dev.* Sep 15 1997;11(18):2323-2334.

[241] Pinson KI, Brennan J, Monkley S, Avery BJ, Skarnes WC. An LDL-receptor-related protein mediates Wnt signalling in mice. *Nature.* Sep 28 2000;407(6803):535-538.

[242] Tamai K, Semenov M, Kato Y, et al. LDL-receptor-related proteins in Wnt signal transduction. *Nature.* Sep 28 2000;407(6803):530-535.
[243] Wehrli M, Dougan ST, Caldwell K, et al. arrow encodes an LDL-receptor-related protein essential for Wingless signalling. *Nature.* Sep 28 2000;407(6803):527-530.
[244] Fodde R, Smits R, Clevers H. APC, signal transduction and genetic instability in colorectal cancer. *Nat Rev Cancer.* Oct 2001;1(1):55-67.
[245] Zeng L, Fagotto F, Zhang T, et al. The mouse Fused locus encodes Axin, an inhibitor of the Wnt signaling pathway that regulates embryonic axis formation. *Cell.* Jul 11 1997;90(1):181-192.
[246] Hart MJ, de los Santos R, Albert IN, Rubinfeld B, Polakis P. Downregulation of beta-catenin by human Axin and its association with the APC tumor suppressor, beta-catenin and GSK3 beta. *Curr Biol.* May 7 1998;8(10):573-581.
[247] Fagotto F, Jho E, Zeng L, et al. Domains of axin involved in protein-protein interactions, Wnt pathway inhibition, and intracellular localization. *J Cell Biol.* May 17 1999;145(4):741-756.
[248] Kishida S, Yamamoto H, Hino S, Ikeda S, Kishida M, Kikuchi A. DIX domains of Dvl and axin are necessary for protein interactions and their ability to regulate beta-catenin stability. *Mol Cell Biol.* Jun 1999;19(6):4414-4422.
[249] Ikeda S, Kishida S, Yamamoto H, Murai H, Koyama S, Kikuchi A. Axin, a negative regulator of the Wnt signaling pathway, forms a complex with GSK-3beta and beta-catenin and promotes GSK-3beta-dependent phosphorylation of beta-catenin. *Embo J.* Mar 2 1998;17(5):1371-1384.
[250] Aberle H, Bauer A, Stappert J, Kispert A, Kemler R. beta-catenin is a target for the ubiquitin-proteasome pathway. *Embo J.* Jul 1 1997;16(13):3797-3804.
[251] Jiang J, Struhl G. Regulation of the Hedgehog and Wingless signalling pathways by the F-box/WD40-repeat protein Slimb. *Nature.* Jan 29 1998;391(6666):493-496.
[252] Marikawa Y, Elinson RP. beta-TrCP is a negative regulator of Wnt/beta-catenin signaling pathway and dorsal axis formation in Xenopus embryos. *Mech Dev.* Sep 1998;77(1):75-80.
[253] Axelrod JD, Miller JR, Shulman JM, Moon RT, Perrimon N. Differential recruitment of Dishevelled provides signaling specificity in the planar cell polarity and Wingless signaling pathways. *Genes Dev.* Aug 15 1998;12(16):2610-2622.
[254] Boutros M, Mihaly J, Bouwmeester T, Mlodzik M. Signaling specificity by Frizzled receptors in Drosophila. *Science.* Jun 9 2000;288(5472):1825-1828.
[255] Smalley MJ, Sara E, Paterson H, et al. Interaction of axin and Dvl-2 proteins regulates Dvl-2-stimulated TCF-dependent transcription. *Embo J.* May 17 1999;18(10):2823-2835.
[256] Molenaar M, Roose J, Peterson J, Venanzi S, Clevers H, Destree O. Differential expression of the HMG box transcription factors XTcf-3 and XLef-1 during early xenopus development. *Mech Dev.* Jul 1998;75(1-2):151-154.
[257] Korinek V, Barker N, Morin PJ, et al. Constitutive transcriptional activation by a beta-catenin-Tcf complex in APC-/- colon carcinoma. *Science.* Mar 21 1997;275(5307):1784-1787.

[258] He TC, Sparks AB, Rago C, et al. Identification of c-MYC as a target of the APC pathway. *Science.* Sep 4 1998;281(5382):1509-1512.

[259] Tetsu O, McCormick F. Beta-catenin regulates expression of cyclin D1 in colon carcinoma cells. *Nature.* Apr 1 1999;398(6726):422-426.

[260] Kim JS, Crooks H, Dracheva T, et al. Oncogenic beta-catenin is required for bone morphogenetic protein 4 expression in human cancer cells. *Cancer Res.* May 15 2002;62(10):2744-2748.

[261] Miwa N, Furuse M, Tsukita S, Niikawa N, Nakamura Y, Furukawa Y. Involvement of claudin-1 in the beta-catenin/Tcf signaling pathway and its frequent upregulation in human colorectal cancers. *Oncol Res.* 2000;12(11-12):469-476.

[262] Wang S, Krinks M, Lin K, Luyten FP, Moos M, Jr. Frzb, a secreted protein expressed in the Spemann organizer, binds and inhibits Wnt-8. *Cell.* Mar 21 1997;88(6):757-766.

[263] Lee CS, Buttitta LA, May NR, Kispert A, Fan CM. SHH-N upregulates Sfrp2 to mediate its competitive interaction with WNT1 and WNT4 in the somitic mesoderm. *Development.* Jan 2000;127(1):109-118.

[264] Bienz M, Clevers H. Linking colorectal cancer to Wnt signaling. *Cell.* Oct 13 2000;103(2):311-320.

[265] Kinzler KW, Vogelstein B. Lessons from hereditary colorectal cancer. *Cell.* Oct 18 1996;87(2):159-170.

[266] Powell SM, Zilz N, Beazer-Barclay Y, et al. APC mutations occur early during colorectal tumorigenesis. *Nature.* Sep 17 1992;359(6392):235-237.

[267] Mei JM, Hord NG, Winterstein DF, Donald SP, Phang JM. Differential expression of prostaglandin endoperoxide H synthase-2 and formation of activated beta-catenin-LEF-1 transcription complex in mouse colonic epithelial cells contrasting in Apc. *Carcinogenesis.* Apr 1999;20(4):737-740.

[268] Wielenga VJ, Smits R, Korinek V, et al. Expression of CD44 in Apc and Tcf mutant mice implies regulation by the WNT pathway. *Am J Pathol.* Feb 1999;154(2):515-523.

[269] Nusse R. Developmental biology. Making head or tail of Dickkopf. *Nature.* May 17 2001;411(6835):255-256.

[270] Pinto D, Gregorieff A, Begthel H, Clevers H. Canonical Wnt signals are essential for homeostasis of the intestinal epithelium. *Genes Dev.* Jul 15 2003;17(14):1709-1713.

[271] van Genderen C, Okamura RM, Farinas I, et al. Development of several organs that require inductive epithelial-mesenchymal interactions is impaired in LEF-1-deficient mice. *Genes Dev.* Nov 15 1994;8(22):2691-2703.

[272] Hovanes K, Li TW, Munguia JE, et al. Beta-catenin-sensitive isoforms of lymphoid enhancer factor-1 are selectively expressed in colon cancer. *Nat Genet.* May 2001;28(1):53-57.

[273] Roose J, Huls G, van Beest M, et al. Synergy between tumor suppressor APC and the beta-catenin-Tcf4 target Tcf1. *Science.* Sep 17 1999;285(5435):1923-1926.

[274] Kaestner KH, Knochel W, Martinez DE. Unified nomenclature for the winged helix/forkhead transcription factors. *Genes Dev.* Jan 15 2000;14(2):142-146.

[275] Kaestner KH, Silberg DG, Traber PG, Schutz G. The mesenchymal winged helix transcription factor Fkh6 is required for the control of gastrointestinal proliferation and differentiation. *Genes Dev.* Jun 15 1997;11(12):1583-1595.

[276] Perreault N, Katz JP, Sackett SD, Kaestner KH. Foxl1 controls the Wnt/beta-catenin pathway by modulating the expression of proteoglycans in the gut. *J Biol Chem.* Nov 16 2001;276(46):43328-43333.
[277] Tsuda M, Kamimura K, Nakato H, et al. The cell-surface proteoglycan Dally regulates Wingless signalling in Drosophila. *Nature.* Jul 15 1999;400(6741):276-280.
[278] Lin X, Perrimon N. Dally cooperates with Drosophila Frizzled 2 to transduce Wingless signalling. *Nature.* Jul 15 1999;400(6741):281-284.
[279] Alexander CM, Reichsman F, Hinkes MT, et al. Syndecan-1 is required for Wnt-1-induced mammary tumorigenesis in mice. *Nat Genet.* Jul 2000;25(3):329-332.
[280] Lickert H, Domon C, Huls G, et al. Wnt/(beta)-catenin signaling regulates the expression of the homeobox gene Cdx1 in embryonic intestine. *Development.* Sep 2000;127(17):3805-3813.
[281] Chawengsaksophak K, James R, Hammond VE, Kontgen F, Beck F. Homeosis and intestinal tumours in Cdx2 mutant mice. *Nature.* Mar 6 1997;386(6620):84-87.
[282] Beck F, Chawengsaksophak K, Waring P, Playford RJ, Furness JB. Reprogramming of intestinal differentiation and intercalary regeneration in Cdx2 mutant mice. *Proc Natl Acad Sci U S A.* Jun 22 1999;96(13):7318-7323.
[283] Theodosiou NA, Tabin CJ. Wnt signaling during development of the gastrointestinal tract. *Dev Biol.* Jul 15 2003;259(2):258-271.
[284] Lickert H, Kispert A, Kutsch S, Kemler R. Expression patterns of Wnt genes in mouse gut development. *Mech Dev.* Jul 2001;105(1-2):181-184.
[285] Katoh M. WNT2 and human gastrointestinal cancer (review). *Int J Mol Med.* Nov 2003;12(5):811-816.
[286] Gregorieff A, Pinto D, Begthel H, Destree O, Kielman M, Clevers H. Expression pattern of Wnt signaling components in the adult intestine. *Gastroenterology.* Aug 2005;129(2):626-638.
[287] Machesky LM, Hall A. Role of actin polymerization and adhesion to extracellular matrix in Rac- and Rho-induced cytoskeletal reorganization. *J Cell Biol.* Aug 25 1997;138(4):913-926.
[288] Van Aelst L, D'Souza-Schorey C. Rho GTPases and signaling networks. *Genes Dev.* Sep 15 1997;11(18):2295-2322.
[289] Stappenbeck TS, Gordon JI. Rac1 mutations produce aberrant epithelial differentiation in the developing and adult mouse small intestine. *Development.* Jun 2000;127(12):2629-2642.
[290] Stappenbeck TS, Gordon JI. Extranuclear sequestration of phospho-Jun N-terminal kinase and distorted villi produced by activated Rac1 in the intestinal epithelium of chimeric mice. *Development.* Jul 2001;128(13):2603-2614.
[291] Hermiston ML, Wong MH, Gordon JI. Forced expression of E-cadherin in the mouse intestinal epithelium slows cell migration and provides evidence for nonautonomous regulation of cell fate in a self-renewing system. *Genes Dev.* Apr 15 1996;10(8):985-996.
[292] Garrett-Sinha LA, Eberspaecher H, Seldin MF, de Crombrugghe B. A gene for a novel zinc-finger protein expressed in differentiated epithelial cells and transiently in certain mesenchymal cells. *J Biol Chem.* Dec 6 1996;271(49):31384-31390.

[293] Shields JM, Christy RJ, Yang VW. Identification and characterization of a gene encoding a gut-enriched Kruppel-like factor expressed during growth arrest. *J Biol Chem.* Aug 16 1996;271(33):20009-20017.

[294] Ton-That H, Kaestner KH, Shields JM, Mahatanankoon CS, Yang VW. Expression of the gut-enriched Kruppel-like factor gene during development and intestinal tumorigenesis. *FEBS Lett.* Dec 15 1997;419(2-3):239-243.

[295] Bieker JJ. Kruppel-like factors: three fingers in many pies. *J Biol Chem.* Sep 14 2001;276(37):34355-34358.

[296] Katz JP, Perreault N, Goldstein BG, et al. The zinc-finger transcription factor Klf4 is required for terminal differentiation of goblet cells in the colon. *Development.* Jun 2002;129(11):2619-2628.

[297] Rempel RE, Saenz-Robles MT, Storms R, et al. Loss of E2F4 activity leads to abnormal development of multiple cellular lineages. *Mol Cell.* Aug 2000;6(2):293-306.

[298] Hemminki A, Markie D, Tomlinson I, et al. A serine/threonine kinase gene defective in Peutz-Jeghers syndrome. *Nature.* Jan 8 1998;391(6663):184-187.

[299] Jenne DE, Reimann H, Nezu J, et al. Peutz-Jeghers syndrome is caused by mutations in a novel serine threonine kinase. *Nat Genet.* Jan 1998;18(1):38-43.

[300] Karuman P, Gozani O, Odze RD, et al. The Peutz-Jegher gene product LKB1 is a mediator of p53-dependent cell death. *Mol Cell.* Jun 2001;7(6):1307-1319.

[301] Olumi AF, Grossfeld GD, Hayward SW, Carroll PR, Tlsty TD, Cunha GR. Carcinoma-associated fibroblasts direct tumor progression of initiated human prostatic epithelium. *Cancer Res.* Oct 1 1999;59(19):5002-5011.

[302] Barcellos-Hoff MH, Ravani SA. Irradiated mammary gland stroma promotes the expression of tumorigenic potential by unirradiated epithelial cells. *Cancer Res.* Mar 1 2000;60(5):1254-1260.

[303] Uchida K, Samma S, Momose H, Kashihara N, Rademaker A, Oyasu R. Stimulation of urinary bladder tumorigenesis by carcinogen-exposed stroma. *J Urol.* Mar 1990;143(3):618-621.

[304] Drakos PE, Nagler A, Or R. Case of Crohn's disease in bone marrow transplantation. *Am J Hematol.* Jun 1993;43(2):157-158.

[305] Kashyap A, Forman SJ. Autologous bone marrow transplantation for non-Hodgkin's lymphoma resulting in long-term remission of coincidental Crohn's disease. *Br J Haematol.* Dec 1998;103(3):651-652.

[306] Lopez-Cubero SO, Sullivan KM, McDonald GB. Course of Crohn's disease after allogeneic marrow transplantation. *Gastroenterology.* Mar 1998;114(3):433-440.

[307] Talbot DC, Montes A, Teh WL, Nandi A, Powles RL. Remission of Crohn's disease following allogeneic bone marrow transplant for acute leukaemia. *Hosp Med.* Jul 1998;59(7):580-581.

[308] Burt RK, Traynor A, Oyama Y, Craig R. High-dose immune suppression and autologous hematopoietic stem cell transplantation in refractory Crohn disease. *Blood.* Mar 1 2003;101(5):2064-2066.

[309] Vaughan D, Drumm B. Treatment of fistulas with granulocyte colony-stimulating factor in a patient with Crohn's disease. *N Engl J Med.* Jan 21 1999;340(3):239-240.

[310] Dieckgraefe BK, Korzenik JR. Treatment of active Crohn's disease with recombinant human granulocyte-macrophage colony-stimulating factor. *Lancet.* Nov 9 2002;360(9344):1478-1480.

In: Stem Cell Research Trends
Editor: Josse R. Braggina, pp. 73-98

ISBN: 978-1-60021-622-0

Chapter II

Pancreatic Stem Cells: Specification Program and Self-Renewal Mechanisms

Francisco Miralles*
Institut Cochin, Dept. de Génétique et Développement,
INSERM U567, CNRS UMR 8104,
Université René Descartes, Paris V, France

Abstract

Optimized islet cell transplantation therapies hold the prospect of providing a curative treatment for diabetes. However, the scarcity of donors prevents the development of these therapies on a large scale. One possible way to reduce the shortage of islets for transplantation would be to produce large amounts of β-cells *in vitro*. Although this approach still remains theoretical, recent advances in the knowledge of embryonic stem cells biology have raised the prospect that these cells may be used for the production of pancreatic β-cells. The experience acquired during the past few years indicates the need to recreate *in vitro* the steps which occur *in vivo*. Stem cells must first be induced to become definitive endoderm, then to differentiate into pancreatic endoderm, and finally to reach the stage of differentiated mature β-cells. Thus, our capacity to drive this process *in vitro* will depend on a thorough understanding of the molecular mechanisms controlling the development of the pancreas. It is now well established, that all the pancreatic cell types are derived from a small pool of apparently identical stem cells, which originate from the early gut endoderm. After a phase of intense proliferation, these progenitor cells differentiate into the pancreatic exocrine and endocrine cell types. Considerable progress has been made in the identification of transcription factors that are required for the differentiation or maintenance of the pancreatic progenitors into the different pancreatic cell types. However less is known about the signalling pathways and the transcription factors controlling the generation of the pancreatic precursors in the primitive endoderm. Also, only very recently have we

* F. Miralles, PhD, Dept. Génétique et Développement, Institut Cochin, 24, rue du Faubourg St. Jacques, 75014 PARIS, France. E-mail: miralles@cochin.iserm.fr; Tel : 33 1 44 41 24 01; Fax : 33 1 44 41 24 21.

started to uncover the mechanisms controlling the maintenance of the pancreatic progenitors in an undifferentiated state. This review will focus on both the signalling pathways and transcription factors, which determine the specification of the pancreatic progenitors, as well as on the mechanisms controlling the maintenance of their self-renewal capacity.

INTRODUCTION

The mammalian pancreas is a mixed exocrine and endocrine gland playing an essential role in the maintenance of nutritional homeostasis, through the synthesis and secretion of enzymes and hormones. The pancreas includes three major cell types : acinar, ductal and endocrine. The acinar and ductal cells form the exocrine component of the pancreas, which represents 90% of the total cells in the adult organ. The acinar cells produce and secrete digestive enzymes such as amylases, lipases, proteases and nucleases, which are transported to the duodenum trough a complex ductal network. The cells forming the ducts are specialized epithelial cells producing a bicarbonate solution which flushes the acinar secretion through the ducts and prevents the premature activation of the digestive enzymes. The endocrine cells are organized into small spherical microorgans, the islets of Langerhans. These cells represent less than 3% of the cells in the adult pancreas. Each endocrine islet is composed by five cell types, α, β, δ, PP, and ε, which produce and secrete, respectively, the hormones glucagon, insulin, somatostatin, pancreatic polypeptide and ghrelin. The mature islets, present a well defined organization, with the β-cells occupying the core of the islet while the other endocrine cells locate at the periphery. In addition to the glandular tissues, the pancreas has a rich blood supply, the arterial blood passing in each lobule first to the islets and then to the adjacent acini. There is also an extensive lymphatic drainage, and a rich sympathetic and parasympathetic nerve supply. Smooth muscle is found around the larger ducts [1, 2].

PANCREATIC MORPHOGENESIS AND DIFFERENTIATION

In mice, the pancreas develops at the junction of the foregut/midgut from a pre-patterned region of the primitive gut endoderm (Figure 1). The first sign of pancreas morphogenesis is the formation of the dorsal pancreatic rudiment anlagen at embryonic day 9.5 (E9.5), followed shortly thereafter by the formation of the ventral anlagen (E10). Subsequently, the epithelial cells forming the dorsal and ventral pancreatic primordia proliferate into the surrounding mesenchyme forming new protusions which will ultimately develop into highly branched ductular structures. Around E13.5 the stomach and duodenum undergoe a rotational mouvement, and, as a consequence, the ventral pancreatic bud comes into contact with the dorsal bud. Both pancreatic rudiments continue to growth independently until they fuse to form a single organ.

Concomitantly to pancreas morphogenesis, the endocrine and exocrine cells differentiate following a well established sequence. Pictet and collaborators proposed a model of pancreatic development describing two major transitions in time [3].

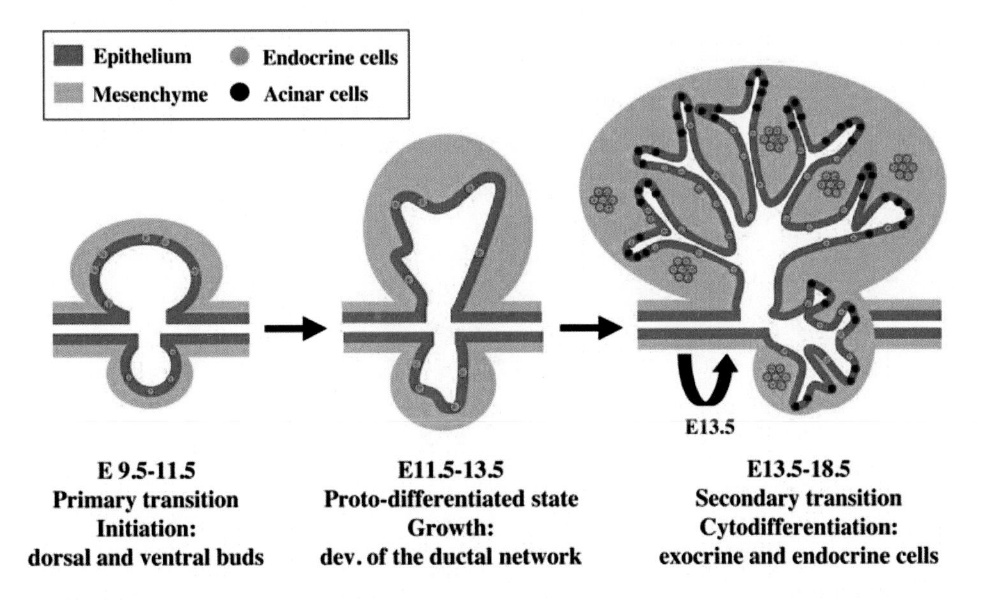

Figure 1. Pancreatic organogenesis. Schematic representation of pancreatic organogenesis in mice from E9 to the end of the gestational period, E18.5. After the formation of the dorsal and ventral buds (primary transition, E9.5-E11.5), the undifferentiated pancreatic epithelium proliferates actively and branches into the surrounding mesenchyme forming a complex ductal network. The cells forming these ducts are precursor cells from which the exocrine and endocrine cells differentiate during the secondary transition (E13.5-E15.5). Later, the endocrine cells migrate into the surrounding stroma and associate to form the islets of Langerhans (Isletogenesis).

During the « primary transition », which parallels the formation of the pancreatic buds, cells of the appropiate region of the gut endoderm become destined to form pancreas. At the begining of this stage, only a few scattered cells expressing glucagon can be detected, and shortly thereafter, a subset of these cells, co-express insulin and the peptide PYY [4-6]. Interestingly, it has been shown that these early glucagon and insulin co-expressing cells do not contribute to the definitive pool of endocrine cells of the mature pancreas [7]. However, their definitive fate and potential function in the developing pancreas remains to be elucidated. This stage is followed by a phase termed « proto-differentiated state » which is characterized by the intense proliferation of the undifferentiated pancreatic epithelium. During this stage low levels of pancreas-specific gene products, such insulin and amylase can be already detected. During the « secondary transition », a period which spans from E13.5 to E15.5, the pancreatic progenitors differentiate massively into acinar, ductal and endocrine cells. After this transition, few undifferentiated cells remain and the subsequent growth of the pancreas results essentially from the proliferation of fully differentiated cells. The proliferation concerns esentially the acinar and ductal cells. The pancreatic endocrine cells present a particularly low rate of proliferation during fetal life. Before birth (E18.5), the endocrine cells which remain closely associated with the ducts during the secondary transition, migrate and aggregate to form clusters in a process termed « isletogenesis ». However, the definitive architecture of the islets is not acquired until 2-3 weeks after birth.

Endoderm Patterning Generates Cells Competent to Form Pancreas

By the end of gastrulation (E6-E7.5), totipotent cells of the early mouse embryo form a sheet of endoderm. Shortly thereafter (E7.5-E9) these cells organize to form the primitive gut tube which becomes divided into several compartments. This regionalization is thought to result from the differential expression of genes. For example, in mammals the first evidence of endoderm regionalization is marked by the expression of the genes *Cerberus-like*, *Otx1* and *Hesx1* in the anterior portion of the endoderm while the *intestinal fatty acid binding protein (IFABP)* and *Cdx2* genes are expressed in the posterior endoderm [8]. Later, the expression of new genes induces more refined territorialization of the endoderm. However, several genes encoding transcription factors are broadly expressed in the primitive gut endoderm including the variant homeodomain hepatocyte nuclear factor *vHnf1* (also called *Tcf2* or *Hnf1β*), *Hnf4*, the forkhead factors *Forkhead box* A1 and A2 (*FoxA1*, *FoxA2*), and the zinc-finger transcription factors *Gata4*, *Gata5*, and *Gata6* [9-12]. These genes constitute a basic endoderm program which requires of the interaction with more restricted genes to generate organ specification. In fine, each compartment differs from the others by a particular combination of common and restricted genes which determines its identity and developmental potential. Thus, differential gene expression along the primitive gut endoderm marks territories for esophagus, lungs, thyroid, thymus, stomach, liver, pancreas and intestines.

Transcription Factors Involved in the Specification of Pancreatic Progenitors

Althougth the first signs of pancreas morphogenesis can be detected only around E9.5, the specification of the foregut/midgut endoderm towards a pancreatic fate occurs as early as E8.5. At this stage, the endodermal cells localized in the foregut/midgut region, initiate the expression of a dynamic transcriptional program commiting them to the adoption of a pancreatic fate. The activation of this transcriptional program results from the interactions between the endoderm and surrounding mesodermal and ectodermal tissues. These interactions involve the concerted activities of members of the Retinoic acid, Fibroblast Growth Factor, Sonic Hedgehog and Bone Morphogenetic Protein signalling pathways [13-16]. Thus, prior to the emergence of the pancreatic buds the foregut/midgut endoderm initiates the expression of several genes which, by interacting with the basic transcriptional endoderm program, specify a group of cells from this region of the endoderm to become pancreatic progenitors. These include the genes encoding the transcription factors *Pdx1*, *Hlxb9*, and *Ptf1α*..

Pdx1: The *Pancreatic duodenal homeobox 1* (*Pdx1*) gene was identified by several laboratories and thus it is also known as *Ipf1*, *Stf1* and *Idx1* [17-21]. *Pdx1* encodes an homeodomain transcription factor which acts as a regulator for the transcriptional expression of the insulin, somatostatin, Glut-2, Glucokinase, and the islet amyloid polypetide genes [22-25]. Besides its expression in β and δ cells of the mature pancreas, *Pdx1* is also expressed in the developing pancreas (Figure 2). Its expression begins at E8.5, in the cells of the endoderm which will develop later into the dorsal and ventral pancreatic rudiments.

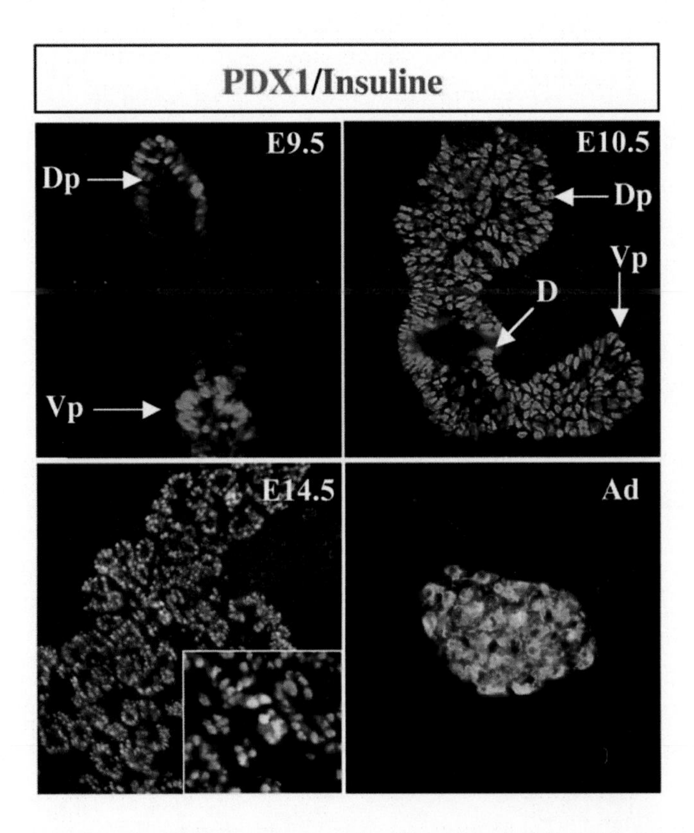

Figure 2. Pdx1 expression throughout pancreas development. At E9.5 Pdx1 is expressed in the emerging dorsal (Dp) and ventral (Vp) pancreatic buds. At E10.5 the pancreatic buds have become much larger and the majority of the epithelial cells in the pancreatic bud express Pdx1. At this stage Pdx1 is also expressed by the cells forming the presumptive duodenum (D). During the protodifferentiated state the undifferentiated cells of the branching epithelium express Pdx1. However, at the onset of the secondary transition (E15.5) the expression of Pdx1 becomes up-regulated in the differentiating β-cells and gradually down-regulated in the exocrine cells. By the end of the gestation period, E18.5, and through adult life (Ad), Pdx1 is expressed only by the β-and δ cells of the islets of Langerhans.

The expression of *Pdx1* is maintained in the majority of the epithelial cells of the rudiments until the end of the protodifferentiated state (E14.5). However, at the onset of the secondary transition the expression of *Pdx1* is strongly down-regulated in the differentiating ductal and acinar cells, although these cells continue to express the protein at low levels. On the contrary, high levels of *Pdx1* persist in the differentiated β and δ cells. Cre-lox lineage tracing studies have revealed that all the epithelial cells of the mature pancreas derive from pancreatic progenitors which express *Pdx1* [26].

Pdx1 null mice (*Pdx1*$^{-/-}$), show strong dorsal and ventral pancreatic hypoplasia, as well as defective development of the rostral duodenal epithelium and enteroendocrine cells [20]. Therefore in these mice pancreatic development is arrested at the primary transition state. Tissue recombination experiments suggest that in *Pdx1*$^{-/-}$ mice, the undifferentiated epithelial cells of the pancreatic buds lack the competence to respond to the mesenchymal signals which ensure their growth [18]. It has been proposed that Pdx1 could be required in a cell-autonomus fashion by the pancreatic precursor cells to maintain their dividing state. Alternatively, Pdx1 could be required for the production of a secreted factor, which would be involved in a feedback signalling mechanism to the adjacent mesenchyme [14].

The transcription factor HNF-6 (also known as Onecut-1) has been identified as an important regulator for *Pdx1* expression in both the dorsal and ventral endoderm [27, 28]. The expression of *Hnf6* precedes that of *Pdx1*, being first expressed at E8.0 in the foregut-midgut region of the endoderm. Later, *Hnf6* becomes restricted to the liver and pancreas. Inactivation of the *Hnf6* gene in mice results in delayed onset of *Pdx1* expression in the endoderm, leading to pancreatic hypoplasia [27]. Recently the vHNF1/TCF2 transcription factor was found to be critical for *Hnf6* expression [29, 30]. Moreover, in *vHnf1*$^{-/-}$ embryos, the dorsal pancreatic bud is hypoplastic and shows delayed *Pdx1* expression, as in *Hnf6*$^{-/-}$ embryos. However, the *vHnf1*$^{-/-}$ mice also show total agenesis of the ventral pancreas. These observations suggest a cascade in which *vHnf1* and *Hnf6* induce *Pdx1* expression in the foregut endoderm. However although delayed, *Pdx1* is still expressed in the dorsal pancreas of the *vHnf1*$^{-/-}$ and *Hnf6*$^{-/-}$. Thus other regulators must control *Pdx1* expression in the dorsal pancreas.

HlxB9: Encodes a homeodomain transcription factor expressed in the early pancreatic progenitors and in mature β cells [31, 32]. Expression of *Hlxb9* is initiated around E8 along a portion of the dorsal endoderm which extends beyond the pancreatic domain and in the ventral pancreatic domain. In addition *Hlxb9* is also expressed in the notochord. *Hlxb9* is expressed in the dorsal and ventral pancreatic buds until E10.5, down-regulated during the protodifferentiation state, and subsequently reactivated during the secondary transition in the newly differentiated β cells. *Hlxb9*$^{-/-}$ mice, display dorsal pancreas agenesis, wereas the ventral pancreas develops normally, although the number of β cells appears significatively reduced. The reason for the differential requierement of Hlxb9 in dorsal and ventral pancreatic rudiments development remains to be elucidated. However, this observation reveals that the developmental program initiating pancreas specification in the ventral and dorsal pancreas could be sligthly different. These differences could be the consequence of the interaction of the dorsal and ventral pro-pancreatic endoderm with their respective surrounding tissues. At the onset of *Hlxb9* expression, the dorsal pro-pancreatic endoderm remains in close contact with the notochord, while the ventral pro-pancreatic is surrounded by mesodermal tissue and very close to the developing liver bud. Thus, the notochord might provide signals which, in the absence of Hlxb9, repress pancreas development in the dorsal pro-pancreatic endoderm. Alternatively, the mesoderm or liver bud could provide signals which would allow the ventral pancreatic bud to bypass the requierement for Hlxb9 to initiate the program of pancreatic specification.

Ptf1α/P48 : The observation of a coordinated regulation of the genes expressing secretory enzymes in the exocrine pancreas led to the identification of a complex cell specific DNA-binding proteins termed PTF1 (pancreas transcription factor-1). PTF1 is an hetero-oligomer of three subunits, including the bHLH (basic helix-loop-helix) homeodomain protein Ptf1α/p48, one of the common class A bHLH proteins, and a subunit recently indentified as the CSL protein RBP-L [33]. The *Ptf1α* gene encodes the pancreas specific component of this complex, the protein P48. In mice, *Ptf1α* is expressed as early as E9.5 by the majority of the cells of the developing pancreatic buds, and at E15 its expression becomes restricted to the acinar cells. *Ptf1α* inactivation results as for *Pdx1* in total pancreatic agenesis [34]. A recombination-based lineage study has shown that at E9.5, *Ptf1α* is expressed by the progenitors of the ductal, acinar and endocrine cells of the pancreas [35]. Moreover, inactivation of *Ptf1α* switches the character of the pancreatic progenitors such that their progeny proliferate and adopt the normal fates of duodenal epithelium. Consistent with the proposal that *Ptf1α* supports the specification of the precursors of all three pancreatic lineages, transgenic expression of *Pdx1* under the control of *Ptf1α* promoter restores pancreatic development in $Pdx1^{-/-}$ mice that otherwise exhibit pancreatic agenesis [35]. These data provide evidence that *Ptf1α* expression is specifically associated with the acquisition of pancreatic fate by the undifferentiated foregut endoderm. In addition, a recent study has shown that in *Xenopus*, ectopic co-expression of *XlHbox8* (the *Xenopus* homologue of *Pdx1*) and *Ptf1α* is sufficient to convert a large portion of non-pancreatic posterior endoderm into endocrine and exocrine tissue [36]. In this study ectopic expression of *Pdx1* alone failed to induce pancreatic differentiation in non-pancreatic endoderm. Also, previous studies had shown that ectopic expression of *Pdx1* in non-pancreatic chicken endoderm was not sufficient to promote differentiation of either endocrine or exocrine cells [37]. Thus, there is now compelling evidence that co-expression of *Ptf1α* and *Pdx1* is both necessary and sufficient for proper pancreatic determination and subsequent proliferation and differentiation.

Tissue Interactions Initiate the Activation of the Pancreatic Specification Program in the Embryonic Endoderm

In mice, the dorsal and ventral pancreatic buds develop from distinct regions of the embryonic gut. The endoderm that will form the dorsal bud lies at the midline of the embryo and has prolonged contact with the notochord initially (E8) and later with the dorsal aorta (E9), both of which have been shown to be sources of signals required for pancreas induction [38, 39]. The endoderm giving rise to the ventral pancreatic buds lies at the lip of the anterior intestinal portal (AIP) and interacts with the lateral plate mesoderm (E7.5-E8) and later, with the cardiac and septum transversum mesoderm (E8.5). (Figure 3).

Signalling from the notochord is necessary for the initiation of pancreatic differentiation in the dorsal bud . Removal of the notochord in chick embryos results in *Sonic hedgehog* (*Shh*) expression in the dorsal pro-pancreatic endoderm and, as a consequence, pancreatic induction is abolished [38]. Similarly, in mice, misexpression of *Shh* under the control of the

Pdx1 promoter severely impairs pancreatic differentiation [40]. Moreover, *in vitro* studies have shown that the inhibition of Hedgehog signalling with cyclopamine leads to ectopic expression of pancreatic markers in the stomach, duodenum and intestine [41]. These results suggest that Hedgehog signalling controls the size of the pancreatic domain by repressing pancreatic differentiation in regions of the embryonic endoderm which do not pertain to the definitive pancreatic domain but which have the potential to differentiate into pancreatic tissue. *In vitro* experiments have shown that FGF2 and activin-βB, can mimic the effects of the notochord by repressing *Shh* expression, while activating *Pdx1* and insulin in isolated dorsal chick endoderm [42].

In mice, the notochord contact with the pro-pancreatic endoderm is interrupted by the interposition of the dorsal aorta at E9.0.

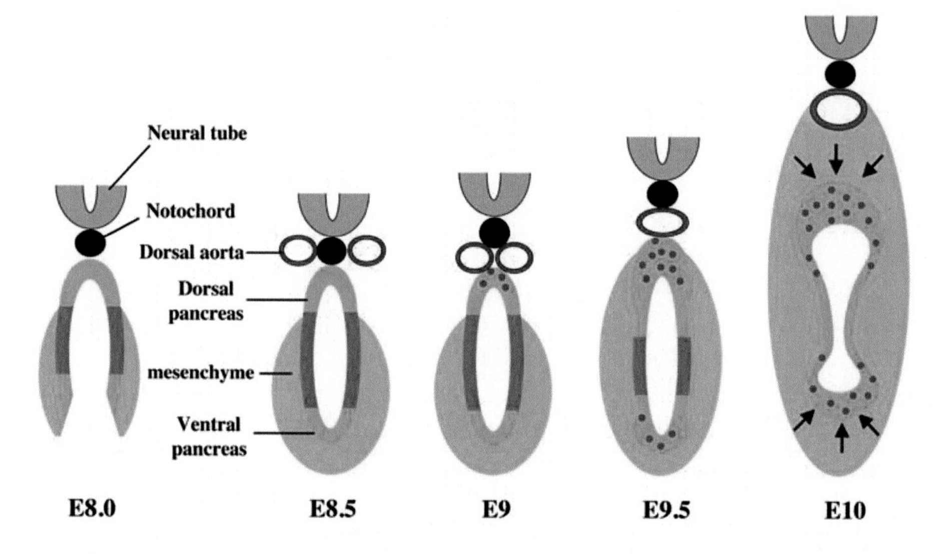

Figure 3. Tissue interactions involved in pancreatic specification. Schematic representations of cross sections of mouse embryos at the level of the developing pancreas. At E8 signals from the notochord repress the expression of *Shh* (blue) in the dorsal pro-pancreatic endoderm allowing *Pdx1* expression (green). Concomitantly, signals from the lateral mesoderm repress *Shh* expression in the presumptive ventral pro-pancreatic endoderm and initiate *Pdx1* expression. By E9.0 the contact of the notochord with the pro-pancreatic endoderm is interrupted by the interposition of the dorsal aorta (red). At this time the endothelial cells induce the expression of *Ptf1a* (purple) in the *Pdx1* cells of the pro-pancreatic endoderm. During the same period the mesenchyme (orange) which initially surrounded the ventral and lateral sides of the endoderm pushes around the dorsal side until it surrounds the entire gut. Signals from the mesenchyme including FGF10 (black arrows) induce the proliferation of the *Pdx1/Ptf1α*-expresing pancreatic precursors and the subsequent growth of the pancreatic buds (E10).

The removal of the dorsal aorta in *Xenopus* embryos leads to the absence of pancreatic markers such glucagon and insulin, suggesting that endothelial signals are requiered for pancreatic differentiation. Moreover explants of pro-pancreatic endoderm cultured with dissected aorta initiate pancreatic differentiation, whereas endoderm cultured with non-endothelial tissues does not. Additionally when the number of blood vessels in the posterior stomach is increased by missexpression of the vascular endothelial growth factor (VEGF) under the control of the *Pdx1* promoter, ectopic insulin-expressing cells are detected [39].

Thus, endothelial cells appear to be the source of pancreas inductive signals. Recent studies have shown that *Flk-1*$^{-/-}$ embryos which lack endothelial cells and vasculature [43], fail to induce *Ptf1α* and lack a dorsal pancreas. The capacity of isolated wild-type aorta to induce *Ptf1α* in co-cultured explants of *Flk-1*$^{-/-}$ dorsal endoderm shows that the endothelial induction of *Ptf1α* in the dorsal endoderm is specific to cell interactions and not due to factors in the bloodstream [44]. Althougth the lack of endothelial cells in the *Flk-1*$^{-/-}$ embryos is associated with a failure in *Ptf1α* induction and a failure in dorsal pancreas buding, the failure in buding is not due to the absence of *Ptf1α*, since *Ptf1α*$^{-/-}$ embryos develop an initial dorsal pancreatic bud [34, 35]. This result suggested that endothelial cells did contribute also to a phase of dorsal pancreas development which follows *Ptf1α* induction. The early studies on pancreas development, some 30 years ago, showed that the buding of the dorsal pancreatic epithelium and subsequent proliferation and branching morphogenesis requires interaction with the surrounding mesenchyme [3, 45-47]. Also, *in vivo* studies have shown that mice deficient for Isl1, a transcription factor expressed by the dorsal pancreatic mesenchyme, lack dorsal pancreatic mesenchyme and fail to develop a dorsal pancreatic bud [48]. The pancreatic mesenchyme has been identified as the source of a number of growth factors which stimulate the growth and morphogenesis of the pancreas, including EGF, HGF, FGF2, FGF10, [49-53]. Interestingly, a recent study has shown that aortic endothelial cells promote the survival of the Isl-1 expressing cells of the dorsal pancreatic mesenchyme. Furthermore, this study reveals that the FGF10 produced by the dorsal mesenchyme cells is requested to maintain *Ptf1α* expression in the dorsal pancreatic bud and can induce the expression of *Pdx1* in *Hnf6*$^{-/-}$ deficient embryos [54]. The later observation suggests that FGF10 can activate pathways and transcription factors within the pro-pancreatic endoderm cells that bypass earlier signals that effect the induction of *Pdx1* in the dorsal endoderm.

Although the ventral pancreatic bud undergoes a developmental program similar to that of the dorsal bud, the ventral bud is not in contact with either the notochord or the dorsal aorta at the time of pancreatic specification. The lateral-plate mesoderm that lies beneath the presumptive ventral pancreatic endoderm has been identified as the source of signals which specify the ventral pancreatic domain. These signals are instructive because they are able to induce pancreatic differentiation in endoderm, anterior to the genuine pancreas domain [55]. The molecular identity of these signals remains to be defined, although members of the activin and BMP family can mimic this pancreatic induction *in vitro* [15]. Since the ventral pancreas and liver bud originate from cells from the same AIP endoderm, Zaret and colleagues suggested the existence of a population of bipotential liver/pancreas precursors in the ventral endoderm [56]. They showed that in the absence of cardiac mesoderm, which is required for hepatic differentiation, the cells of the AIP endoderm, instead of differentiating into hepatic tissue, initiate *Pdx1* expression and downregulate *Shh*. Thus, pancreas might be the default identity of the AIP endoderm at this stage. Some FGFs and BMPs which are expressed within the cardiac mesoderm and septum trasversum mesenchyme were able to mimic the effect of the cardiac mesoderm on the AIP endoderm by inducing liver differentiation and repressing pancreas induction. This suggests that FGFs and BMPs could be responsible for directing the bipotential precursors towards a liver fate. However the existence of a bipotential precursor for liver and ventral pancreas has not been demonstrated formally, and it has been proposed that these results could be alternatively explained by the coexistence of committed liver and pancreas precursors in the AIP endoderm. FGFs and BMPs could preferentially allow the survival or differentiation of the liver precursors [15].

SIGNALLING PATHWAYS CONTROLLING PANCREATIC PROGENITORS PROLIFERATION AND SELF-RENEWAL CAPACITY

After pancreas specification and buds formation, the pancreatic progenitors undergo a phase of intense proliferation and morphogenesis (E11-E13.5). Although during this period a few of these cells differentiate into endocrine cells, most remain in an undifferentiated state and proliferate actively. This leads to a considerable expansion of the initially small pool of pancreatic progenitors. Therefore, the control of the pancreatic progenitors differentiation and self-renewal capacity is a critical step in the development of the pancreas. The maintenance of the self-renewal capacity of the pancreatic progenitors is temporally and spatially controlled, by means of several signalling pathways including Notch, FGFs and Wnt/β-catenin signalling.

Notch Signalling

The Notch pathway has been identified as the mechanism controlling the self-renewal of the pancreatic precursors by preventing premature differentiation. In this pathway, the transmembrane ligands of the DSL family (Delta, Jagged/Serrate and Lag-2) activate the Notch receptors expressed in adjacent cells (Figure 4). Ligand activation leads to the cleavage of the intracellular domain of Notch (Notch-IC), which interacts with the DNA binding CSL factors (CBF1/RBP-J/RBPSUH in mammals, Su(H) in *Drosophila*, and Lag-1 in nematodes). This complex transactivates target genes including the Hairy Enhancer of Split (HES) family of bHLH repressors [57]. The HES repressors generally act by down-regulating pro-differentiation factors such as Achaete and Scute in *Drosophila* proneurons [58], MyoD in mammalian myoblasts [59] and Neurogenin 3 (Ngn3) in a neuroendocrine context [60]. The studies of the genetic inactivation of different components of this pathway, *Hes1* [61], *Ngn3* [62], *Delta* and *RBP-J* [63] as well as mice over-expressing Ngn3 in early pancreatic progenitors [63], collectively show that Notch signalling controls the choice between differentiated endocrine cell and progenitor cell fates.

These studies demonstrated also that Ngn3 not only promotes, but is also indispensable for ensuring the differentiation of the pancreatic precursors into endocrine cells [62]. According to this model, the inactivation of the Notch receptor or signalling, results in Ngn3 activation, leading to premature endocrine cell differentiation at the expense of pancreatic progenitors expansion and exocrine cell differentiation. In contrast, cells with active Notch signalling remain as undifferentiated progenitor cells that would proliferate, undergo morphogenesis and late differentiation. Recent studies have shown that Notch activation prevents not only endocrine differentiation but also acinar cell differentiation [64-66].

In several developmental systems including neuronal diferentiation, Notch signalling operates in a mode known as « lateral inhibition » [67]. In this situation a precursor cell expresses extracellular ligands, such as Delta or Jagged which activate Notch receptors in an adjacent cell. Notch activation then leads to activation of the bHLH Hes genes in that adjacent cell. The *Hes* genes act as repressors of other target genes, including *Ngn3* which is required for *Delta* activation. As a consequence, Notch receptors in the adjacent precursor cell are not activated and this cell can initiate a differentiation program. Several observations indicate that the lateral inhibition operates during pancreas development including the patterns of expression of Hes1 and Ngn3 during the primary transition as well as the upregulation of Ngn3 in null-mice for *Hes1*, *Delta-1* and *RBP-J* [63]. However, other observations suggest that in addition to lateral inhibition Notch signalling could also operate in the pancreas by other modes.

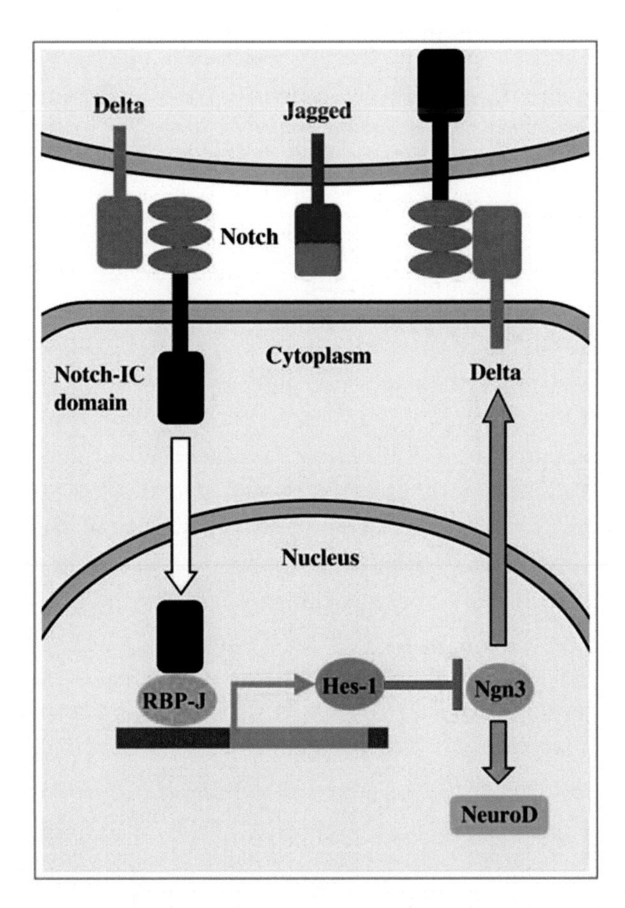

Figure 4. The Notch signalling pathway. The transmembrane ligands Delta or Jagged activate Notch receptors in adjacent cells. Ligand activation leads to the cleavage of the intracellular domain of Notch (Notch-IC) which interacts with the DNA-bound factor RBP-J. This complex transactivates several target genes including Hes1. In the pancreatic progenitors, Hes1 represses the expression of the pro-endocrine gene Ngn3. The transcription factor Ngn3, in addition to controlling the expression of Delta, is also required for activation of NeuroD, a bHLH factor crucial for the maintenance of endocrine differentiation.

In particular, it has been noted that during the phase of protodifferentiation, Notch1 and Notch2 are ubiquitously expressed in the pancreatic precursors as well as their ligands Jagged1 and Jagged 2, and their target gene *Hes1* [61, 68]. This second mode of Notch signalling has been linked to the maintenance of an expanding population of pancreatic precursors during the protodifferentiated state. Thus, to distinguish between the two modes of Notch signalling, Jensen and collaborators refered to this mechanism as « suppressive maintenace » [14, 68].

FGF Signalling

The early studies on pancreas development revealed the importance of epithelial-mensenchymal interactions in the control of pancreas development. These studies showed that in the absence of mesenchyme, the isolated pancreatic epithelium fails to grow and the pancreatic progenitors differentiate only into endocrine cells. Moreover, in the absence of mesenchyme, morphogenesis is also abolished [45-47]. More recent studies have shown that mesenchyme signals also regulate the ratio between endocrine and exocrine cell differentiation [69-71].

The mesenchyme produces and secretes a large variety of growth factors which act as mediators of epithelial-mesenchymal interactions in many organs. Some of them, i.e. EGF, HGF, TGFβ and different FGFs, have been implicated in pancreatic growth and differentiation [16]. In particular, in vitro studies have shown that several FGFs (FGF1, FGF2, FGF4, FGF7 and FGF10) induce the proliferation of the isolated rat pancreatic epithelium depleted of mesenchyme [52, 53]. Among these, the FGFs signalling through the receptor FGFR2b seems to play a critical role in pancreatic organogenesis. Indeed, mice expressing a dominant negative form of FGFR2b, and mutant null-mice for *FGF1R2b* show severe pancreatic hypoplasia [72, 73]. Moreover, mice in which the gene encoding FGF10, a high affinity ligand of FGFR2b, has been inactivated show severe pancreatic hypoplasia [74]. In these *FGF10* $^{-/-}$ mice pancreatic specification and bud formation occurs normally but the progenitor cells show a very low index of proliferation. This suggest that FGF10 acts as a trophic factor necessary to ensure the expansion of the pancreatic progenitors pool. Interestingly, in the developing pancreas, FGF10 is produced essentially by the pancreatic mesenchyme, while its receptor FGFR2b is expressed only in the epithelial cells [53, 74, 75]. However, FGF10 seems to act not only as a mere trophic factor for pancreatic precursors. Two independent studies have shown that mice expressing FGF10 under the control of the promoter of *Pdx1* (*pPdx1-FGF10*) show enhanced and prolonged proliferation of pancreatic progenitors, pancreatic hyperplasia and impaired pancreatic cell differentiation [68, 76]. In these mice most epithelial cells co-expressed pancreatic progenitor markers such Pdx1, Ptf1α/P48 and Nkx6.1 and failed to express markers of endocrine or exocrine differentiation. Moreover, most of these cells expressed Notch1 and Hes1, suggesting that Notch activation is maintained througout the pancreatic epithelium of the *pPdx1-FGF10* mice. Thus, both studies provided evidence that FGF10 acts not only as a trophic factor for pancreatic progenitors but also maintains active Notch signalling, and thereby prevents differentiation. The activation of Notch signalling by FGF10 in the pancreatic precursors confirms the existence of the

« suppressive maintenance » mode of Notch signalling postulated by Jensen and collaborators [14]. *In vitro* cultures of isolated mouse embryonic pancreatic epithelium have confirmed the effects of FGF10 on the pancreatic progenitors. In isolated E10 mice pancreatic epithelium, FGF10 induces the proliferation of the pancreatic progenitors and blocks their differentiation. Moreover, by using specific Notch signalling inhibitors these studies provided experimental evidence demonstrating that the effect of FGF10 on the self-renewal of the pancreatic progenitors requires active Notch-signalling [77]. Therefore, the Notch pathway is required as a downstream mediator of the FGF10 signalling in pancreatic precursors. Thus, the pancreatic mesenchyme controls the self-renewal capacity of the pancreatic progenitors througth the interplay between FGF10 and Notch signalling. How FGF10 can maintain Notch activation in the pancreatic precursors remains to be elucidated. It has been suggested, based on the expression of the Notch ligand genes *Jagged-1* and *Jagged-2* in the undifferentiated pancreatic epithelium of the *pPdx1-FGF10* mice, that FGF10 might induce the expression of these ligands [68]. FGF10 might also down-regulate repressors of Notch activity such as Sel1 [76] and mNumb [78] or up-regulate Notch expression. In this regard, it has been shown that the FGF1 and FGF2 induce the proliferation and inhibit the differentiation of neuroepithelial precursors through Notch signalling. Both FGFs efficiently up-regulate the expression of Notch-1 in these neuronal precursors [79]. It has been shown also that *Lunatic fringe*, an enhancer of Notch activity, is expressed throughout pancreatic development, and its maximal expression coincides with the expansion of the pancreatic precursors. Moreover, FGF10 induces *Lunatic fringe* expression in the E10.5 isolated dorsal pancreatic epithelium. Thus, *in vivo* FGF10 could maintain Notch activity in the pancreatic precursors by inducing *Lunatic fringe* [77]. The study of the interactions between the Notch pathway and other signalling cascades implicated in pancreas development will be crucial to further unravel the mechanisms controlling the self-renewal of the pancreatic precursors.

Wnt/β-Catenin Signalling

Wnt proteins are a large family of signalling molecules with well established roles in regulating embryonic patterning, cell proliferation and cell determination [80-82]. In particular, Wnt signalling has been found to be required for the maintenance of the self-renewal capacity of several progenitors cell types including hematopietic, intestine and neural progenitors [83-85]. In these instances, the different Wnt proteins activate gene transcription through a pathway controlled by β-catenin. Wnt binding to Frizzled (Frz) receptors activates the intracellular protein dishevelled. This results in the inhibition of the glycogen synthase kinase-3β (GSK-3β) which results, in turn, in the stabilization of free cytosolic β-catenin. Free β-catenin enters the nucleus to activate target genes in collaboration with members of the Lef/TCF family of transcription factors. In the absence of Wnt signalling, β-catenin is associated with a complex including GSK-3β, Axin, and the adenomatous polyposis coli tumor protein (APC). In this complex, the β-catenin is phosphorylated by the GSK-3β and subsequently degraded by the proteasome [81, 86].

Several members of the Wnt family as well as their regulators, the sFRP extracellular proteins, and Frz receptors have been found to be expressed in the epithelium and mesenchyme of the developing pancreas [87, 88]. Moreover, expression of *Wnt1* and *Wnt5a* under the control of the *Pdx1* promoter resulted in agenesis or hypoplasia of the pancreas [87, 89]. A recent study has shown that the early pancreatic epithelium (E11-E13.5) exhibits robust expression of unphosphorylated β-catenin, which indicates activation of the Wnt signalling pathway in the developing pancreas [90]. To determine whether this pathway is required for pancreas development, *β-catenin* was specifically deleted in the pancreas. The pancreas developing in the absence of β-catenin where dramatically smaller than the wild type, and contained very few acinar cells. In these mice, the specification of the early pancreatic progenitors occurs normally and they do not exhibit premature differentiation or death. However, the authors of this study noticed an important decrease in the proliferation of the pancreatic progenitor cells which paralleled a significant downregulation of the proto-oncogene Myc. The decrease in the proliferation rate of the pancreatic progenitors, however, could not explain the paucity of acinar cells in these animals. Thus, in addition to maintaining robust progenitors proliferation, β-catenin could be required for their subsequent differentiation into acinar cells. Another study has shown that overexpression of a dominant-negative form of mouse Frz8 under the control of the *Pdx1* promoter (*pPdx1-Frz8CRD*) severely pertubs pancreatic growth [91]. At birth, these mice show a 75% reduction in overall pancreatic mass. The detailed analysis of the development of the pancreas of the *pPdx1-Frz8CRD* mice ruled-out the existence of any defects in pancreatic specification. Premature differentiation or increased apoptosis were also ruled-out. However, a 25% reduction in the number of proliferating epithelial cells was observed at E10, and a 50% reduction at E13. At E15, when the proliferation of the pancreatic epithelium declines, the difference in the proliferation rate was of merely 20%. Thus, this study corroborates that Wnt/β-catenin signalling is required to maintain a high rate of proliferation in the pancreatic progenitors.

Cross-Talk between Signalling Pathways in Pancreas Development

Notch, FGF, EGF and Wnt signalling pathways cross-talk during embryogenesis, tissue regeneration and carcinogenesis [92, 93]. As discussed above, all these pathways are involved in pancreas development. Thus, cross-talk between these pathways is most likely essential to ensure the specification of the pancreatic progenitors, their expansion and later differentiation. However, to date, such interactions remain poorly explored. As mentioned above, interplay between Notch and FGF signalling is required to maintain the self-renewal capacity of the pancreatic progenitors and to avoid their premature differentiation [68, 76, 77]. It has been also shown that the TGFα can induce Notch activation in explant cultures of pancreatic acinar cells. Upon treatment with TGFα, these cells expressed high levels of Hes1 and Pdx1 and underwent acinar to ductal metaplasia [94]. In addition, the fact that the FGF and Wnt signalling pathways are both required to stimulate the proliferation of the pancreatic progenitors suggests that these pathways act together to ensure maximal growth of the pancreatic progenitors [90, 91]. Recent studies have shown that *Jagged-1*, *mNumb*, *Spry*, and several *FGFs* are target genes of the canonical Wnt pathway [95-97]. Such results suggest that Wnt signalling could be involved in the control of Notch activity, in particular by controlling the expression of *Jagged-1* and *mNumb*. mNumb is an adaptator molecule which binds to the intracellular domain of Notch (Noth-IC). In addition, mNumb has a proline-rich-domain which interacts with the E3 ligase, Itch. mNumb and Itch act cooperatively to promote the ubiquitination of Notch at the cell membrane, which, in turn, promotes the degradation of Notch-IC and, as a consequence, circunvents the activation of Notch target genes [98]. mNumb has four splicing isoforms which can be divided into two types based on the presence (PRRL) or absence (PRRS) of an aminoacid insert in the proline-rich-region (PRR). It has been proposed that distinct functions may be atributable to these two different types of isoforms. For example, in the optical anlage of the *Drosophila* larval brain, the PRRL-isoform promotes the proliferation of the neuroepithelial stem cells and post-embryonic neuroblasts, while the PRRS-isoform inhibits the proliferation of the stem cells and promotes neural differentiation. Moreover, it has been shown that the PRRS-isoform decreases more effectively the amount of nuclear Notch than the PRRL-isoform [99].

From E10.5 to E14.5, the PRRL-isoform of mNumb is expressed in the majority of the undifferentiated cells of the mouse pancreatic rudiments. However, at E14.5 the PRRL-isoform of mNumb is down-regulated in the endocrine progenitors expressing Ngn3. By contrast, at this stage, the pro-endocrine precursors express the PRRS-isoform. The mature β-cells do also express the PRRS-isoform of Numb. On the contrary, differentiated acinar cells mantain the expression of both the PRRL and PRRS-isoforms of mNumb [78]. These observations suggest that the maintenance of Notch signalling in the pancreatic progenitors could be regulated by mNumb. In the pancreatic endocrine progenitors at least, the down-regulation of PRRL mNumb and the activation of the PRRS-isoform could lead to the ubiquitination and degradation of Notch-IC, thus interferring with the activation of Notch target genes, including Hes1, and therefore allowing the expression of Ngn3. Although, mNumb is a target gene of Wnt signalling, our current knowledge suggests that, in the pancreatic progenitors, mNumb expression is probably not exclusively under the control of the Wnt pathway. Indeed, mice with pancreas specific inactivation of *β-catenin* or *pPdx1-Frz8CRD* mice overexpressing a dominant negative form of the Frz8 receptor do not show premature endocrine or exocrine differentiation, which would be expected if mNumb were exclusively under the control of the Wnt pathway. Alternatively, mNumbs could be only part of a more complex mechanism controlling the maintenance of Notch signalling in the pancreatic progenitors, which would include other Notch regulators, such Sel1, Lunatic fringe, etc… and several signalling pathways. Further studies will be required to establish the role of these modulators of Notch signalling in the pancreatic progenitors and to identify the signalling pathways involved in their regulation.

Direct Interactions between Notch Signalling Components and P48 Are Involved in the Maintenance of the Self-Renewal Capacity of Pancreatic Progenitors

As previously mentioned, *Ptf1α* encodes the transcription factor P48, a bHLH homeodomain protein essential for the specification and expansion of the pancreatic progenitors. Later in development, P48 is also required to ensure the differentiation of the acinar cells. The late effects of P48 are known to occur through the PTF1 transcriptional complex. This complex includes the transcription factor P48, one of the common class A bHLH proteins (either, HEB, E2-2, E12 or E47) and the CSL protein RBP-L. The PTF1 complex binds to common promoter elements in acinar cell-specific genes, resulting in acinar cell differentiation and cell cycle exit [100, 101]. Experimental evidence suggests that Notch signalling modulates the switch between the early and late functions of P48 [64-66]. Although the early effects of P48 on pancreas specification occur in the setting of an active Notch pathway, Notch signalling is silenced with the onset of acinar cell differentiation. Moreover, Notch has been shown to actively inhibit acinar differentiation by preventing the binding of the PTF1 complex to DNA [66]. This effect would be mediated by the Notch target proteins HES (Hes1, Hey1 and Hey2). Thus, active Notch signalling would delay the onset of acinar differentiation, and thereby allow the expansion of the pancreatic progenitors pool. It has been demonstrated that the capacity of Hes1 to inhibit the transcriptional activity of the PTF1 complex is dependent on its capacity to bind to P48 [102]. Hes1 interacts with the C-terminal domain of P48 through both its C-terminal and bHLH domains. Thus, the physical interactions between Hes1 and P48 provide a simple mechanism to explain the inhibitory effect of Notch on acinar differentiation. However, further complexity has been added by the recent discovery that P48 can bind to the CSL factor RBP-J, the DNA-bound transcriptional mediator of Notch signalling (Figure 4). This study has shown that the binding of P48 and the Notch-IC to RBP-J are mutually exclusive. According to that, both P48 and Notch-IC are competing for the same binding site on RBP-J [33]. Thus, binding of Notch-IC to RBP-J could interfere with the formation of the PTF1 complex and prevent the differentiation of the pancreatic progenitors into acinar cells. Interestingly, the authors reported that RBP-J can replace its Notch insensitive paralogue RBP-L, in the PTF1 complex and activate a luciferase reporter driven by six copies of the PTF1 binding site in the *Ela1* promoter. However, PTF1 complexes containing RBP-L provided highest transcriptional activity. Adult acinar cells express both the RBP-J and RBP-L forms, but the PTF1 complex contains only the RBP-L form. Thus, it has been proposed that in mature acinar cells there should be a selective mechanism which excludes RBP-J from the PTF1 complex. The relevance of P48 binding to RBPs for pancreas development is further stressed by the fact that mutations which delete one or both of the RBP-interacting motifs of P48 are associated with a human genetic disorder characterized by pancreatic agenesis, indicating that the association of P48 and the RBPs is required for proper embryonic development [103].

CONCLUSION

Over the last 10 years our understanding of pancreas development has evolved spectacularly. In this relatively short period of time we have progressed from an essentially descriptive knowledgement of the morphological aspects of pancreas organogenesis to gain considerable insight into the molecular mechanisms directing the development of this organ (Figure 5). Much of this progress has been triggered by the conviction that a sound knowledge of pancreas development will allow us to develop more effective therapies to fight the pathologies which grieve this organ, in particular diabetes. However, several important aspects of pancreas development remain to be elucidated.

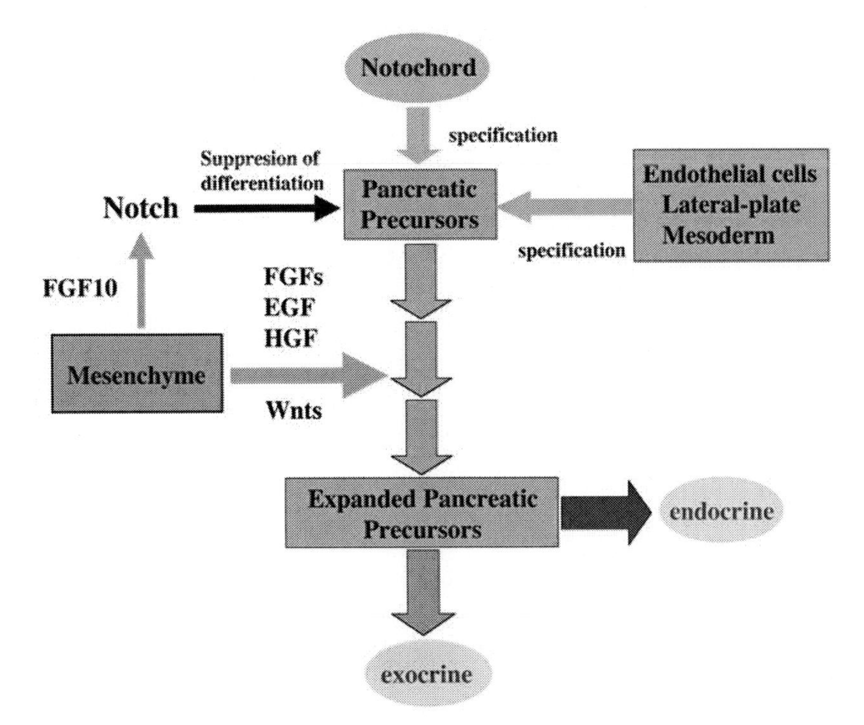

Figure 5. Schematic representation of pancreas development. A small pool of cells from the pro-pancreatic dorsal and ventral endoderm become specified as pancreatic progenitors by signals originating from the notochord, dorsal aorta and lateral-plate mesoderm. The nature of these signals remains elusive. Once specified, the pancreatic progenitors initiate a phase of intense proliferation. The pancreatic mesenchyme secretes growth factors such FGFs, HGF, EGF and Wnts which stimulate the proliferation of the pancreatic progenitors. The capacity of the pancreatic progenitors to proliferate without differentiating (self-renewal) is dependent upon the maintenance of active Notch signalling in these cells. It has been demonstrated that the FGF10, secreted by the pancreatic mesenchyme is required to maintain active Notch signalling in the pancreatic progenitors. Therefore signals from the pancreatic mesenchyme ensure the expansion of the initially small pool of pancreatic progenitors and prevent their premature differentiation. Later, the pancreatic progenitors differentiate into exocrine and endocrine cells. The origin and identity of the signals instructing the differentiation of the pancreatic progenitors towards a particular cell type remains to be elucidated.

Much remains to be discovered about the formation of the endoderm, its regionalization in organ domains and the initiation of pancreas specification. Although in *Xenopus* ectopic co-expression of *Pdx1* and *Ptf1α* is sufficient to convert non-pancreatic endodermal cells into pancreatic precursor cells, we still do not know how the activation of these two factors modifies the basic transcriptional program of the endoderm. The notochord, vascular tissue and embryonic mesoderm have been identified as the source of signals initiating pancreatic induction in the adjacent endoderm. However, such signals remain to be identified. Moreover, although we know that the initiation of the specification of the precursors in the dorsal and ventral pancreatic buds has different developmental requirements we still do not know their precise nature. Another crucial aspect of pancreas development concerns the mechanisms orchestrating the necessary coordination between the growth, morphogenesis and cytodifferentiation of the organ.

The central role of the Notch pathway in the maintenance of the self-renewal capacity of the pancreatic precursors and its interplay with the FGF10 signalling have been recently discovered. These interactions constitute a plausible mechanism for the regulation of growth and differentiation. However, the details of these interactions remain to be elucidated. The interactions between other pathways involved in pancreas development, TGF-β, Wnt/β-catenin, Hedgehog, EGF, etc… must also to be explored. Another prominent area in the field of pancreas development concerns the specification of the different endocrine cell types. The Ngn3-expressing cells in the developing pancreas have been identified as endocrine precursors. However we still do not know the origin of the signals and signalling pathways instructing these precursors to differentiate towards a particular endocrine cell type.

Great expectations have been placed in the conviction that our knowledge of the mechanisms of pancreas development could allow the emergence of new therapies for diabetes. Optimized islet cell transplantation therapies hold the prospect to provide a curative treatment for diabetes [104, 105]. However, the scarcity of donor supply prevents the development of these therapies on a large scale. One way to reduce the shortage of islets for transplantation would be to produce large amounts of β-cells *in vitro*. However, β-cells have a very low replicative capacity thus alternative sources must be found. Several strategies are currently investigated including the use of xenogenic islets, of transformed cells (insulinomas), the transdifferentiation of liver cells, and the differentiation of embryonic stem cells, pluripotent mesenchymal cells, or hematopoietic stem cells [106, 107]. Recent advances in the knowledge of embryonic stem (ES) cells biology and their potential have considerably encouraged the development of the strategies based on the use of these cells for the production of pancreatic β-cells. Several reports have appeared on attempts to differentiate mouse ES into functional β-cells by means of empirical protocols [108-110]. However, to date these protocols have generated scarce insulin-expressing cells, which express and secrete very low levels of insulin and often lack the expression of key components of the β–cells [106]. Thus, the experience acquired during the past few years indicates the need to develop rational protocols, which recreate *in vitro* what happens *in vivo*. Stem cells must first be induced to become definitive endoderm, then differentiate into pancreatic endoderm, and finally reach the stage of differentiated mature β-cells. Although this approach still remains theoretical, an important step forward has been recently made. D'Amour and collaborators have developed an efficient protocol allowing the generation of definitive endoderm from human ES cells [111]. Certainly, many difficulties would need to be overcome before we find a way to induce the expression in ES cells of the transcription factors initiating the pancreatic specification program. Other difficulties will need to be resolved to differentiate the pancreatic precursors generated *in vitro* into functional β-cells. No doubt however, that a thorough understanding of the molecular mechanisms controlling pancreas development will provide the keys to overcome all these difficulties and open the way to innovative and efficient therapies for diabetes.

ACKNOWLEDGEMENTS

The author thanks Drs. Jean-Paul Concordet, Francina Langa, Pascal Maire and Suzan Saint-Just Elsevier for critical comments and valuable suggestions on this manuscript.

REFERENCES

[1] Bockman, D. E. (1993). Anatomy of the pancreas. In V. Go, E. DiMagno, J. Gardner, E. Lebenthal, H. Reber, and G. Scheele, (Eds.), *The pancreas: biology, pathophysiology, and disease.* (2nd, ed. pp. 1-8). New York, Raven Press Ltd.

[2] Slack, J. (1995). Developmental biology of the pancreas. *Development, 121,* 1569-1580.

[3] Pictet, R. and Rutter, W. (1972). Development of the embryonic endocine pancreas. In D. Steiner and N. Freinkel, (Eds.), *Handbook of Physiology* (pp. 25-66). Washington, Williams and Wilkins.

[4] Herrera, P., Huarte, J., Sanvito, F., Meda, P., Orci, L. and Vassali, J. (1991). Embryogenesis of the murine pancreas; early expression of pancreatic polypeptide gene. *Development, 113,* 1257-1265.

[5] Teitelman, G., Alpert, S., Polak, J. M., Martinez, A. and Hanahan, D. (1993). Precursor cells of mouse endocrine pancreas coexpress insulin, glucagon and the neuronal proteins tyrosine hydroxylase and neuropeptide Y, but not pancreatic polypeptide. *Development, 118,* 1031-9.

[6] Upchurch, B. H., Aponte, G. W. and Leiter, A. B. (1994). Expression of peptide YY in all four islet cell types in the developing mouse pancreas suggests a common peptide YY-producing progenitor. *Development, 120,* 245-52.

[7] Herrera, P. L. (2000). Adult insulin- and glucagon-producing cells differentiate from two independent cell lineages. *Development, 127,* 2317-22.

[8] Wells, J. M. and Melton, D. A. (1999). Vertebrate endoderm development. *Annu Rev Cell Dev Biol, 15,* 393-410.

[9] Barbacci, E., Reber, M., Ott, M. O., Breillat, C., Huetz, F. and Cereghini, S. (1999). Variant hepatocyte nuclear factor 1 is required for visceral endoderm specification. *Development, 126,* 4795-805.

[10] Monaghan, A. P., Kaestner, K. H., Grau, E. and Schutz, G. (1993). Postimplantation expression patterns indicate a role for the mouse forkhead/HNF-3 alpha, beta and gamma genes in determination of the definitive endoderm, chordamesoderm and neuroectoderm. *Development, 119,* 567-78.

[11] Zaret, K. (1999). Developmental competence of the gut endoderm: genetic potentiation by GATA and HNF3/fork head proteins. *Dev. Biol, 209,* 1-10.

[12] Laverriere, A. C., MacNeill, C., Mueller, C., Poelmann, R. E., Burch, J. B. and Evans, T. (1994). GATA-4/5/6, a subfamily of three transcription factors transcribed in developing heart and gut. *J. Biol. Chem., 269,* 23177-84.

[13] Hebrok, M. (2003). Hedgehog signaling in pancreas development. *Mech. Dev., 120,* 45-57.

[14] Jensen, J. (2004). Gene regulatory factors in pancreatic development. *Dev. Dyn., 229,* 176-200.

[15] Kumar, M. and Melton, D. (2003). Pancreas specification: a budding question. *Curr. Opin. Genet Dev., 13,* 401-7.

[16] Kim, S. K. and MacDonald, R. J. (2002). Signaling and transcriptional control of pancreatic organogenesis. *Curr. Opin. Genet Dev., 12,* 540-7.

[17] Ahlgren, U., Jonsson, J., Jonsson, L., Simu, K. and Edlund, H. (1998). beta-cell-specific inactivation of the mouse Ipfl/Pdx1 gene results in loss of the beta-cell phenotype and maturity onset diabetes. *Genes Dev., 12,* 1763-8.

[18] Ahlgren, U., Jonsson, J. and Edlund, H. (1996). The morphogenesis of the pancreatic mesenchyme is uncoupled from that of the pancreatic epithelium in IPF1/PDX1-deficient mice. *Development, 122,* 1409-1416.

[19] Jonsson J, Carlsson L, Edlund T and Edlund H. (1994). Insulin-promoter-factor I is required for pancreas development in mice. *Nature, 371,* 606-609.

[20] Offield MF, Jetton TL, Laborsky PA, Ray M, Stein RW, Magnuson MA, Hogan BLM and Wright CVE. (1996). PDX-1 is required for pancreatic outgrowth and differentiation of the rostral duodenum. *Development, 122,* 983-995.

[21] Stoffers, D. A., Heller, R. S., Miller, C. P. and Habener, J. F. (1999). Developmental expression of the homeodomain protein IDX-1 in mice transgenic for an IDX-1 promoter/lacZ transcriptional reporter. *Endocrinology, 140,* 5374-81.

[22] Petersen, H. V., Serup, P., Leonard, J., Michelsen, B. K. and Madsen, O. D. (1994). Transcriptional regulation of the human insulin gene is dependent on the homeodomain protein STF1/IPF1 acting through the CT boxes. *Proc. Natl. Acad. Sci USA, 91,* 10465-9.

[23] Bonny, C., Thompson, N., Nicod, P. and Waeber, G. (1995). Pancreatic-specific expression of the glucose transporter type 2 gene: identification of cis-elements and islet-specific trans-acting factors. *Mol. Endocrinol., 9,* 1413-26.

[24] Watada, H., Kajimoto, Y., Miyagawa, J., Hanafusa, T., Hamaguchi, K., Matsuoka, T., Yamamoto, K., Matsuzawa, Y., Kawamori, R. and Yamasaki, Y. (1996). PDX-1 induces insulin and glucokinase gene expressions in alphaTC1 clone 6 cells in the presence of betacellulin. *Diabetes, 45,* 1826-31.

[25] Serup, P., Jensen, J., Andersen, F. G., Jorgensen, M. C., Blume, N., Holst, J. J. and Madsen, O. D. (1996). Induction of insulin and islet amyloid polypeptide production in pancreatic islet glucagonoma cells by insulin promoter factor 1. *Proc. Natl. Acad. Sci. USA, 93,* 9015-20.

[26] Gu, G., Dubauskaite, J. and Melton, D. A. (2002). Direct evidence for the pancreatic lineage: NGN3+ cells are islet progenitors and are distinct from duct progenitors. *Development, 129,* 2447-57.
[27] Jacquemin, P., Lemaigre, F. P. and Rousseau, G. G. (2003). The Onecut transcription factor HNF-6 (OC-1) is required for timely specification of the pancreas and acts upstream of Pdx-1 in the specification cascade. *Dev. Biol, 258,* 105-16.
[28] Rausa, F., Samadani, U., Ye, H., Lim, L., Fletcher, C. F., Jenkins, N. A., Copeland, N. G. and Costa, R. H. (1997). The cut-homeodomain transcriptional activator HNF-6 is coexpressed with its target gene HNF-3 beta in the developing murine liver and pancreas. *Dev. Biol, 192,* 228-46.
[29] Poll, A. V., Pierreux, C. E., Lokmane, L., Haumaitre, C., Achouri, Y., Jacquemin, P., Rousseau, G. G., Cereghini, S. and Lemaigre, F. P. (2006). A vHNF1/TCF2-HNF6 cascade regulates the transcription factor network that controls generation of pancreatic precursor cells. *Diabetes, 55,* 61-9.
[30] Haumaitre, C., Barbacci, E., Jenny, M., Ott, M. O., Gradwohl, G. and Cereghini, S. (2005). Lack of TCF2/vHNF1 in mice leads to pancreas agenesis. *Proc. Natl. Acad. Sci. USA, 102,* 1490-5.
[31] Harrison, K. A., Thaler, J., Pfaff, S. L., Gu, H. and Kehrl, J. H. (1999). Pancreas dorsal lobe agenesis and abnormal islets of Langerhans in Hlxb9-deficient mice. *Nat. Genet, 23,* 71-5.
[32] Li, H., Arber, S., Jessell, T. M. and Edlund, H. (1999). Selective agenesis of the dorsal pancreas in mice lacking homeobox gene Hlxb9. *Nat. Genet, 23,* 67-70.
[33] Beres, T. M., Masui, T., Swift, G. H., Shi, L., Henke, R. M. and MacDonald, R. J. (2006). PTF1 is an organ-specific and Notch-independent basic helix-loop-helix complex containing the mammalian Suppressor of Hairless (RBP-J) or its paralogue, RBP-L. *Mol. Cell Biol., 26,* 117-30.
[34] Krapp, A., Knofler, M., Ledermann, B., Burki, K., Berney, C., Zoerkler, N., Hagenbuchle, O. and Wellauer, P. K. (1998). The bHLH protein PTF1-p48 is essential for the formation of the exocrine and the correct spatial organization of the endocrine pancreas. *Genes Dev., 12,* 3752-63.
[35] Kawaguchi, Y., Cooper, B., Gannon, M., Ray, M., MacDonald, R. J. and Wright, C. V. (2002). The role of the transcriptional regulator Ptf1a in converting intestinal to pancreatic progenitors. *Nat. Genet, 32,* 128-34.
[36] Afelik, S., Chen, Y. and Pieler, T. (2006). Combined ectopic expression of Pdx1 and Ptf1a/p48 results in the stable conversion of posterior endoderm into endocrine and exocrine pancreatic tissue. *Genes Dev., 20,* 1441-6.
[37] Grapin-Botton, A., Majithia, A. R. and Melton, D. A. (2001). Key events of pancreas formation are triggered in gut endoderm by ectopic expression of pancreatic regulatory genes. *Genes Dev., 15,* 444-54.
[38] Kim, S. K., Hebrok, M. and Melton, D. A. (1997). Notochord to endoderm signaling is required for pancreas development. *Development, 124,* 4243-52.
[39] Lammert, E., Cleaver, O. and Melton, D. (2001). Induction of pancreatic differentiation by signals from blood vessels. *Science, 294,* 564-7.

[40] Apelqvist, A., Ahlgren, U. and Edlund, H. (1997). Sonic hedgehog directs specialised mesoderm differentiation in the intestine and pancreas [published erratum appears in Curr Biol 1997 Dec 1;7(12):R809]. *Curr. Biol., 7,* 801-4.

[41] Kim, S. K. and Melton, D. A. (1998). Pancreas development is promoted by cyclopamine, a hedgehog signaling inhibitor. *Proc. Natl. Acad. Sci. USA, 95,* 13036-41.

[42] Hebrok, M., Kim, S. K. and Melton, D. A. (1998). Notochord repression of endodermal Sonic hedgehog permits pancreas development. *Genes Dev., 12,* 1705-13.

[43] Shalaby, F., Rossant, J., Yamaguchi, T. P., Gertsenstein, M., Wu, X. F., Breitman, M. L. and Schuh, A. C. (1995). Failure of blood-island formation and vasculogenesis in Flk-1-deficient mice. *Nature, 376,* 62-6.

[44] Yoshitomi, H. and Zaret, K. S. (2004). Endothelial cell interactions initiate dorsal pancreas development by selectively inducing the transcription factor Ptf1a. *Development, 131,* 807-17.

[45] Golosow, N. and Grobstein, C. (1962). Epitheliomesenchymal interaction in pancreatic morphogenesis. *Dev. Biol, 4,* 242-255.

[46] Wessels, N. and Cohen, J. (1967). Early pancreas organogenesis: morphogenesis, tissue interactions, and mass effects. *Dev. Biol, 15,* 237-270.

[47] Spooner, B., Cohen, H. and Faubion, J. (1977). Development of the embryonic mammalian pancreas: the relationship between morphogenesis and cytodifferentiation. *Dev. Biol, 61,* 119-130.

[48] Ahlgren, U., Pfaff, S., Jessel, T., Edlund, T. and Edlund, H. (1997). Independent requirement for ISL1 in formation of pancreatic mesenchyme and islet cells. *Nature, 385,* 257-260.

[49] Miettinen, P. J., Huotari, M., Koivisto, T., Ustinov, J., Palgi, J., Rasilainen, S., Lehtonen, E., Keski-Oja, J. and Otonkoski, T. (2000). Impaired migration and delayed differentiation of pancreatic islet cells in mice lacking EGF-receptors. *Development, 127,* 2617-27.

[50] Sanvito F, Herrera PL, Huarte J, Nichols A, Montesano R, Orci L and Vassali JD. (1994). TGF-ß1 influences the relative development of the exocrine and endocrine pancreas in vitro. *Development, 120,* 3451-3462.

[51] Otonkoski, T., Cirulli, V., Beattie, G., Mally, M., Soto, G., Rubin, J. and Hayek, A. (1996). A role for hepatocyte growth factor/scatter factor in fetal mesenchyme-induced pancreatic ß-cell growth. *Endocrinology, 137,* 3131-3139.

[52] Le Bras, S., Miralles, F., Basmaciogullari, A., Czernichow, P. and Scharfmann, R. (1998). Fibroblast growth factor 2 promotes pancreatic epithelial cell proliferation via functional fibroblast growth factor receptors during embryonic life. *Diabetes, 47,* 1236-42.

[53] Miralles, F., Czernichow, P., Ozaki, K., Itoh, N. and Scharfmann, R. (1999). Signaling through fibroblast growth factor receptor 2b plays a key role in the development of the exocrine pancreas. *Proc. Natl. Acad. Sci. USA, 96,* 6267-72.

[54] Jacquemin, P., Yoshitomi, H., Kashima, Y., Rousseau, G. G., Lemaigre, F. P. and Zaret, K. S. (2006). An endothelial-mesenchymal relay pathway regulates early phases of pancreas development. *Dev. Biol, 290,* 189-99.

[55] Kumar, M., Jordan, N., Melton, D. and Grapin-Botton, A. (2003). Signals from lateral plate mesoderm instruct endoderm toward a pancreatic fate. *Dev. Biol, 259,* 109-22.

[56] Deutsch, G., Jung, J., Zheng, M., Lora, J. and Zaret, K. S. (2001). A bipotential precursor population for pancreas and liver within the embryonic endoderm. *Development, 128,* 871-81.

[57] Beatus, P. and Lendahl, U. (1998). Notch and neurogenesis. *J. Neurosci. Res., 54,* 125-36.

[58] Fisher, A. and Caudy, M. (1998). The function of hairy-related bHLH repressor proteins in cell fate decisions. *Bioessays, 20,* 298-306.

[59] Kuroda, K., Tani, S., Tamura, K., Minoguchi, S., Kurooka, H. and Honjo, T. (1999). Delta-induced Notch signaling mediated by RBP-J inhibits MyoD expression and myogenesis. *J. Biol. Chem., 274,* 7238-44.

[60] Lee, J. C., Smith, S. B., Watada, H., Lin, J., Scheel, D., Wang, J., Mirmira, R. G. and German, M. S. (2001). Regulation of the pancreatic pro-endocrine gene neurogenin3. *Diabetes, 50,* 928-36.

[61] Jensen, J., Pedersen, E. E., Galante, P., Hald, J., Heller, R. S., Ishibashi, M., Kageyama, R., Guillemot, F., Serup, P. and Madsen, O. D. (2000). Control of endodermal endocrine development by Hes-1. *Nat. Genet, 24,* 36-44.

[62] Gradwohl, G., Dierich, A., LeMeur, M. and Guillemot, F. (2000). neurogenin3 is required for the development of the four endocrine cell lineages of the pancreas. *Proc. Natl. Acad. Sci. USA, 97,* 1607-11.

[63] Apelqvist, A., Li, H., Sommer, L., Beatus, P., Anderson, D. J., Honjo, T., Hrabe de Angelis, M., Lendahl, U. and Edlund, H. (1999). Notch signalling controls pancreatic cell differentiation. *Nature, 400,* 877-81.

[64] Hald, J., Hjorth, J. P., German, M. S., Madsen, O. D., Serup, P. and Jensen, J. (2003). Activated Notch1 prevents differentiation of pancreatic acinar cells and attenuate endocrine development. *Dev. Biol, 260,* 426-37.

[65] Murtaugh, L. C., Stanger, B. Z., Kwan, K. M. and Melton, D. A. (2003). Notch signaling controls multiple steps of pancreatic differentiation. *Proc. Natl. Acad. Sci. USA, 100,* 14920-5.

[66] Esni, F., Ghosh, B., Biankin, A. V., Lin, J. W., Albert, M. A., Yu, X., MacDonald, R. J., Civin, C. I., Real, F. X., Pack, M. A., Ball, D. W. and Leach, S. D. (2004). Notch inhibits Ptf1 function and acinar cell differentiation in developing mouse and zebrafish pancreas. *Development, 131,* 4213-24.

[67] Katsube, K. and Sakamoto, K. (2005). Notch in vertebrates--molecular aspects of the signal. *Int. J. Dev. Biol., 49,* 369-74.

[68] Norgaard, G. A., Jensen, J. N. and Jensen, J. (2003). FGF10 signaling maintains the pancreatic progenitor cell state revealing a novel role of Notch in organ development. *Dev. Biol., 264,* 323-38.

[69] Rutter, W., Pictet, R., Harding, J., Chirgwin, J., MacDonald, R. and Przybyla, A. (1978). An analysis of pancreattic development: role of mesenchymal factor and other extracellular factors. *Symp. Soc. Dev. Biol., 35,* 205-227.

[70] Gittes, G., Galante, P., Hanahan, D., Rutter, W. and Debas, H. (1996). Lineage specific morphogenesis in the developing pancreas: role of mesenchymal factors. *Development, 122,* 439-447.

[71] Miralles, F., Czernichow, P. and Scharfmann, R. (1997). Follistatin regulates the relative proportions of endocrine versus exocrine tissue during pancreatic development. *Development,125,* 1017-1024.

[72] Celli, G., LaRochelle, W. J., Mackem, S., Sharp, R. and Merlino, G. (1998). Soluble dominant-negative receptor uncovers essential roles for fibroblast growth factors in multi-organ induction and patterning. *Embo. J., 17,* 1642-55.

[73] Revest, J. M., Spencer-Dene, B., Kerr, K., De Moerlooze, L., Rosewell, I. and Dickson, C. (2001). Fibroblast growth factor receptor 2-IIIb acts upstream of Shh and Fgf4 and is required for limb bud maintenance but not for the induction of Fgf8, Fgf10, Msx1, or Bmp4. *Dev. Biol., 231,* 47-62.

[74] Bhushan, A., Itoh, N., Kato, S., Thiery, J. P., Czernichow, P., Bellusci, S. and Scharfmann, R. (2001). Fgf10 is essential for maintaining the proliferative capacity of epithelial progenitor cells during early pancreatic organogenesis. *Development, 128,* 5109-17.

[75] Ohuchi, H. (2000). [Roles for FGF-FGFR signaling during vertebrate development]. *Hum. Cell, 13,* 169-75.

[76] Hart, A., Papadopoulou, S. and Edlund, H. (2003). Fgf10 maintains notch activation, stimulates proliferation, and blocks differentiation of pancreatic epithelial cells. *Dev. Dyn., 228,* 185-93.

[77] Miralles, F., Lamotte, L., Couton, D. and Joshi, R. L. (2006). Interplay between FGF10 and Notch signalling is required for the self-renewal of pancreatic progenitors. *Int. J. Dev. Biol, 50,* 17-26.

[78] Yoshida, T., Tokunaga, A., Nakao, K. and Okano, H. (2003). Distinct expression patterns of splicing isoforms of mNumb in the endocrine lineage of developing pancreas. *Differentiation, 71,* 486-95.

[79] Faux, C. H., Turnley, A. M., Epa, R., Cappai, R. and Bartlett, P. F. (2001). Interactions between fibroblast growth factors and Notch regulate neuronal differentiation. *J. Neurosci., 21,* 5587-96.

[80] Karner, C., Wharton, K. A. and Carroll, T. J. (2006). Apical-basal polarity, Wnt signaling and vertebrate organogenesis. *Semin. Cell Dev. Biol., 17,* 214-22.

[81] Widelitz, R. (2005). Wnt signaling through canonical and non-canonical pathways: recent progress. *Growth Factors, 23,* 111-6.

[82] Wodarz, A. and Nusse, R. (1998). Mechanisms of Wnt signaling in development. *Annu Rev. Cell Dev. Biol., 14,* 59-88.

[83] Khan, N. I. and Bendall, L. J. (2006). Role of WNT signaling in normal and malignant hematopoiesis. *Histol. Histopathol, 21,* 761-74.

[84] Ross, J. and Li, L. (2006). Recent advances in understanding extrinsic control of hematopoietic stem cell fate. *Curr. Opin. Hematol., 13,* 237-42.

[85] Ille, F. and Sommer, L. (2005). Wnt signaling: multiple functions in neural development. *Cell Mol. Life Sci., 62,* 1100-8.

[86] Patapoutian, A. and Reichardt, L. F. (2000). Roles of Wnt proteins in neural development and maintenance. *Curr. Opin. Neurobiol., 10,* 392-9.

[87] Heller, R. S., Dichmann, D. S., Jensen, J., Miller, C., Wong, G., Madsen, O. D. and Serup, P. (2002). Expression patterns of Wnts, Frizzleds, sFRPs, and misexpression in transgenic mice suggesting a role for Wnts in pancreas and foregut pattern formation. *Dev. Dyn, 225,* 260-70.

[88] Gu, G., Wells, J. M., Dombkowski, D., Preffer, F., Aronow, B. and Melton, D. A. (2004). Global expression analysis of gene regulatory pathways during endocrine pancreatic development. *Development, 131,* 165-79.

[89] Lin, Y., Liu, A., Zhang, S., Ruusunen, T., Kreidberg, J. A., Peltoketo, H., Drummond, I. and Vainio, S. (2001). Induction of ureter branching as a response to Wnt-2b signaling during early kidney organogenesis. *Dev. Dyn., 222,* 26-39.

[90] Murtaugh, L. C., Law, A. C., Dor, Y. and Melton, D. A. (2005). Beta-catenin is essential for pancreatic acinar but not islet development. *Development, 132,* 4663-74.

[91] Papadopoulou, S. and Edlund, H. (2005). Attenuated Wnt signaling perturbs pancreatic growth but not pancreatic function. *Diabetes, 54,* 2844-51.

[92] Artavanis-Tsakonas, S., Rand, M. D. and Lake, R. J. (1999). Notch signaling: cell fate control and signal integration in development. *Science, 284,* 770-6.

[93] Radtke, F. and Raj, K. (2003). The role of Notch in tumorigenesis: oncogene or tumour suppressor? *Nat. Rev. Cancer, 3,* 756-67.

[94] Miyamoto, Y., Maitra, A., Ghosh, B., Zechner, U., Argani, P., Iacobuzio-Donahue, C. A., Sriuranpong, V., Iso, T., Meszoely, I. M., Wolfe, M. S., Hruban, R. H., Ball, D. W., Schmid, R. M. and Leach, S. D. (2003). Notch mediates TGF alpha-induced changes in epithelial differentiation during pancreatic tumorigenesis. *Cancer Cell, 3,* 565-76.

[95] Katoh, M. (2006). Notch ligand, JAG1, is evolutionarily conserved target of canonical WNT signaling pathway in progenitor cells. *Int. J. Mol. Med., 17,* 681-5.

[96] Katoh, M. (2006). NUMB is a break of WNT-Notch signaling cycle. *Int. J. Mol. Med., 18,* 517-21.

[97] Katoh, M. (2005). Comparative genomics on FGF8, FGF17, and FGF18 orthologs. *Int. J. Mol. Med, 16,* 493-6.

[98] McGill, M. A. and McGlade, C. J. (2003). Mammalian numb proteins promote Notch1 receptor ubiquitination and degradation of the Notch1 intracellular domain. *J. Biol. Chem., 278,* 23196-203.

[99] Toriya, M., Tokunaga, A., Sawamoto, K., Nakao, K. and Okano, H. (2006). Distinct functions of human numb isoforms revealed by misexpression in the neural stem cell lineage in the Drosophila larval brain. *Dev. Neurosci., 28,* 142-55.

[100] Rose, S. D., Swift, G. H., Peyton, M. J., Hammer, R. E. and MacDonald, R. J. (2001). The role of PTF1-P48 in pancreatic acinar gene expression. *J. Biol. Chem., 276,* 44018-26.

[101] Rodolosse, A., Chalaux, E., Adell, T., Hagege, H., Skoudy, A. and Real, F. X. (2004). PTF1alpha/p48 transcription factor couples proliferation and differentiation in the exocrine pancreas [corrected]. *Gastroenterology, 127,* 937-49.

[102] Ghosh, B. and Leach, S. D. (2006). Interactions between hairy/enhancer of split-related proteins and the pancreatic transcription factor Ptf1-p48 modulate function of the PTF1 transcriptional complex. *Biochem J., 393,* 679-85.

[103] Sellick, G. S., Barker, K. T., Stolte-Dijkstra, I., Fleischmann, C., Coleman, R. J., Garrett, C., Gloyn, A. L., Edghill, E. L., Hattersley, A. T., Wellauer, P. K., Goodwin, G. and Houlston, R. S. (2004). Mutations in PTF1A cause pancreatic and cerebellar agenesis. *Nat. Genet., 36,* 1301-5.

[104] Bertuzzi, F., Marzorati, S. and Secchi, A. (2006). Islet cell transplantation. *Curr. Mol. Med., 6,* 369-74.

[105] Shapiro, A. M., Lakey, J. R., Ryan, E. A., Korbutt, G. S., Toth, E., Warnock, G. L., Kneteman, N. M. and Rajotte, R. V. (2000). Islet transplantation in seven patients with type 1 diabetes mellitus using a glucocorticoid-free immunosuppressive regimen. *N. Engl. J. Med., 343,* 230-8.

[106] Stoffel, M., Vallier, L. and Pedersen, R. A. (2004). Navigating the pathway from embryonic stem cells to beta cells. *Semin. Cell Dev. Biol., 15,* 327-36.

[107] Colman, A. (2004). Making new beta cells from stem cells. *Semin Cell Dev Biol, 15,* 337-45.

[108] Lumelsky, N., Blondel, O., Laeng, P., Velasco, I., Ravin, R. and McKay, R. (2001). Differentiation of embryonic stem cells to insulin-secreting structures similar to pancreatic islets. *Science, 292,* 1389-94.

[109] Blyszczuk, P., Czyz, J., Kania, G., Wagner, M., Roll, U., St-Onge, L. and Wobus, A. M. (2003). Expression of Pax4 in embryonic stem cells promotes differentiation of nestin-positive progenitor and insulin-producing cells. *Proc Natl Acad Sci U S A, 100,* 998-1003.

[110] Moritoh, Y., Yamato, E., Yasui, Y., Miyazaki, S. and Miyazaki, J. (2003). Analysis of insulin-producing cells during in vitro differentiation from feeder-free embryonic stem cells. *Diabetes, 52,* 1163-8.

[111] D'Amour, K. A., Agulnick, A. D., Eliazer, S., Kelly, O. G., Kroon, E. and Baetge, E. E. (2005). Efficient differentiation of human embryonic stem cells to definitive endoderm. *Nat Biotechnol, 23,* 1534-41.

In: Stem Cell Research Trends
Editor: Josse R. Braggina, pp. 99-116
ISBN: 978-1-60021-622-0

Chapter III

STEM CELLS IN GASTROINTESTINAL CANCER – COMBINING EMBRYOGENESIS AND CARCINOGENESIS?

Daniel Neureiter[1], Christoph Herold[2] and Matthias Ocker[2*]
[1] Institute of Pathology, Salzburger Landeskliniken,
Paracelsus Private Medical University, Müllnerstrasse 48,
A-5020 Salzburg, Austria
[2] Department of Medicine 1, University Hospital Erlangen,
Ulmenweg 18, D-91054 Erlangen, Germany

ABSTRACT

Current concepts of gastrointestinal tumor formation favor a multi-step model with the acquisition of several targeted mutations in oncogenes and tumor suppressor genes that finally lead to the acquisition of different "hallmarks of cancer" like self-sufficiency in growth or anti-growth signals, evasion of apoptosis, immortality and angiogenesis as has been highlighted by Hanahan and Weinberg in their outstanding review [1]. Recent evidences indicate that targeted genetic events alone (e.g. point mutations in oncogenes) are not sufficient to provide tumor cells with these growth advantages but that the process of carcinogenesis closely resembles and involves processes of trans- and dedifferentiation as have been observed during embryonic development. Comparative analyses of embryogenesis and tumorigenesis, especially of the gastrointestinal tract, revealed immense analogies of morphological patterning as well as of expression of differentiation markers. Both processes of development and carcinogenesis as well as integral processes of regeneration like reparation after acute or chronic inflammation are characterized by pattern maintenance which is essentially maintained by different gradients of markers of embryonic differentiation. So called epigenetic modulations of chromatin and DNA elements lead to changes in phenotypic properties that can also

* Correspondence to: Dr. Matthias Ocker; Assistant Professor of Experimental Medicine; Department of Medicine 1, University Hospital Erlangen; Ulmenweg 18, D-91054 Erlangen, Germany; Phone: ++49-9131-8535057, Fax: ++49-9131-8535058; Matthias.Ocker@med1.imed.uni-erlangen.de

favor the malignant conversion of progenitor cells. Furthermore, the role of cell-cell and cell-matrix contacts has gathered new attention as these signaling pathways strongly influence survival and differentiation of embryonic and tumorigenic cells and is one of the key features of the still to be defined (cancer) stem cell niche, especially in the gastrointestinal tract. The so far unclear factors that contribute to the maintenance and longevity of tumor stem cells may also contribute to relapsing tumor diseases after successful first-line treatment and the outbreak of disseminated micro metastases. This novel concept yields great impact for the development of new diagnostic and therapeutic approaches, as highly chemo- and radioresistant long-living tumor progenitor or stem cells will enter the focus of future cancer medicine and research.

Introduction

Cancer diseases are a major healthcare and socioeconomic burden worldwide with global incidences and mortality increasing constantly during the past decades [2]. Although some progresses have been made in treatment of distinct cancers, overall prognosis still remains poor and improved means of diagnosis and therapy are urgently needed. Here, the current concepts focus on already established, commonly even already metastasized, tumor masses and aim at reducing the tumor load with curative or palliative intention. Present diagnostic and therapeutic tools mainly focus on genetic alterations in advanced cancer cell generations, e.g. mutations in oncogenes or activation of growth factor receptors, and several concepts have been postulated for the evolution of these mutations that finally provide the tumor cell with what has been defined as "the hallmarks of cancer" [1]. Although many, if not all, cancers exhibit these growth-promoting properties, these features and models do not sufficiently explain commonly observed phenomena of cancer development, treatment resistance and recurrence. Novel concepts of tumorigenesis thus adopted knowledge from the immune system and from embryonic development and evidence is accumulating that not already differentiated cells acquire the initiating genetic or epigenetic pro-carcinogenic events, but that peripheral, disseminated or bone-marrow related cells with a still stem cell like phenotype are the culprit of cancer development (see figure 1).

Multistep Carcinogenesis Models

Today, several models have been proposed how solid malignancies develop and acquire key features of cancer [1], i.e. independence to growth-regulating signals (e.g. activating point mutations in *K-ras*, mutations in *p53* or *Rb*) [3-9], deficiency in apoptosis inducing or executing pathways (e.g. overexpression of bcl-2) [10-15], tissue remodeling and metastasis (e.g. dysregulation in the β-catenin/WNT pathway) [16-20], unrestricted proliferation (e.g. by activation of telomerase) [21-23] and induction of neo-angiogenesis, e.g. by upregulation of VEGF-related signaling pathways [24-26]. These alterations are commonly found in a variety of human cancers and have been summed up to so-called multistep models of carcinogenesis that finally lead to a malignant phenotype.

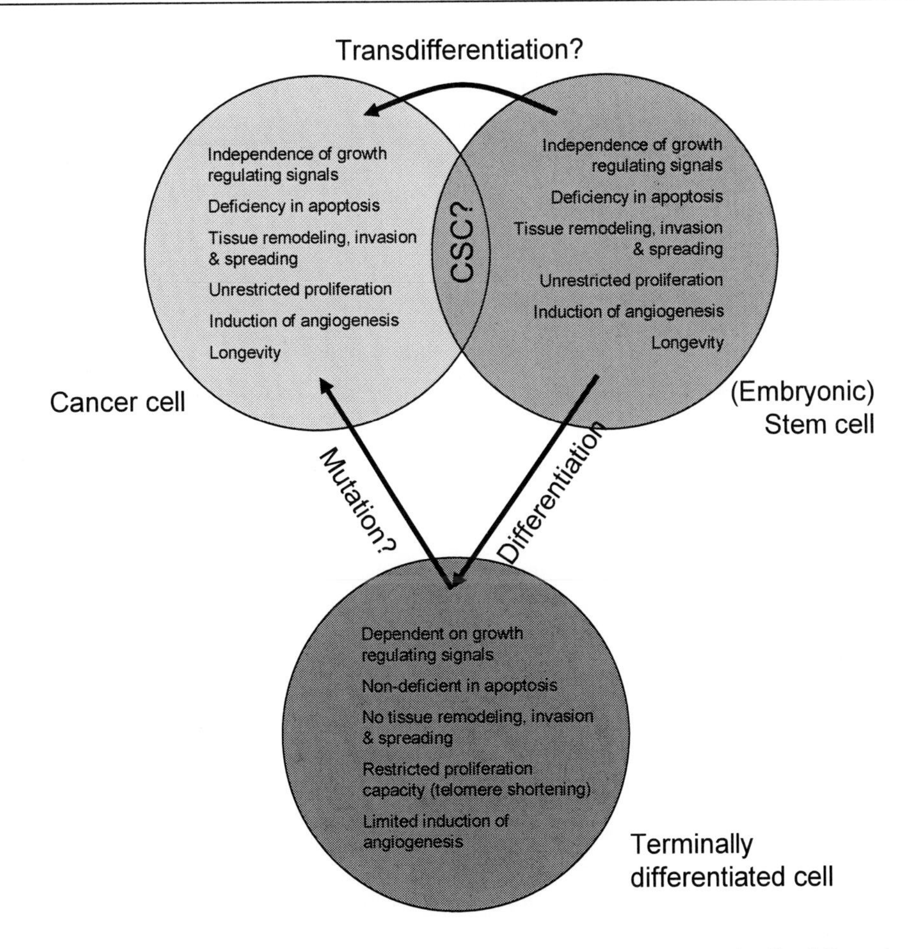

Figure 1. Schematic comparison of (embryonic) stem cells, cancer cells, terminally differentiated cells and possible pathways of converting these cell types. The intersection of embryonic stem cells and cancer cells represents a putative cancer stem cell (CSC) compartment that can directly maturate to cancer cells without previous terminal differentiation.

Especially for colorectal [6] and pancreatic cancer [27, 28], these models are widely used for staging and diagnosis of the disease and have helped to develop novel therapeutic algorithms. Yet, the accumulation of these growth and survival promoting events is not a linear process and some steps might be omitted while others may be repeated during carcinogenesis and should therefore be regarded as a network of intertwining but independent steps [1, 29].

Although these models are attractive tools for a static view from the endpoint of the carcinogenic process, there are several constraints to functional aspects in these settings, too. Cancer is a dysregulation of cell division and cell death that is crucial for maintaining tissue homeostasis, organization and function. Proliferation and apoptosis are therefore redundantly controlled by several pathways, e.g. cell cycle check points (p53, Rb), cyclin dependent kinases (CDKs, p21), DNA integrity sensors (p53) and repair systems as well as executioners (caspases) and inhibitors of apoptosis (survivin, bcl-2). According to the multistep model, targeted genetic mutations or epigenetic inactivation of several genes are necessary to provide cells with the capability to circumvent these controls. As most of the genetic material of quiescent, i.e. terminally differentiated parenchyma cells in the G_0 phase, is tightly packed in the center of the nucleus, it is of low statistic probability that various mutational events can be acquired in a coordinated manner by carcinogenic agents or events [30, 31]. Moreover, the overall lifespan of parenchyma cells in the gastrointestinal tract is limited to a maximum of 7 to 10 days, which further decreases the statistical probability for acquiring targeted mutations [32-35].

Cancer Development as a Repetition of Embryogensis in Adult Organisms?

Embryogenesis is characterized by the development of complex organs by correct pattern formation integrating cell division and differentiation. In adult organisms, pattern formation is changed to the phenomenon of pattern maintenance mainly characterized by spatio-temporal and iterative renewal and repair of differentiated cells of complex organs (see table 1). Comparing patterning during embryogenesis and in carcinogenesis in adult organisms, the same basic structures and morphogenes could be observed in both systems, indicating that patterning is an evolutionary highly conserved process [36]. But which factors are known to be essentially involved in these complex processes of embryonic and adult morphogenesis?

Early during the development of vertebrate embryos, the process of gastrulation forms three distinct layers, i.e. ectoderm, endoderm and mesoderm, being responsible for further embryonic organ development (e.g. ectoderm: skin and nervous systems; endoderm: respiratory and digestive tract; mesoderm: blood, bone and muscle) [37]. Especially, the development of the gastrointestinal tract form the primitive gut tube is under the control of so called morphogenic factors which are secreted by a specific set of cells distributed in a gradient fashion and thus influence cell fate along this gradient [38-41]. Among these morphogens, WNT and Hedgehog (Hh) signaling pathways play a key role for growth and patterning control [42, 43]. In detail, Hh proteins essentially influence positional identity of embryonic segments as well as imaginal disc-derived structures (e.g. eye, appendages) in Drosophila [44].

Table 1. Similarities and differences of embryogenesis in comparison to regeneration and cancer in adult organisms

	Embryogenesis/ development	**Regeneration**	**Cancer**
Morphology			
1. Pattern formation	+++	+	+ ↔ +++ §
2. Pattern mantainance	+	+++	+
Morphogens			
1. Expression of morphogens	+++	+	+ ↔ +++ §
2. Building of morphogen gradients	+++	+++	+ ↔ +++ §

+: weak, ++: moderate, +++ strong.

§: Heterogeneous distribution.

On the other hand, the WNT signaling pathway is integrally involved in building up embryonic axis in different metazoans such as Hydra or Xenopus [45]. The key role of the well moderated signaling-transducing components of WNT and Hh signaling pathways is emphasized by the generation of different genetic mutants with gain or loss of function which lead to major and lethal morphological disturbances of embryogenesis [45-47]. Interestingly, gradients of proteins of WNT and Hh signaling pathways proteins are found in adult organisms, especially in the gastrointestinal tract regulating stem-cell self-renewal and differentiation [42, 48, 49]. Finally, mutations or amplifications of down-stream proteins of both pathways are found in different human cancers (e.g. the gastrointestinal tract) indicating that WNT and Hh signaling pathways are not only involved in pattern formation and maintenance, but also in human cancerogenesis [50-52] (see table 2).

Other factors identified in both stem cell differentiation and carcinogenesis are for example bone morphogenic protein (BMP), fibroblast growth factors (FGFs), Notch/Delta pathways as well as the janus family kinase (JAK) and its downstream mediators (e.g. STAT and SMAD proteins) in different anatomic gastrointestinal sites indicating the close relationship between the tightly regulated physiologic process of stem cell maintenance and differentiation and the dysregulated malignant transformation [53, 54].

Interestingly, using gene expression analysis via DNA microarrays, Lee et al. found that a subtype of individuals with hepatocellular carcinoma has a poor prognosis indicated by gene expression pattern of fetal hepatoblasts [55, 56].

Table 2. Involvement of WNT- and Hh pathways in gastrointestinal cancer adapted from: OMIM (http://www.ncbi.nlm.nih.gov/omim/) and [72, 125]

Tissue	Pathway	Tumor-Type	Occurrence of mutations in sporadic cases	Familial syndrome, tumor incidence
Oesophagus	Hh (SHh, IHh, PTCH)	Adenocarcinoma	*	*
Stomach	Hh (SHh, IHh, PTCH)	Adenocarcinoma	*	*
Colon	WNT, β-Catenin (inactivation of APC)	Adenocarcinoma	About 85%	FAP, very high
Pancreas	Hh (SHh)	Adenocarcinoma	*	*
Biliary tract	Hh (SHh, IHh, PTCH)	Adenocarcinoma	*	*
Liver	WNT, β-Catenin (inactivation of APC and by stabilization of β-Catenin)	Hepatoblastoma	About 67%	FAP, <1%

*: until now no valid data available.
SHh: sonic hedgehog, IHh : indian hedgehog, PTCH : patched.
APC: adenomatous polyposis coli.

Inflammation, Repair and Cancer

According to the review of Balkwil et al. many links between cancer and inflammation are known [57]: (i) cancers arise at sites of inflammation, (ii) immune cells of chronic inflammation are found in cancer and could promote tumor growth in cell transfer experiments, (iii) cancers can produce mediators that regulate inflammation, (iv) deletion or inhibition of inflammatory mediators inhibits development of cancer in experimental settings, (v) changes in inflammatory genes influence susceptibility to and severity of cancer and (vi) nonsteroidal anti-inflammatory agents could reduce risk of cancer.

An association between chronic inflammation and malignancy was expressed by Virchow already in 1863 [58, 59]: Acute injury and inflammation are self-limited processes, whereas chronic injury and inflammation could induce expansion of tissue proliferation zones associated with initiation and propagation of tumorigenesis. Additionally, observed interaction of inflammation, connective tissue and tumorigenesis lead to the statement of Dvorak of tumors as "wounds that do not heal" indicating that the peritumorous stroma with new blood vessels and modified connective tissue represents a significant part of tumorigenesis [60].

A number of chronic inflammatory diseases have been identified as risk factors for tumor development in the gastrointestinal tract [59, 61], e.g. ulcerative colitis [62], Crohn's disease [63], chronic pancreatitis [64] or chronic forms of viral and non-viral hepatitis with progredient liver cirrhosis [65-67]. It is estimated that infections and associated inflammatory conditions are responsible for 10 to 15% of all cancer diseases [68]. Yet, the exact pathomechanism of inflammation-driven carcinogenesis is still unclear. Possibly, cancerous genetic events on "regenerative proliferating cells" in inflammation are caused through reactive oxygen and nitrogen species produced by inflammatory cells [69]. In inflammatory bowel disease and especially in ulcerative colitis, enhanced numbers of p53 mutations of inflammatory damaged colon epithelial cells were observed. In hepatocellular carcinoma, interactions of the HCV core protein and an transcription factor of cytokine signaling (signal transducer and activator of transcription 3) is found [70]. A similar pathway is involved in H. pylori associated generation of stomach cancer [71].

Comparing regenerative processes in other organisms could provide insight into this association of inflammation and carcinogenesis: Experimental manipulations, especially in amphibians, induced impressive re-organizations of certain organs which were accompanied by up regulation of specific proteins, especially proteins of the WNT and Hh signaling family resembling embryonic pattern formation (see table 3) [72]. Interestingly, these regenerative processes were mediated by expression of β-catenin or Hh-proteins and could be influenced by pharmacological interventions like cylcopamine treatment, a specific inhibitor of the Hh signaling pathway. In another experimental setting, sonic hedgehog was upregulated in human gastric cell lines under acidic conditions [73]. Using the duct ligature animal model of rat pancreas, a time dependant upregulation of different transcription factors (e.g. PDX1, pbx1, Meis2, Nkx2.2 as well as c-kit and nestin) could be observed during remodeling [74]. Reviewing the experimental hepatocarcinogenesis literature, different markers (e.g. like OV-6, Thy-1, c-kit, CD34, CK7, CK19, CK14 and AFP) of hepatocellular stem cells (so called oval cells) are differentially upregulated through the influence of inflammatory cells, growth factors and DNA-damaging agents (like reactive oxygen and nitrogen species) [75], which could also be observed in human specimens of chronic hepatitis B and C [76, 77]. Taken together, the pool of activated stem-cells is increased in dependence of intensity and time of injury whereby the influence of additional genetic and epigenetic events (as discussed above) on these activated stem-cells is raised and therefore the possibility of cancer development increases (see figure 2) [78].

Table 3. WNT- and Hh pathway in regeneration indicating that WNT and Hh pathway is widespread involved in different organisms and organs (modified according to [72])

Organism	Organ/Tissue (injury)	Associated signaling pathway	Experimental pathway modulations
Newt	Lens (ectomy), Limb (amputation)	Hh	regeneration blocked by cycopamine or HIP-transfection
Zebrafish	Fin (amputation)	Hh	regeneration blocked by cycopamine
Hydra	Axis (dissociation)	WNT	
Mouse	Vasculature (ischaemia), Bile duct (immune), Lung (chemical), Bone (fracture), Liver (Hepatectomy)	Hh	regeneration blocked by cycopamine or anti-Hh antibody
Human	Bile duct (immune) Kidney (ureteral obstruction)	WNT	no data available

HIP: Hedgehog inhibitory protein

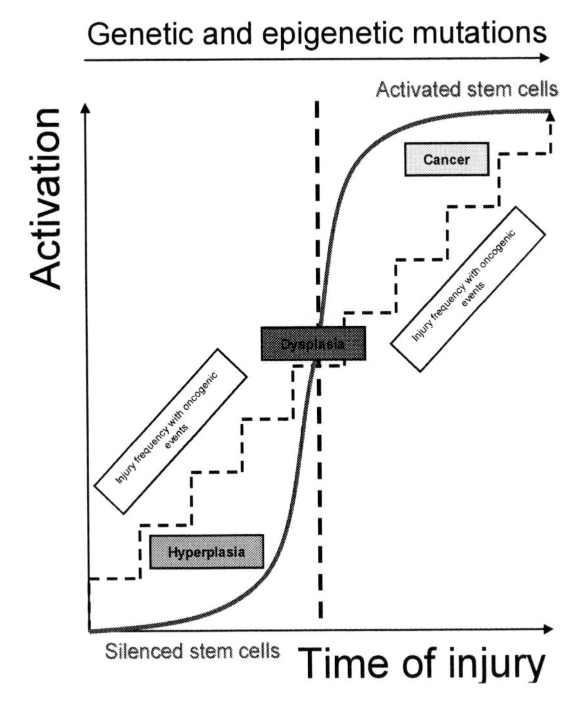

Figure 2. The concept for carcinogenesis based on persistence of injury and inflammation. During chronic tissue injury silenced stem cells or stem cell-like cells are activated. Time dependant accumulation of oncogenic events on these activated stem cells lead to genetic and epigenetic mutations via reactive oxygen and nitrogen species. Finally these genetically mutated activated stem cells gain the different “hallmarks of cancer” like self-sufficiency in growth or anti-growth signals, evasion of apoptosis, immortality and angiogenesis as has been highlighted by Hanahan and Weinberg [1] and can be described morphologically by invasiveness and metastases.

Chronic inflammation leads to repetitive tissue repair [58, 60] and recently, these conditions have been shown to attract bone marrow derived stem cells to facilitate tissue regeneration [79]. In humans, this concept has been verified in patients receiving mixed-sex tissue or bone marrow transplantations. Especially during phases of acute tissue renewal, e.g. after graft-versus-host disease, a high number of bone marrow derived cells has been found as differentiated epithelial cells in various organs, also in the gastrointestinal tract [80-82]. These findings indicate that tissue homeostasis is highly dependent on migrating stem cells that can adapt a terminally differentiated phenotype in various organs and that peripheral tissue specific stem or precursor cells are not sufficient to maintain organ integrity in phases of acute demand or for a whole life-time of several decades.

Stem Cells and Cancer Stem Cells

In parallel to differentiated parenchyma, the vast majority of cells in a malignant tumor resembles terminally differentiated cells lacking the high proliferating or adaptive capacity of stem cells [83-85]. The concept of stem cells is commonly accepted in the field of non-solid, i.e. hematological, cancers where bone-marrow cells with an immature phenotype represent the majority of tumor cells. Yet, experiences from solid as well as hematological tumors showed that only a limited number of primary cancer cells is capable of establishing xenografts or stable cell lines *in vitro* [86]. These cells have been attributed to be cancer stem cells (CSC). As in normal organs, this stem cell pool was believed to represent only a small fraction of the whole cell number [87] and recently these CSC have been identified in a variety of human solid tumors like breast, gastric or liver cancer [88-93]. These cells can be identified by flow cytometry by their increased drug efflux capacities (e.g. via mdr-1) and are designated as the so-called "side population" [94, 95]. Especially in gastrointestinal cancers, high expression of these p-glycoproteins has been shown to be a common phenomenon associated with poor prognosis and resistance to chemotherapy [96].

Several surface markers of hematopoietic and solid stem cells as well as of various differentiated epithelia have been identified on cancer stem cells. Besides this increased resistance against cytotoxic agents, side population cells show further stem cell properties, like the capability to unlimited replication, apoptosis deficiency and independence of growth-promoting signals. Xenograft experiments with thus identified mammary cancer stem cells have shown that the small number of about 200 $CD44^{+}CD24^{-}$ stem cells is capable of forming a new tumor in NOD-SCID mice, while cells isolated from the same tumor without these properties do not induce reproducible tumors and similar results were reported for leukemia ($CD34^{+}CD38^{-}$), glioblastoma, medulloblastoma ($CD133^{+}$) or liver cancer [83, 90, 93, 97-100]. Interestingly, markers of embryonic development (e.g. integrin β_1 or receptors of the WNT signaling pathway like FZD7) and differentiation (e.g. cytokeratins) are commonly upregulated in side population cells [93, 101, 102] indicating a close resemblance between controlled embryonic growth and differentiation with dysregulated cancer development from putative long living progenitor cells. In concordance to this concept, we have shown previously that an up-regulation or *de novo* expression of regulators of embryonic pancreas development can be observed in different human pancreatic cancer cell lines when provided with a distinct environment as compared to cell culture conditions [103]. Especially the pancreatic and duodenal homeobox gene 1 (PDX-1) and members of the hedgehog signal transduction pathway (SHh, PTC) are of great importance in pancreatic cancer development and embryonic organogenesis [52, 103-105]. Furthermore, the expression of PDX-1 seems to be correlated with the degree of PanIN (pancreatic intraepithelial neoplasia) development, which is considered as an early event during pancreatic carcinogenesis [106].

During embryongenesis as well as for tissue homeostasis, signals from the extracellular matrix and other environmental factors are crucial for cell survival and differentiation [107]. In stem cells and other progenitor cells, this signaling is mediated by a number of genes that have oncogenic functions in tumors, e.g. WNT, SHh or integrins [29]. This indicates that not only intracellular changes, i.e. genetic or epigenetic events, but also the extracelular environment contributes to tumor formation. For physiologic stem cell maintenance the so called "stem cell niche" has attracted significant levels of attention recently and several factors contributing to the "stemness" of progenitor cells have been identified [85, 108, 109]. The normal stem cell niche harbors quiescent and only slowly dividing stem cell populations that are under control of suppressive signals to regulate proliferation and differentiation. The cancer stem cell niche is considered to be constitutively active with uncontrolled proliferation and impaired differentiation in the stem cell population. Again, several factors identified in terminally differentiated cancers are involved in this signaling like WNT or SHh [108] indicating a close resemblance between these phenomena [110, 111]. Besides maintaining stem cell function and differentiation, cell-matrix interactions regulate survival, migration and angiogenesis during development and carcinogenesis [112].

The stem cell niche as well as the contribution of bone marrow derived cells to tumorigenesis has been investigated in several human cancers. Especially under inflammatory conditions, e.g. after acute graft-versus-host disease, or other processes associated with high tissue turnover, a significant proportion of novel epithelial cells possesses stem cell properties [79, 80, 113, 114].

These findings indicate that the crosstalk and interdependency of tumor cells, stem cells and the surrounding environment is of great importance for further understanding of cancer biology and the development of novel treatment options.

Implications of Therapeutic Consequences

Presuming that human gastrointestinal tumors resemble de-differentiated states of embryonic or adult stem cells, it will be an interesting therapeutic approach to induce differentiation of these tumor cells into normal resting adult cells or to reduce the malignant potency of de-differentiation (see figure 3).

Presently, cancer therapy aims at reducing the tumor cell mass, either by surgical interventions or systemic or loco-regional treatment strategies like radio- or chemotherapy approaches. The main target of this strategy is the terminally differentiated tumor cell that has only limited replicative potency and ignores the role of tumor-stroma interactions and tumor stem cells that might be recruited from the bone marrow or the periphery.

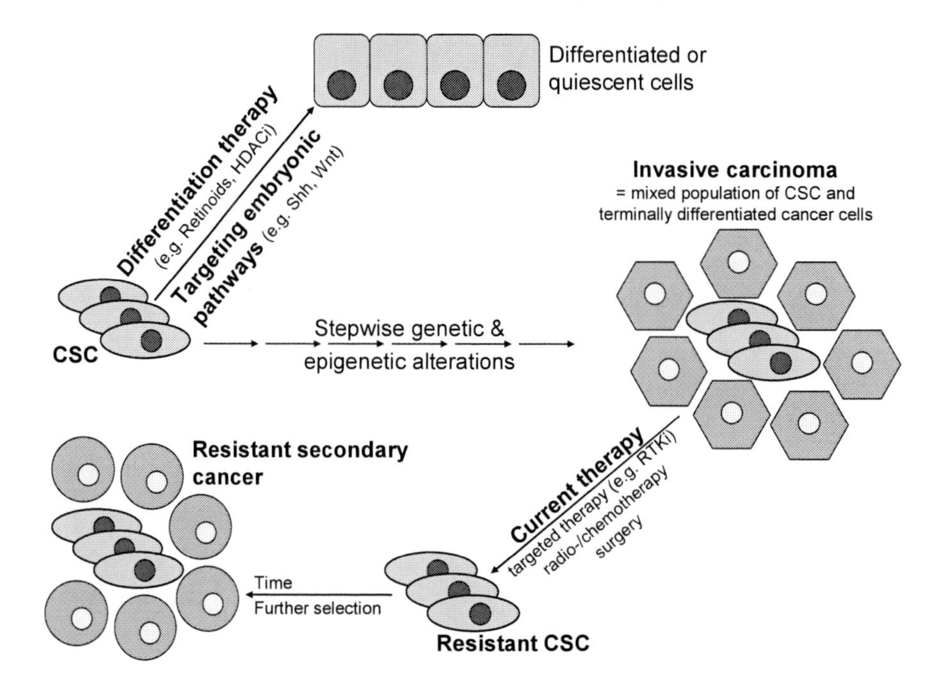

Figure 3. Paradigm shift in cancer treatment. Current treatment strategies aim at reducing the mass of already established cancer, which represent a mixed population of cancer stem cells (CSC) and terminally differentiated tumor cells lacking a high replicative capacity. Resistant CSC (either locally or disseminated in the body) can survive this treatment and give rise to a secondary cancer with different properties and resistance to conventional treatment. Novel treatment strategies aim at CSC at the beginning of the malignant transformation either by targeting specific embryonic signaling pathways or by inducing differentiation to phenotypically normal and quiescent cell populations.

Yet, this view explains the failure of the currently available treatment options as the stem cell population has a high intrinsic resistance to radio- or chemotherapy and is capable of repopulating the "diseased" organ, i.e. the attacked primary tumor. Furthermore, this cell population does not necessarily reside at the location of the treatment, but may be hiding in the bone marrow or may even circulate throughout the whole body. Therefore, a paradigm shift in cancer treatment has to evolve that targets not the end of the carcinogenesis process, but tries to target stem cell (like) cell types and structures at the beginning of the malignant transformation. To this end either a differentiation modulating therapy or targeting embryonic signaling pathways seem to be promising.

Until now, several classes of differentiation modulating agents have been examined and tested in pre-clinical or clinical settings [115]. Among these, natural or synthetic derivatives of retinoic acids (e.g. all-*trans* retinoic acid, ATRA), epigenetic modulators like the DNA methyltransferase inhibitor Zebularine, inhibitors of histone deacetylases like suberoylanilide hydroxamic acid (SAHA) or Trichostatin A (TSA) as well as "specific" inhibitors of WNT/β-catenin or hedgehog signaling like cyclopamine are the most prominent. Besides inhibition of proliferation and induction of apoptosis, retinoids interact with nuclear receptors forcing differentiation of cells in several non-gastrointestinal malignancies like acute promyelocytic leukemia, teratocarcinomas and different solid tumors (e.g. squamous cell carcinoma and breast carcinoma) [116]. Several natural and synthetic derivatives are currently tested in clinical trials. Additionally, retinoids have the potential of chemoprevention [117]. Epigenetic modulators (e.g. Zebularine or SAHA) regulate gene transcription via inhibition of DNA methylation or deacetylation of lysine residues in core histones. Hypermethylation and hypoacetylation are observed in many solid tumors, especially in gastrointestinal tumors. Here, these phenomena have been linked to the inactivation of tumor suppressor genes (e.g. transcriptional repression of $p16^{ink4a}$ by promoter hypermethylation), while the inactivation of genes by these processes is also commonly observed during embryogenesis and cellular differentiation. *In vitro* and *in vivo* experiments confirmed that these two classes of drugs (inhibitors of DNA methylation and histone deacetylation) have anti-proliferative and pro-apoptotic capabilities as well as pro-differentiation potency [102, 118-120]. Our own experiences with Zebularine and SAHA confirmed earlier findings, that these compounds have anti-proliferative and pro-apoptotic effects [121, 122]. Additionally, pancreatic carcinoma xenografts in nude mice show a morphological and molecular stability after this treatment, esp. regarding the expression of different cytokeratins or PDX-1 [103]. Finally, different agents (e.g. cyclopamine or sulindac) have been identified which selectively inhibit components of the WNT/β-catenin or hedgehog signaling pathways, thus inhibiting proliferation and induce apoptosis and differentiation [123]. These results indicate that interference with these embryonic pathways, which represent early changes during the process of carcinogenesis, might be promising approaches for the development of future therapies [115].

As indicated previously, a strong association between cancer development and various inflammatory conditions has been described and connected to high cellular turnover. Besides inducing cell proliferation, several inflammatory mediators (e.g. interleukins, reactive oxygen species, NO, TGFβ, prostaglandins etc.) have been shown to induce genetic or epigenetic changes in evolving tumor cell populations and to interfere with known embryonic differentiation pathways like NFκB [124]. Anti-inflammtory drugs, e.g. conventional NSAID, NO-donating NSAID, COX-2 inhibitors, proteasome inhibitors, have been shown to have beneficial effects in cancer chemoprevention and therapy and could be useful new targets as an adjunctive stem cell targeting therapy, too.

Conclusions

The future challenge is to gain better understanding and deeper knowledge on mechanisms and impact of differentiation and its dysregulation in the process of malignant transformation of embryonal or adult stem cells. As current therapy options aim at tumor cells at the end of a differentiation process, novel therapies should concentrate on the other side of this scale, i.e. early changes in stem cell like tumor progenitor cells that could revert the instability of the differentiation status and lead to a phenotypic stabilization.

References

[1] Hanahan, D. and R.A. Weinberg, The hallmarks of cancer. *Cell,* 2000. 100(1): p. 57-70.

[2] Thun, M.J. and A. Jemal, *Cancer Epidemiology, in Cancer Medicine*, D.W. Kufe, et al., Editors. 2006, BC Decker Inc.: Hamilton, London. p. 339-53.

[3] Bos, J.L., ras oncogenes in human cancer: a review. Cancer Res, 1989. 49(17): p. 4682-9.

[4] Chau, B.N. and J.Y. Wang, Coordinated regulation of life and death by RB. *Nat. Rev. Cancer,* 2003. 3(2): p. 130-8.

[5] Hruban, R.H., et al., K-ras oncogene activation in adenocarcinoma of the human pancreas. A study of 82 carcinomas using a combination of mutant-enriched polymerase chain reaction analysis and allele-specific oligonucleotide hybridization. *Am. J. Pathol,* 1993. 143(2): p. 545-54.

[6] Kinzler, K.W. and B. Vogelstein, Lessons from hereditary colorectal cancer. *Cell,* 1996. 87(2): p. 159-70.

[7] Levine, A.J., p53, the cellular gatekeeper for growth and division. *Cell,* 1997. 88(3): p. 323-31.

[8] Sherr, C.J., Principles of tumor suppression. *Cell,* 2004. 116(2): p. 235-46.

[9] Sherr, C.J. and F. McCormick, The RB and p53 pathways in cancer. *Cancer Cell,* 2002. 2(2): p. 103-12.

[10] Cory, S., D.C. Huang, and J.M. Adams, The Bcl-2 family: roles in cell survival and oncogenesis. *Oncogene,* 2003. 22(53): p. 8590-607.

[11] Ghobrial, I.M., T.E. Witzig, and A.A. Adjei, Targeting apoptosis pathways in cancer therapy. *CA Cancer J. Clin,* 2005. 55(3): p. 178-94.

[12] Gross, A., J.M. McDonnell, and S.J. Korsmeyer, BCL-2 family members and the mitochondria in apoptosis. *Genes Dev,* 1999. 13(15): p. 1899-911.

[13] Kim, R., et al., Therapeutic potential of antisense Bcl-2 as a chemosensitizer for cancer therapy. *Cancer,* 2004. 101(11): p. 2491-502.

[14] Kountouras, J., C. Zavos, and D. Chatzopoulos, Apoptotic and anti-angiogenic strategies in liver and gastrointestinal malignancies. *J. Surg. Oncol,* 2005. 90(4): p. 249-59.

[15] Piro, L.D., Apoptosis, Bcl-2 antisense, and cancer therapy. Oncology (Williston Park), 2004. 18(13 Suppl 10): p. 5-10.

[16] Behrens, J. and B. Lustig, The Wnt connection to tumorigenesis. *Int. J. Dev. Biol,* 2004. 48(5-6): p. 477-87.

[17] Christofori, G. and H. Semb, The role of the cell-adhesion molecule E-cadherin as a tumour-suppressor gene. *Trends Biochem. Sci,* 1999. 24(2): p. 73-6.

[18] Moon, R.T., et al., WNT and beta-catenin signalling: diseases and therapies. *Nat. Rev. Genet,* 2004. 5(9): p. 691-701.

[19] Perl, A.K., et al., A causal role for E-cadherin in the transition from adenoma to carcinoma. *Nature,* 1998. 392(6672): p. 190-3.

[20] Reya, T. and H. Clevers, Wnt signalling in stem cells and cancer. *Nature,* 2005. 434(7035): p. 843-50.

[21] Artandi, S.E. and L.D. Attardi, Pathways connecting telomeres and p53 in senescence, apoptosis, and cancer. *Biochem. Biophys. Res. Commun,* 2005. 331(3): p. 881-90.

[22] Rudolph, K.L., et al., Telomere dysfunction and evolution of intestinal carcinoma in mice and humans. *Nat. Genet,* 2001. 28(2): p. 155-9.

[23] Sharpless, N.E. and R.A. DePinho, Telomeres, stem cells, senescence, and cancer. *J. Clin. Invest,* 2004. 113(2): p. 160-8.

[24] Folkman, J., Tumor angiogenesis: therapeutic implications. *N. Engl. J. Med,* 1971. 285(21): p. 1182-6.

[25] Hicklin, D.J. and L.M. Ellis, Role of the vascular endothelial growth factor pathway in tumor growth and angiogenesis. *J. Clin. Oncol,* 2005. 23(5): p. 1011-27.

[26] von Marschall, Z., et al., De novo expression of vascular endothelial growth factor in human pancreatic cancer: evidence for an autocrine mitogenic loop. *Gastroenterology,* 2000. 119(5): p. 1358-72.

[27] Hruban, R.H., et al., Progression model for pancreatic cancer. *Clin. Cancer Res,* 2000. 6(8): p. 2969-72.

[28] Wilentz, R.E., et al., Loss of expression of Dpc4 in pancreatic intraepithelial neoplasia: evidence that DPC4 inactivation occurs late in neoplastic progression. *Cancer Res,* 2000. 60(7): p. 2002-6.

[29] Neureiter, D., C. Herold, and M. Ocker, Gastrointestinal cancer - only a deregulation of stem cell differentiation? (Review). *Int. J. Mol. Med,* 2006. 17(3): p. 483-9.

[30] Cremer, T. and C. Cremer, Chromosome territories, nuclear architecture and gene regulation in mammalian cells. *Nat. Rev. Genet,* 2001. 2(4): p. 292-301.

[31] Fisher, A.G. and M. Merkenschlager, Gene silencing, cell fate and nuclear organisation. *Curr. Opin. Genet. Dev,* 2002. 12(2): p. 193-7.

[32] Heath, J.P., Epithelial cell migration in the intestine. *Cell Biol. Int,* 1996. 20(2): p. 139-46.

[33] Okamoto, R. and M. Watanabe, Molecular and clinical basis for the regeneration of human gastrointestinal epithelia. *J. Gastroenterol,* 2004. 39(1): p. 1-6.

[34] Radtke, F. and H. Clevers, Self-renewal and cancer of the gut: two sides of a coin. *Science,* 2005. 307(5717): p. 1904-9.

[35] McDonald, S.A., et al., Mechanisms of disease: from stem cells to colorectal cancer. *Nat. Clin. Pract. Gastroenterol. Hepatol,* 2006. 3(5): p. 267-74.

[36] Kirchner, T. and T. Brabletz, Patterning and nuclear beta-catenin expression in the colonic adenoma-carcinoma sequence. Analogies with embryonic gastrulation. *Am. J. Pathol,* 2000. 157(4): p. 1113-21.

[37] Thomson, J.A., et al., Embryonic stem cell lines derived from human blastocysts. *Science,* 1998. 282(5391): p. 1145-7.

[38] Cummings, F.W., The interaction of surface geometry with morphogens. *J. Theor. Biol,* 2001. 212(3): p. 303-13.

[39] McHale, P., W.J. Rappel, and H. Levine, Embryonic pattern scaling achieved by oppositely directed morphogen gradients. *Phys. Biol,* 2006. 3(2): p. 107-20.

[40] Mishra, L., et al., The role of TGF-beta and Wnt signaling in gastrointestinal stem cells and cancer. *Oncogene,* 2005. 24(37): p. 5775-89.

[41] Mehlen, P., F. Mille, and C. Thibert, Morphogens and cell survival during development. *J. Neurobiol,* 2005. 64(4): p. 357-66.
[42] Leedham, S.J., et al., Intestinal stem cells. *J. Cell Mol. Med,* 2005. 9(1): p. 11-24.
[43] Walters, J.R., Recent findings in the cell and molecular biology of the small intestine. *Curr. Opin. Gastroenterol,* 2005. 21(2): p. 135-40.
[44] Ingham, P.W. and A.P. McMahon, Hedgehog signaling in animal development: paradigms and principles. *Genes Dev.,* 2001. 15(23): p. 3059-87.
[45] Logan, C.Y. and R. Nusse, The Wnt signaling pathway in development and disease. *Annu. Rev. Cell Dev. Biol,* 2004. 20: p. 781-810.
[46] Wetmore, C., Sonic hedgehog in normal and neoplastic proliferation: insight gained from human tumors and animal models. *Curr. Opin. Genet. Dev,* 2003. 13(1): p. 34-42.
[47] Lau, J., H. Kawahira, and M. Hebrok, Hedgehog signaling in pancreas development and disease. *Cell Mol. Life Sci,* 2006. 63(6): p. 642-52.
[48] van den Brink, G.R., et al., Indian Hedgehog is an antagonist of Wnt signaling in colonic epithelial cell differentiation. *Nat. Genet,* 2004. 36(3): p. 277-82.
[49] Watt, F.M., Unexpected Hedgehog-Wnt interactions in epithelial differentiation. *Trends Mol. Med,* 2004. 10(12): p. 577-80.
[50] Clements, W.M., A.M. Lowy, and J. Groden, Adenomatous polyposis coli/beta-catenin interaction and downstream targets: altered gene expression in gastrointestinal tumors. *Clin. Colorectal Cancer,* 2003. 3(2): p. 113-20.
[51] Berman, D.M., et al., Widespread requirement for Hedgehog ligand stimulation in growth of digestive tract tumours. *Nature,* 2003. 425(6960): p. 846-51.
[52] Thayer, S.P., et al., Hedgehog is an early and late mediator of pancreatic cancer tumorigenesis. *Nature,* 2003. 425(6960): p. 851-6.
[53] Nicolas, M., et al., Notch1 functions as a tumor suppressor in mouse skin. *Nat. Genet,* 2003. 33(3): p. 416-21.
[54] Sell, S., Stem cell origin of cancer and differentiation therapy. *Crit. Rev. Oncol. Hematol,* 2004. 51(1): p. 1-28.
[55] Roberts, L.R. and G.J. Gores, Hepatocellular carcinoma: molecular pathways and new therapeutic targets. *Semin. Liver Dis,* 2005. 25(2): p. 212-25.
[56] Lee, J.S., et al., A novel prognostic subtype of human hepatocellular carcinoma derived from hepatic progenitor cells. *Nat. Med,* 2006. 12(4): p. 410-6.
[57] Balkwill, F., K.A. Charles, and A. Mantovani, Smoldering and polarized inflammation in the initiation and promotion of malignant disease. *Cancer Cell,* 2005. 7(3): p. 211-7.
[58] Balkwill, F. and A. Mantovani, Inflammation and cancer: back to Virchow? *Lancet,* 2001. 357(9255): p. 539-45.
[59] Coussens, L.M. and Z. Werb, Inflammation and cancer. *Nature,* 2002. 420(6917): p. 860-7.
[60] Dvorak, H.F., Tumors: wounds that do not heal. Similarities between tumor stroma generation and wound healing. *N. Engl. J. Med,* 1986. 315(26): p. 1650-9.
[61] Moss, S.F. and M.J. Blaser, Mechanisms of disease: Inflammation and the origins of cancer. *Nat. Clin. Pract. Oncol,* 2005. 2(2): p. 90-7; quiz 1 p following 113.
[62] Eaden, J.A., K.R. Abrams, and J.F. Mayberry, The risk of colorectal cancer in ulcerative colitis: a meta-analysis. *Gut,* 2001. 48(4): p. 526-35.

[63] Jess, T., et al., Increased risk of intestinal cancer in Crohn's disease: a meta-analysis of population-based cohort studies. *Am. J. Gastroenterol,* 2005. 100(12): p. 2724-9.

[64] Jura, N., H. Archer, and D. Bar-Sagi, Chronic pancreatitis, pancreatic adenocarcinoma and the black box in-between. *Cell Res,* 2005. 15(1): p. 72-7.

[65] Sherman, M., Hepatocellular carcinoma: epidemiology, risk factors, and screening. *Semin. Liver Dis,* 2005. 25(2): p. 143-54.

[66] Blum, H.E., Hepatocellular carcinoma: therapy and prevention. *World J. Gastroenterol,* 2005. 11(47): p. 7391-400.

[67] Kuper, H.E., et al., Hepatitis B and C viruses in the etiology of hepatocellular carcinoma; a study in Greece using third-generation assays. *Cancer Causes Control,* 2000. 11(2): p. 171-5.

[68] Kuper, H., H.O. Adami, and D. Trichopoulos, Infections as a major preventable cause of human cancer. *J. Intern. Med,* 2000. 248(3): p. 171-83.

[69] Maeda, H. and T. Akaike, Nitric oxide and oxygen radicals in infection, inflammation, and cancer. *Biochemistry.* (Mosc), 1998. 63(7): p. 854-65.

[70] Bromberg, J. and J.E. Darnell, Jr., The role of STATs in transcriptional control and their impact on cellular function. *Oncogene,* 2000. 19(21): p. 2468-73.

[71] Tebbutt, N.C., et al., Reciprocal regulation of gastrointestinal homeostasis by SHP2 and STAT-mediated trefoil gene activation in gp130 mutant mice. *Nat. Med,* 2002. 8(10): p. 1089-97.

[72] Salo, E., The power of regeneration and the stem-cell kingdom: freshwater planarians (Platyhelminthes). *Bioessays,* 2006. 28(5): p. 546-59.

[73] Dimmler, A., et al., Transcription of sonic hedgehog, a potential factor for gastric morphogenesis and gastric mucosa maintenance, is up-regulated in acidic conditions. *Lab. Invest,* 2003. 83(12): p. 1829-37.

[74] Peters, K., et al., Expression of stem cell markers and transcription factors during the remodeling of the rat pancreas after duct ligation. *Virchows Arch,* 2005. 446(1): p. 56-63.

[75] Alison, M.R. and M.J. Lovell, Liver cancer: the role of stem cells. *Cell Prolif,* 2005. 38(6): p. 407-21.

[76] Sun, C., X.L. Jin, and J.C. Xiao, Oval cells in hepatitis B virus-positive and hepatitis C virus-positive liver cirrhosis: histological and ultrastructural study. *Histopathology,* 2006. 48(5): p. 546-55.

[77] Libbrecht, L. and T. Roskams, Hepatic progenitor cells in human liver diseases. *Semin. Cell Dev. Biol,* 2002. 13(6): p. 389-96.

[78] Beachy, P.A., S.S. Karhadkar, and D.M. Berman, Tissue repair and stem cell renewal in carcinogenesis. *Nature,* 2004. 432(7015): p. 324-31.

[79] Houghton, J., et al., Gastric cancer originating from bone marrow-derived cells. *Science,* 2004. 306(5701): p. 1568-71.

[80] Okamoto, R., et al., Damaged epithelia regenerated by bone marrow-derived cells in the human gastrointestinal tract. *Nat. Med,* 2002. 8(9): p. 1011-7.

[81] Korbling, M., et al., Hepatocytes and epithelial cells of donor origin in recipients of peripheral-blood stem cells. *N. Engl. J. Med,* 2002. 346(10): p. 738-46.

[82] Forbes, S.J., et al., A significant proportion of myofibroblasts are of bone marrow origin in human liver fibrosis. *Gastroenterology,* 2004. 126(4): p. 955-63.
[83] Reya, T., et al., Stem cells, cancer, and cancer stem cells. *Nature,* 2001. 414(6859): p. 105-11.
[84] Al-Hajj, M., et al., Therapeutic implications of cancer stem cells. *Curr. Opin. Genet Dev,* 2004. 14(1): p. 43-7.
[85] Li, L. and W.B. Neaves, Normal stem cells and cancer stem cells: the niche matters. *Cancer Res,* 2006. 66(9): p. 4553-7.
[86] Masters, J.R., Human cancer cell lines: fact and fantasy. *Nat. Rev. Mol. Cell Biol,* 2000. 1(3): p. 233-6.
[87] Setoguchi, T., T. Taga, and T. Kondo, Cancer stem cells persist in many cancer cell lines. *Cell Cycle,* 2004. 3(4): p. 414-5.
[88] Haraguchi, N., et al., Characterization of a side population of cancer cells from human gastrointestinal system. *Stem Cells,* 2006. 24(3): p. 506-13.
[89] Patrawala, L., et al., Side population is enriched in tumorigenic, stem-like cancer cells, whereas ABCG2+ and ABCG2- cancer cells are similarly tumorigenic. *Cancer Res,* 2005. 65(14): p. 6207-19.
[90] Al-Hajj, M., et al., Prospective identification of tumorigenic breast cancer cells. *Proc. Natl. Acad. Sci. U S A,* 2003. 100(7): p. 3983-8.
[91] Dontu, G., et al., Stem cells in normal breast development and breast cancer. *Cell Prolif,* 2003. 36 Suppl 1: p. 59-72.
[92] Hirschmann-Jax, C., et al., A distinct "side population" of cells with high drug efflux capacity in human tumor cells. *Proc. Natl. Acad. Sci. U S A,* 2004. 101(39): p. 14228-33.
[93] Chiba, T., et al., Side population purified from hepatocellular carcinoma cells harbors cancer stem cell-like properties. *Hepatology,* 2006. 44(1): p. 240-251.
[94] Bunting, K.D., ABC transporters as phenotypic markers and functional regulators of stem cells. *Stem Cells,* 2002. 20(1): p. 11-20.
[95] Goodell, M.A., Multipotential stem cells and 'side population' cells. *Cytotherapy,* 2002. 4(6): p. 507-8.
[96] Ho, G.T., F.M. Moodie, and J. Satsangi, Multidrug resistance 1 gene (P-glycoprotein 170): an important determinant in gastrointestinal disease? *Gut,* 2003. 52(5): p. 759-66.
[97] Bhatia, M., et al., A newly discovered class of human hematopoietic cells with SCID-repopulating activity. *Nat. Med,* 1998. 4(9): p. 1038-45.
[98] Bonnet, D. and J.E. Dick, Human acute myeloid leukemia is organized as a hierarchy that originates from a primitive hematopoietic cell. *Nat. Med,* 1997. 3(7): p. 730-7.
[99] Lessard, J. and G. Sauvageau, Bmi-1 determines the proliferative capacity of normal and leukaemic stem cells. *Nature,* 2003. 423(6937): p. 255-60.
[100] Singh, S.K., et al., Identification of human brain tumour initiating cells. *Nature,* 2004. 432(7015): p. 396-401.
[101] Ivanova, N.B., et al., A stem cell molecular signature. *Science,* 2002. 298(5593): p. 601-4.
[102] Ramalho-Santos, M., et al., "Stemness": transcriptional profiling of embryonic and adult stem cells. *Science,* 2002. 298(5593): p. 597-600.

[103] Neureiter, D., et al., Different capabilities of morphological pattern formation and its association with the expression of differentiation markers in a xenograft model of human pancreatic cancer cell lines. *Pancreatology,* 2005. 5(4-5): p. 387-97.
[104] Kayed, H., et al., Hedgehog signaling in the normal and diseased pancreas. *Pancreas,* 2006. 32(2): p. 119-29.
[105] Ohuchida, K., et al., Sonic hedgehog is an early developmental marker of intraductal papillary mucinous neoplasms: clinical implications of mRNA levels in pancreatic juice. *J. Pathol,* 2006.
[106] Stintzing, S., et al., Morphological Expression Pattern of PDX-1 and SHH in Human Pancreatic Cancer. *Pancreatology,* 2006. 6: p. P29.
[107] Schuppan, D. and M. Ocker, Integrin-mediated control of cell growth. *Hepatology,* 2003. 38(2): p. 289-91.
[108] Li, L. and T. Xie, Stem cell niche: structure and function. *Annu. Rev. Cell Dev. Biol,* 2005. 21: p. 605-31.
[109] Naveiras, O. and G.Q. Daley, Stem cells and their niche: a matter of fate. *Cell Mol. Life Sci,* 2006. 63(7-8): p. 760-6.
[110] Proia, D.A. and C. Kuperwasser, Stroma: tumor agonist or antagonist. *Cell Cycle,* 2005. 4(8): p. 1022-5.
[111] Kopfstein, L. and G. Christofori, Metastasis: cell-autonomous mechanisms versus contributions by the tumor microenvironment. *Cell Mol. Life Sci,* 2006. 63(4): p. 449-68.
[112] Brabletz, T., et al., Opinion: migrating cancer stem cells - an integrated concept of malignant tumour progression. *Nat. Rev. Cancer,* 2005. 5(9): p. 744-9.
[113] Haura, E.B., Is repetitive wounding and bone marrow-derived stem cell mediated-repair an etiology of lung cancer development and dissemination? *Med. Hypotheses,* 2006.
[114] Li, H.C., et al., Stem cells and cancer: evidence for bone marrow stem cells in epithelial cancers. *World J. Gastroenterol,* 2006. 12(3): p. 363-71.
[115] Massard, C., E. Deutsch, and J.C. Soria, Tumour stem cell-targeted treatment: elimination or differentiation. *Ann. Oncol,* 2006.
[116] Niles, R.M., Recent advances in the use of vitamin A (retinoids) in the prevention and treatment of cancer. *Nutrition,* 2000. 16(11-12): p. 1084-9.
[117] Okuno, M., et al., Retinoids in cancer chemoprevention. *Curr. Cancer Drug Targets,* 2004. 4(3): p. 285-98.
[118] Fang, J.Y., Histone deacetylase inhibitors, anticancerous mechanism and therapy for gastrointestinal cancers. *J. Gastroenterol. Hepatol,* 2005. 20(7): p. 988-94.
[119] Munster, P.N., et al., The histone deacetylase inhibitor suberoylanilide hydroxamic acid induces differentiation of human breast cancer cells. *Cancer Res,* 2001. 61(23): p. 8492-7.
[120] Cheng, J.C., et al., Preferential response of cancer cells to zebularine. *Cancer Cell,* 2004. 6(2): p. 151-8.
[121] Neureiter, D., et al., Apoptosis, proliferation and differentiation patterns are influenced by Zebularine and SAHA in pancreatic cancer models. *Scand. J. Gastroenterol,* 2007. DOI: 10.1080/00365520600874198. In press.

[122] Ocker, M., et al., The histone-deacetylase inhibitor SAHA potentiates proapoptotic effects of 5-fluorouracil and irinotecan in hepatoma cells. *J. Cancer Res. Clin. Oncol,* 2005. 131(6): p. 385-94.

[123] Li, H., R. Pamukcu, and W.J. Thompson, beta-Catenin signaling: therapeutic strategies in oncology. *Cancer Biol. Ther,* 2002. 1(6): p. 621-5.

[124] Yoshimura, A., Signal transduction of inflammatory cytokines and tumor development. *Cancer Sci,* 2006. 97(6): p. 439-47.

[125] Taipale, J. and P.A. Beachy, The Hedgehog and Wnt signalling pathways in cancer. *Nature,* 2001. 411(6835): p. 349-54.

In: Stem Cell Research Trends
Editor: Josse R. Braggina, pp. 117-154
ISBN: 978-1-60021-622-0

Chapter IV

DIFFERENTIATION AND LINEAGE-COMMITMENT OF STEM CELLS FOR THERAPEUTIC APPLICATIONS

Boon Chin Heng[1], Wei Seong Toh[1], Hua Liu[1], Abdul Jalil Rufaihah[2] and Tong Cao[1*]

[1]Stem Cell Program, Faculty of Dentistry, National University of Singapore, 5 Lower Kent Ridge Road, 119074 Singapore

[2]Department of Surgery, Yong Loo Lin School of Medicine, National University of Singapore
5 Lower Kent Ridge Road, 119074 Singapore

ABSTRACT

An essential prerequisite for the application of adult and embryonic stem cells in clinical therapy is the development of efficient protocols for directing their differentiation or commitment to specific lineages. It must be noted that there probably exist a subtle balance somewhere between the undifferentiated and terminally-differentiated state that would be optimal for transplantation/transfusion therapy, which is likely to differ for different cell/tissue/organ models of regeneration. For example, differentiated/committed stem cell progenies may express specific cell-surface markers or extracellular matrix that could facilitate their rapid engraftment and integration within the recipient tissue or organ. Moreover, the commitment/differentiation of stem cells prior to clinical therapy, would also avoid spontaneous differentiation into undesired lineages that can compromise tissue/organ regeneration i.e. teratoma and fibroblastic scar tissue formation. On the other hand, differentiated/committed stem cell progenies have reduced proliferative capacity and possess complex nutritional requirements that could compromise their survival within the 'hostile' pathological environment at the

* Correspondence and reprint requests: Dr. Tong Cao, Stem Cell Laboratory, Faculty of Dentistry, National University of Singapore, 5 Lower Kent Ridge Road, Singapore 119074, e-mail : dencaot@nus.edu.sg; Tel : +65-6516-4630; Fax : +65-6774-5701.

transplantation/transfusion site. There are two major approaches for directing lineage-specific commitment/differentiation of stem cells. Firstly, a 'milieu-based' approach would involve exposing stem cells to a cocktail of exogenous chemicals, cytokines, growth factors or extracellular matrix substratum, over an extended duration of in vitro culture. Nevertheless, the non-specific pleiotropic effects exerted by various cytokines, growth factors, and extracellular matrix would make this a relatively inefficient approach. Moreover, a "milieu-based" approach is likely to require extended durations of in vitro culture, which might delay autologous transplantation of adult stem cells to the patient and might alter their immunogenicity through prolonged exposure to xenogenic proteins within the culture milieu. Secondly, a 'gene-modulatory' approach would involve transfection of stem cells with either recombinant DNA or iRNA, to directly influence gene regulatory networks implicated in cellular fate and lineage determination. However, overwhelming safety and ethical concerns are likely to preclude the application of genetically modified stem cells in human clinical therapy for the foreseeable near future. To avoid permanent genetic modification, new technology platforms have been developed for the modulation of gene expression without recombinant DNA. This includes the use of protein transduction domain (PTD) – fusion transcription factors, PTD-PNA (peptide nucleic acid) analogs, as well as immunoliposome-mediated delivery of proteins or iRNA directly into the cytosol. Nevertheless, because such molecules have a limited active half-life in the cytosol and are obviously not incorporated into the genetic code of the cell, these would only exert a transient modulatory effect on gene expression.

Keywords: Commitment, Differentiation, Lineage, Stem cells, Therapy.

1. INTRODUCTION

Stem cells are undifferentiated progenitor cells that are capable of both self-renewal and multilineage differentiation [1, 2]. They can be broadly classified into two major categories, depending on whether they are of embryonic or postnatal origin. Stem cells of embryonic origin include embryonic stem (ES) cells [3, 4] that are derived from the inner cell mass of blastocyst-stage embryos, as well as embryonic germ (EG) cells [5] that are derived from the developing fetal gonadal ridge. Additionally, embryonic carcinoma (EC) cells [6] that develop from testicular tumour can also be considered stem cells of embryonic origin, even though this is controversial. Stem cells originating from differentiated postnatal tissues are referred to as adult stem cells [7, 8], and these are believed to be intimately involved in tissue/ organ regeneration and repair during injury and ageing. More often than not, stem cells originating from foetal tissues [9] are also classified as adult stem cells.

Only ES cells are considered to be pluripotent, since these have been shown to be capable of giving rise to differentiated cell lineages of all three embryonic germ layers: endoderm, mesoderm and ectoderm [10]. Because adult stem cells have a much lower degree of plasticity compared to ES cells, they are deemed to be only multipotent. Previously, it was believed that adult stem cells were capable of differentiation into lineages that are characteristic of the tissue or organ from which they originated. However, recent evidence from the scientific literature would suggest that adult stem cells could possess a much higher degree of plasticity than previously thought [11]. For example, hematopoietic stem cells derived from either the bone marrow or peripheral blood have been reported to differentiate into lineages possessing the characteristics of cardiac [12, 13] and skeletal muscles [14, 15], endothelium [16], neuroectoderm [17], skin epithelium [18] and endodermal lineages such as hepatocytes [19] and gastrointestinal epithelium [18]. However, at this point in time, adult stem cell plasticity and transdifferentiation are highly controversial issues [20, 21] due to evidence of cell fusion and heterokaryon formation [22, 23]. In any case, evidence from the scientific literature suggests that adult stem cell plasticity and transdifferentiation are relatively rare and sporadic events for most type of adult stem cells, with the possible exception of mesenchymal stromal cells [24].

2. Adult Versus Embryonic Stem Cells for Therapeutic Applications

A major issue of contention in regenerative medicine is the choice of utilising either adult or embryonic stem cells for transplantation. Both of these have their inherent advantages and shortcomings in clinical therapy. Currently, there is much ethical dispute on utilizing embryonic stem cells for either therapeutic or non-therapeutic applications [25, 26], which arises mainly from the controversy on when life actually begins and the destruction of a potential human life [27]. There is however no such ethical or moral inhibition over the use of autologous or allogenic adult stem cells in either therapy or research.

Nevertheless, adult stem cells are often difficult to isolate and their self-renewal and proliferative capacity are usually very much limited and appears to decrease with age [28–30]. This would obviously limit their usefulness in autologous cell-based therapy for the treatment of age-related degenerative diseases. If the capacity for continuous self renewal is taken strictly as the defining criteria for stem cells, then an extremely low proportion (perhaps less than 1 in 100,000) of most putative adult stem cell populations can be considered as true stem cells [31, 32]. This would pose a major challenge in the development of isolation and purification protocols. Furthermore, most putative adult stem cell populations are in fact highly heterogenous, with only a limited proportion of cells being capable of differentiation into a desired lineage, which in turn poses another major challenge to their application in clinical therapy. Additionally, adult stem cells may also contain more genetic abnormalities than ES cells, caused by exposure to metabolic toxins and errors in DNA replication accumulated during the course of a lifetime [33].

By contrast, the unlimited proliferative capacity of ES cells can provide an almost endless supply of cells for transplantation therapy. However, the major challenge is to overcome immunological rejection from the transplant recipient unless an isogenic source of ES cells is derived from therapeutic cloning. Although this has already been achieved in a number of animal models [34, 35], this is a remote option, given the technical difficulties and ethical problems faced in trying to replicate this in the human model [36]. A more viable alternative is to create a bank of ES cells with different major histocompatibilty complex (MHC) genotypes for matching with the transplant recipient [37]. It could also be possible to downregulate the antigenicity of ES cells through suppression of MHC gene expression. By contrast, autologous adult stem cells derived from the patient's own tissues would not face any immunological barriers in transplantation, even though there is a minor problem of donor site morbidity [38]. Another major challenge is the risk of teratoma formation by ES cells upon transplantation in situ [39], which is not the case for adult stem cells.

3. The Differentiation State of Transplanted Stem Cells Is a Critical Factor in Determining the Therapeutic Efficacy of Tissue/Organ Regeneration

A key consideration in stem cell transplantation therapy would be the differentiation status of the stem cells being transplanted. This issue has largely been ignored in the majority of transplantation studies in animal models and phase I human clinical trials. The differentiation status of transplanted stem cells or their progenies is likely to profoundly influence several factors that determine their ability to contribute to tissue/organ regeneration within the recipient.

It is a well-known fact that undifferentiated human embryonic stem (hES) cells form teratomas (Figure 1) upon transplantation into immunologically compromised animal models [40, 41]. Teratoma formation would obviously be detrimental to the transplant recipient. Hence for human clinical therapy, it is imperative that hES cells are differentiated to some degree prior to transplantation, so as to reduce the likelihood of teratoma formation in situ [42, 43]. Undifferentiated adult stem cells lack the ability to form teratomas, but could instead undergo non-specific multi-lineage differentiation in vitro [44, 45]. This is likely to reduce the clinical efficacy of transplantation therapy, since only a sub-fraction of the transplanted stem cells would have differentiated into the required tissue lineage.

Further complications may arise from the pathological state of damaged or diseased tissues/organs providing an 'abnormal' milieu for undifferentiated stem cells. The presence of necrotic/apoptotic cells [46], free radicals [47] and inflammatory cytokines [48] within the pathological environment, may very well exert an adverse effect on the differentiation pathway of the transplanted stem cells. This could further reduce the proportion of transplanted cells giving rise to the required lineage of interest, as well as result in the formation of undesired tissue lineages at the transplantation site, which could impair tissue/organ regeneration. For example, in the study of Wang et al. [49], the transplantation of undifferentiated mesenchymal stem cells into the pathological environment of damaged cardiac muscles, resulted in the differentiation of some of these cells into fibroblastic scar tissue, which could in turn impair recovery of heart function after myocardial infarction.

Undifferentiated stem cells may also lack expression of specific cell-surface markers that could facilitate their rapid engraftment and integration within the recipient tissue or organ, i.e. gap-junction connexin proteins, surface ligands for interaction with extracellular matrix. Hence, there is a strong likelihood that some degree of differentiation of stem cells in vitro, to allow expression of such surface markers may enhance their engraftment efficiency and subsequent integration in situ, thereby improving the clinical efficacy of transplantation therapy.

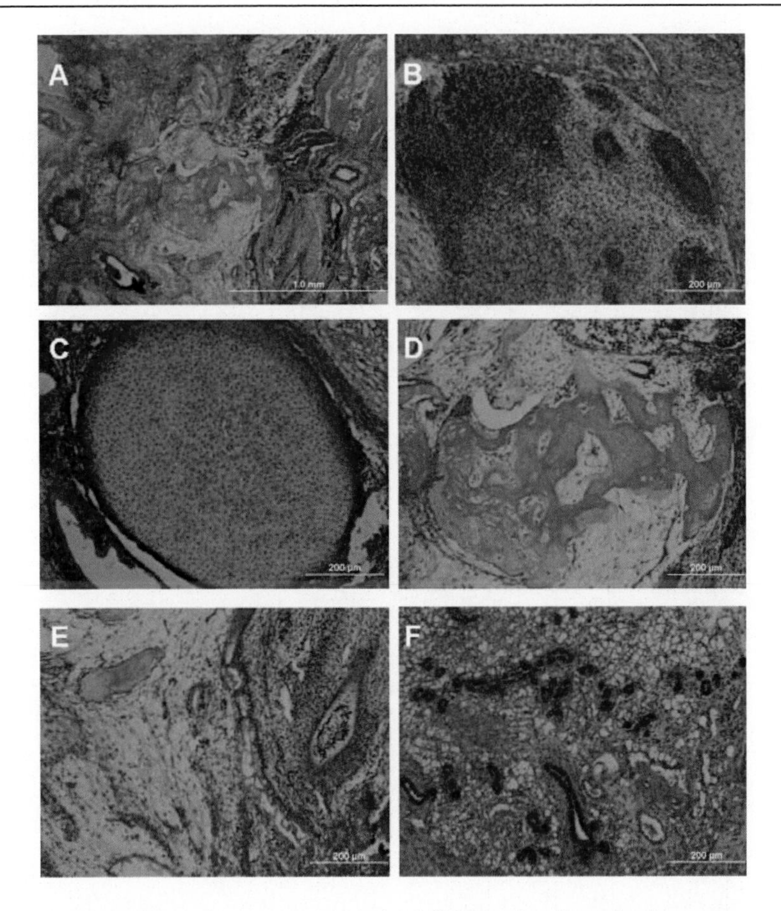

Figure 1. HESCs injected into immunocompromised SCID mice form benign teratomas five weeks later. Present within these teratomas (A) are advanced derivatives of ectoderm, such as neural rosettes (B), of mesoderm, such as cartilage (C) and bone (D) and of endoderm, such as gut-like structures (E and F). All photomicrographs are of haematoxylin and eosin-stained sections.

Nevertheless, there are a number of factors which would discourage the use of differentiated stem cell progenies in transplantation therapy. In the case of autologous adult stem cells, directed differentiation into specific lineages is likely to require prolonged durations of ex vivo culture. This would obviously delay treatment to the patient who is likely to be in dire need of therapeutic intervention. There is also evidence that prolonged durations of ex vivo culture could somehow alter the immunogenicity of cultured autologous cells, which may lead to immuno-rejection upon transplantation [50, 51]. It can certainly be argued that genetic modulation with recombinant DNA technology could potentially offer a 'rapid' and efficient means of directing autologous adult stem cell commitment and differentiation into a required lineage [52, 53], without the need for prolonged ex vivo culture. There are however overwhelming safety concerns with regards to the application of genetic manipulation in human clinical therapy [54].

Differentiated stem cells may also have reduced proliferative capacity, which could impair their ability to contribute to tissue/organ regeneration. By committing or differentiating stem cells into specific lineages prior to transplantation, there could be a loss in their ability to give rise to other lineages that might also be beneficial to tissue/organ regeneration, in particular the vascular endothelial lineage. This is best illustrated by the transplantation of undifferentiated mesenchymal stem cells for the repair of myocardially infarcted hearts. The transplanted stem cells give rise to not only the cardiomyogenic lineage [55], but also to the vascular endothelial lineage [56]. The resultant increased angiogenesis appears to play a crucial role in myocardial regeneration [57].

Differentiated progenies of stem cells are also likely to be much more fastidious in their nutritional and oxygen requirement, which could compromise their ability to survive in the adverse pathological environment of damaged/diseased tissues or organs. This is particularly the case with cell lineages of highly vascularized tissues/organs, such as the myocardium and liver. Indeed, the transplantation of fully differentiated cardiomyocytes and hepatocytes is severely hindered by the ischemic environment present within pathological heart and liver models, which often leads to an extremely low survival and engraftment rate of the transplanted cells [58, 59].

Another major factor that would discourage the use of differentiated stem cell progenies is their higher degree of immunogenicity, in the case of allogenic models of transplantation. It is well-established that undifferentiated hES cells have low levels of expression of major histocompatibility complex (MHC) class I and II antigens [60]. There is a gradual upregulation in the expression of MHC antigens by hES cells, with increased degree of differentiation [60, 61]. Hence, transplantation of hES progenies at early stages of differentiation, would be expected to provoke much less immunological reaction, as compared to hES progenies at later stages of differentiation. In the case of undifferentiated bone marrow-derived mesenchymal stem cells, these have been reported to be both immuno-privileged [62, 63], as well as immuno-suppressive [64, 65], which would favor the use of these cells in allogenic transplantation. As with ES cells, there has also been reported to be a gradual upregulation of MHC expression by mesenchymal stem cells, with increasing degree of differentiation [66].

Hence, it would appear that a major issue of contention in stem cell transplantation therapy is the choice of transplanting undifferentiated stem cells, or their differentiated progenies. Each of these has their advantages and disadvantages. Probably, there exists a subtle balance somewhere between the undifferentiated and fully differentiated state, which would be optimal for achieving maximal efficacy of stem cell transplantation therapy. This would be expected to differ for different tissue and organ models of transplantation. At present, this particular aspect of stem cell transplantation has not been well-studied. More exhaustive investigations should be carried out in this area of research, in view of its potentially important clinical implications.

4. Milieu-Based Versus a Gene-Modulatory Approach for Directing Stem Cell Differentiation and Lineage Commitment

There are two major approaches for committing or differentiating stem cells into a required lineage. Firstly, a 'milieu-based' approach would involve exposing stem cells to a cocktail of exogenous chemicals, cytokines, growth factors or extracellular matrix substratum, over an extended duration of in vitro culture. This approach is given further credence by the 'niche' microenvironment theory of stem cell regulation, as proposed by Spradling et al. [67]; who hypothesized that stem cells are strongly dependent on environmental cues for control of cellular fate and lineage determination, rather than evolve complex internal regulatory mechanisms to achieve that purpose. Secondly, a 'gene-modulatory' approach would involve directly influencing gene regulatory networks implicated in cellular fate and lineage determination, through transfection with recombinant DNA, or delivery of proteins, iRNA and their synthetic analogues directly into the cell.

First and foremost, a prime consideration would be which of these two approaches offer a more 'efficient' means of committing/directing stem cells to a specific lineage. With the 'milieu-based' approach, only a limited fraction of stem cells would be expected to commit and subsequently differentiate into a particular lineage, over an extended duration of in vitro culture. This is because the various chemicals, cytokines, growth factors and ECM substrata within the culture milieu, are likely to exert non-specific pleiotropic effects on stem cell differentiation into multiple lineages. At best, the dosage and combination of various chemicals, cytokines, growth factors and ECM substrata can be 'optimized' by trial and error, to maximize the proportion of stem cells committing and differentiating into a specific lineage, while at the same time yielding a large number of unrequired lineages. Hence, extensive selection, purification and proliferation of the cellular lineage of interest would be necessary, in order to achieve the level of purity and cell numbers required for transplantation therapy. This would certainly pose a formidable technical challenge with relatively scarce adult stem cells, which are likely to have limited proliferative capacity in vitro [68].

By contrast, a 'gene-modulatory' approach through recombinant DNA transfection, or delivery of proteins iRNA and their synthetic analogues into the cell would be expected to have a more specific effect on committing/directing stem cell to a required lineage, particularly if there is a resultant upregulation of genes specific to that particular lineage. An added advantage is that gene modulatory strategies could incorporate the use of marker genes to facilitate the selection and purification of specific cell lineages. Commonly used marker genes include green fluorescent protein (GFP) and the neomycin resistance cassette [69, 70].

Of particular interest for genetic modulation, would be transcription factors that exert upstream control of somatic differentiation [71-73]. These can be thought of as 'master switches' that control the expression of the entire array of proteins that are characteristic or specific to a particular lineage. Indeed, numerous transcription factors have been implicated in early lineage commitment of both adult and embryonic stem cells. Some prominent examples include the role of *Sox9* in chondrogenesis [71], *Runx2* in osteogenesis [72] and *MyoD* in myogenesis [73]. However, a note of caution is that there is still limited knowledge on gene regulatory networks implicated in stem cell differentiation and cell-fate decisions. No doubt, the upregulation of one master regulatory transcription factor specific to a particular lineage would apparently result in the differentiating stem cells acquiring some of the molecular markers and morphological characteristics of that lineage. However, the pertinent question is whether such differentiated cells can display the full array of biological functions normally associated with that particular lineage upon transplantation in situ. Would the upregulation of several transcription factors be necessary to achieve this purpose? If so, is a hierarchical order of transcription factor activation necessary? Moreover, it is also possible that some transcription factors may act in a pleiotropic manner just like the majority of cytokines and growth factors. Certainly, we all dream of a 'one gene determining one phenotype' scenario, but this may actually be far from the truth in biological systems.

Another drawback of a 'milieu-based' approach for committing/directing stem cells into a specific lineage is the requirement for prolonged durations of in vitro culture. This has two major disadvantages if autologous adult stem cells are to be utilized for transplantation therapy. Firstly, this would delay treatment to the patient who is likely to be in dire need of therapeutic intervention. Secondly, there is evidence that prolonged durations of ex vivo culture could somehow alter the immunogenicity of cultured autologous cells, which could in turn lead to immuno-rejection upon transplantation. This could arise from the presence of xenogenic proteins within the in vitro culture milieu, which could potentially adhere to the surface of cultured cells, thereby conferring some degree of antigenicity to the cells [74, 75]. Additionally, the study of Hodgetts et al. [76] on autologous myoblasts (which are progenitor rather than true stem cells), would suggest that some changes also occur in the expression of endogenous cell-surface proteins during extended durations of in vitro culture, which could in turn trigger an immunogenic response from T-lymphocytes.

By contrast, genetic modulation through recombinant DNA transfection or delivery of iRNA, proteins and their synthetic analogues could offer a 'rapid' and almost 'instantaneous' means of committing stem cells into a required lineage, thereby obviating the need for prolonged durations of in vitro culture. In the case of iRNA, these can either be delivered directly into the cell [77], or be encoded by recombinant transfected DNA [78]. To avoid long-term genetic modification to the cell, direct delivery of iRNA may be preferable. Nevertheless, a major technical challenge is the relatively low efficiency of nucleic acid delivery into living cells, in the absence of a viral-based vector. Commonly-used physical techniques such as lipofection [79] and electroporation [80] may also severely compromise cellular viability, in addition to being relatively inefficient in nucleic acid delivery. The use of viral-based vectors could dramatically enhance transfection efficiency, particularly in the case of adenoviral vectors, where transfection efficiencies of > 95% are routinely reported [81].

There are however overwhelming safety concerns with regards to both the use of viral-based vectors, as well as the application of genetic manipulation in human clinical therapy. For example, the constitutive overexpression of any one single protein by transfected stem cells would certainly have unpredictable physiological consequences upon transplantation in situ. Genetically-manipulated stem cells also run the risk of becoming malignant within the transplant recipient. In a recent study [82] on gene therapy for X-linked severe combined immunodeficiency (SCID-X1), it was reported that retroviral vector insertion triggered malignant cell proliferation, which was driven by retrovirus enhancer activity on the LMO2 promoter. Hence, these various safety concerns have to be rigorously addressed before genetically manipulated stem cells are acceptable for use in human clinical therapy.

In conclusion, both the 'milieu-based' and 'gene-modulatory' approach for committing/directing stem cells into a required lineage have their associated advantages and disadvantages, which would turn out to be a major issue of contention in transplantation medicine. The choice between these two approaches would ultimately depend on the particular model of organ/tissue transplantation concerned, as well as further technical developments in this field.

5. MILIEU-BASED APPROACHES

5.1. Development of Defined Culture Milieu for Directing Stem Cell Differentiation and Lineage Commitment *In Vitro*

5.1.1. Advantages of Defined Culture Milieu

For clinical applications of stem cell transplantation therapy, it is imperative that in vitro culture protocols should be devoid of animal or human products to avoid potential contamination with pathogens. The avoidance of products of animal or human origin would also reduce variability within the culture milieu and provide a more stringent level of quality control. Moreover, supplemented animal or human proteins may adhere onto the surface of cultured stem cells, which could possibly enhance their antigenicity upon transplantation. Hence, the ideal culture milieu for promoting the lineage commitment and differentiation of stem cells in vitro should be chemically defined and either be serum-free or use synthetic serum replacements [83, 84], with the possible supplementation of specific recombinant cytokines and growth factors if so required.

5.1.2. Serum-Free Culture Conditions

The major problems with culturing stem cells under serum-free conditions is that cells generally tend to have a lower mitotic index, become apoptotic, and display poor adhesion, in the absence of serum [83]. Serum supplementation is usually required for most types of adult stem cells. There are, however, exceptions. Most notably, neural stem cells are routinely cultured under serum-free conditions [85]. For clinical applications, there are a number of reasons why the use of serum should be completely eliminated from the in vitro culture milieu developed for stem cells. First and foremost, the composition of serum is poorly defined, with a considerable degree of inter-batch variation, even when obtained from the same manufacturer. This would impede good quality control in the laboratory. Serum is also not completely physiological, since it is essentially a pathological fluid formed in response to blood clotting. Additionally, it also contains uncharacterized growth and differentiation factors, which may confound the controlled differentiation of stem cells into specific well-defined lineages in vitro.

A step towards serum-free culture conditions is the development of chemically defined synthetic serum substitutes. At present, there are a number of such commercially available synthetic serum substitutes [83, 84]. Most notable of these is Knockout Serum Replacement (KSR), which was specifically developed for the maintenance of ES cells in an undifferentiated state within in vitro culture [84]. The exact chemical composition of KSR is not available, since it is a protected by trade secret. However, it has been reported to be completely devoid of any undefined growth factors or differentiation-promoting factors [84]. This would be extremely useful for achieving controlled differentiation of stem cells into specific well-defined lineages in vitro. In the case of adult stem cells, culture conditions devoid of either serum or synthetic serum replacement have been developed for hematopoietic [86, 87], mesenchymal [88] and neural [85, 89] stem cells. However, the lower mitotic index in the absence of serum 87, 90] poses a formidable challenge for the expansion and proliferation of differentiated stem cells. This may be overcome by the supplementation of various combinations of recombinant cytokines and growth factors in the culture milieu that could stimulate the proliferation of adult stem cells under serum-free conditions.

5.1.3. Exogenous Cytokines and Growth Factors

The use of recombinant cytokines is another step forward towards the development of defined culture conditions for stem cell differentiation. In the case of ES cells, they have the ability to spontaneously differentiate into multiple lineages in vitro [91], even in the absence of cytokines and growth factors. This is manifested by the formation of an embryoid body. Hence, a possible strategy for ES cell transplantation therapy would be to initially allow spontaneous differentiation of ES cells into multiple lineages in vitro, in the absence of cytokines and growth factors. This would be followed by selective purification and expansion of the differentiated subpopulation of interest. However, spontaneous differentiation within an embryoid body is a relatively inefficient and haphazard process, in which only a very small fraction of the ES cells would have differentiated into the lineage of interest. It is likely that the use of recombinant cytokines and growth factors would enhance the differentiation of ES cells into desired lineages. Unlike ES cells, adult stem cells do not appear to either spontaneously differentiate or proliferate well in the absence of cytokines and growth factors. The in vitro culture milieu of adult stem cells must therefore be supplemented with cytokines and growth factors. At present, rapid progress is being made in this area, and there are several studies which have successfully used recombinant cytokines to induce differentiation of adult stem cells in vitro. In an interesting study by Hall et al. [92], it was reported that recombinant TGFb1-VWF fusion protein bound to collagen matrices could promote the differentiation of bone marrow progenitor cells into several lineages including fibrogenic, osteogenic, chondrogenic and adipogenic lineages. In another study, Bernhard et al. [93] reported that Hyper-IL6, a recombinant fusion protein of interleukin-6 with its specific receptor, promoted the differentiation of human hematopoietic stem cells (derived from either peripheral blood or bone marrow) into functional dendritic cells. In the case of umbilical cord blood, Flores-Guzman et al. [94] used a combination of recombinant cytokines to induce differentiation of the CD34+ enriched population into myeloid progenitors, while Yoshikawa et al. [95] reported differentiation into both myeloid and lymphoid lineages through the use of a recombinant cytokine cocktail. In addition to bone marrow and hematopoietic stem cells, the use of recombinant cytokines for the differentiation of hepatic stem cells [96] has also been reported.

5.1.4. Non-Proteinaceous Chemical Compounds

Besides protein-based cytokines and growth factors, a number of synthetic chemical compounds have also been shown to promote stem cell differentiation in vitro [97-107]. Synthetic chemicals tend to be less labile, with a longer active half-life in solution, compared to protein-based cytokines and growth factors. This would be advantageous for prolonged in vitro cell culture over several days or even weeks. Moreover, unlike proteins that have to be synthesized in living organisms and subjected to complex posttranslational modifications (i.e. glycosylation, peptidesplicing, conformational folding), synthetic chemical compounds are manufactured by chemical reactions in the laboratory, and hence are more structurally and chemically defined compared to proteins.

Among the synthetic chemicals that are known to possess morphogenic and teratogenic properties are retinoic acid, dexamethasone and 5-azacytidine. The effect of retinoic acid (a derivative of vitamin A) on neurogenic differentiation is particularly well documented [97-100]. Additionally, retinoic acid has also been reported to induce differentiation of ES cells into the cardiomyogenic [101] and adipogenic [102] lineages. Dexamethasone, a synthetic steroid, has been shown to promote osteogenic differentiation of both embryonic [103] and adult [104] stem cells. 5-Azacytidine is a synthetic nucleoside that is commonly used as an inhibitor of DNA methylation. It has been reported to enhance the maturation of neural stem cells [105], as well as induce mesenchymal stem cells to differentiate into the cardiomyogenic lineage [106, 107].

5.1.5. Naturally-Occurring Extracellular Matrix and Novel Biomaterials as Culture Substratum

The three-dimensional acellular extracellular matrix (ECM) in living tissues is primarily composed of high molecular weight molecules such as collagen and proteoglycans, and makes up approximately 75% of the biological mass [108]. Besides maintaining the structural integrity of organs and tissues, the ECM also has diverse physiological functions. For example, the ECM controls the passage of nutrients to the cells, and acts as the reservoir where growth factors and other physiological mediators are concentrated. Additionally, the interaction between extracellular matrix components such as collagen and proteoglycans with specific cellsurface receptors mediates cellular function. Hence, the physiological function of organs and tissues is in fact maintained by the complex interactions between cells and their surrounding ECM [108].

One strategy to achieve controlled differentiation of stem cells into well-defined lineages in vitro would be to introduce extracellular matrix molecules within the culture milieu. Type I collagen, one of the most commonly occurring extracellular matrix molecules, has been reported to slow down cellular senescence within in vitro culture [109], and has therefore been proposed for use in reducing the rate of ageing during the in vitro expansion of stem cells for tissue engineering applications. Additionally, the use of collagen gels for ES cell differentiation into multiple lineages has also been reported [110]. Integrins are a major class of extracellular matrix receptor molecules that have an important role in cellular differentiation and tissue development [111]. Integrin-activated signaling pathways have been implicated in the differentiation of ES cells into the mesodermal and neuroectodermal lineages [112], as well as in the differentiation of bone marrow stromal precursors into functional osteoblasts [113]. Laminin, an integrin ligand, has been shown to direct osteogenic differentiation in vitro [114]. Another major class of extracellular matrix molecules are the glycosaminoglycans (GAGs), which are long-chain sugar molecules. These have been reported to maintain CD34+ expression in vitro during the proliferation and expansion of CD34+ enriched cord blood cells [115]. Heparan sulfate, another member of the GAG family, appears to have an important role in directing the phenotypic maturation of neural stem cells

[116]. One commonly occurring adhesion molecule within ECM is fibronectin, which is involved in the homing of CD34+ cells into the bone marrow during transplantation [117], as well as in the differentiation of neural crest stem cells into melanocytes [118].

Besides naturally occurring ECM, much progress has also been made in synthetic artificial matrices for stem cell differentiation in vitro. In an interesting study [119], a recombinant extracellular matrix protein incorporating multiple active domains of different proteins on an elastin backbone was used to direct the differentiation of neural progenitor cells into the glial lineage. A collagen-based biopolymer scaffold for corneal stem cells has been successfully tested on a rabbit model [120], while an artificial thymic organoid was tissue engineered by seeding thymic stroma on a tantulum-coated carbon matrix [121]. Because the in vivo hematopoietic environment within bone marrow is composed of stromal cells and extracellular matrix in a three-dimensional configuration, it has been proposed that the in vitro culture of hematopoietic progenitors should be carried out in an artificial matrix embedded with stromal cells. Non-woven synthetic three-dimensional fibrous matrices seeded with stromal cells were reported to be superior to two-dimensional culture for the propagation of hematopoietic stem cells [122, 123]. The use of three-dimensional polyvinyl formal (PVF) resin, together with growth factors, has been successfully used for the propagation of hematopoietic stem cells, in the absence of coculture with stromal cells [124]. Moreover, PVF resin also achieved host-derived hematopoiesis within the artificial matrix when it was transplanted in vivo [125]. Another promising avenue of development is in bone tissue-engineering. The differentiation of mesenchymal stem cells into the osteogenic lineage has been reported with the use of macroporous calcium phosphate ceramics [126], hyaluronan-gelatin composite sponge [127], and fibrin microbeads [128].

At present, the development of artificially synthesized matrices for tissue engineering applications is progressing rapidly. It is anticipated that more novel types of artificial matrices would be developed in the near future for the differentiation of stem cells into various lineages for ultimate use in transplantation therapy.

5.2. Application of Biophysical Stimuli

A possible avenue for stem cell differentiation that has remained less well explored is the application of biophysical stimuli, i.e. in the form of mechanical forces, heat treatment, as well as magnetic and electrical fields. Indeed, there is much evidence in the literature to show that the physiological function of in vitro cultured cells could be radically altered in response to such physical stimuli.

Osteogenic differentiation induced by mechanical stimulation has been widely documented. Matsuda et al. [129] reported that the differentiation of human osteoblastic periodontal ligament cells was enhanced in response to mechanical stress, while Yoshikawa et al. [130] reported a similar result with bone marrow osteogenic progenitors. In the absence of exogenous growth and differentiation factors, mechanical stimuli alone could induce the differentiation of mesenchymal stem cells into the osteogenic lineage [131]. Further investigations by Kapur et al. [132] showed that there existed multiple competing signalling pathways for osteogenic differentiation, in response to mechanical stimuli.

The activation of heat shock protein expression, in response to elevated temperatures, has also been shown to promote cellular differentiation. With human embryonal carcinoma cells, heat treatment induced differentiation into the trophectodermal [133] and neuronal [134] lineages. Additionally, heat shock proteins have also been implicated in erythroid differentiation [135], glial differentiation [136] and neural plate induction in early mammalian central nervous system and brain development [137].

Another interesting area of research is the application of electrical and magnetic fields to stimulate cellular differentiation in vitro. It was reported that exposure to magnetic fields could increase the mitotic index of erythroleukemia cells [138] and induce neurite outgrowth in PC12 cells [139]. With applied electrical fields, cardiomyogenic differentiation of ES cells was enhanced [140], while neonatal cardiomyocytes were stimulated to undergo hypertrophy and cellular maturation, through an increase in mitochondrial content and activity [141]. Additionally, neuronal [142] and osteogenic [143] differentiation were also reported to be stimulated by applied electrical fields.

At the present moment, there is a paucity of information on the effects of physical stimuli on stem cell differentiation. Hopefully, more investigations will be conducted in this potentially interesting area of research in the near future.

5.3. Co-Culture with Differentiated Somatic Cells

Another strategy to direct the differentiation of stem cells into well-defined lineages in vitro is to coculture with fully differentiated somatic cell types. In fact, prior to differentiation studies, cocultures with somatic cells were used for the maintenance and proliferation of both embryonic [144] and adult [145] stem cells in vitro. In the case of ES cells, the most commonly used feeder cells are mitotically inactivated murine fibroblasts, even though human fetal fibroblasts [146] and bone marrow cells [147] as feeders have also been reported. In the case of hematopoietic progenitors derived from bone marrow or cord blood, the most commonly used feeder cells are bone marrow stromal cells [148], even though the use of other cell types such as brain endothelial cells has also been reported [149].

At present, there are many studies that have directed embryonic stem cell differentiation by coculture with various somatic cell types. Differentiation of ES cells into the hematopoietic and osteogenic lineage has been achieved through coculture with stromal cells [150-152] and fetal osteoblasts [153] respectively. In an interesting study, Mummery et al. [154] managed to differentiate ES cells into beating cardiomyocytes within in vitro culture, by coculture with visceral endoderm-like cells.

Besides ES cells, cocultures have also been used to differentiate adult stem cells in vitro. It was reported that neuronal differentiation was promoted in the presence of coculture with chromaffin cells [155] and Schwann cells [156, 157], as well as with primary neurons and astrocytes derived from neonatal hippocampus [158]. When cocultured with myoblasts, neural stem cells formed neuromuscular synaptic junctions [159], and trans-differentiated into the myogenic lineage [160]. Trans-differentiation of mesencymal stem cells into the cardiomyogenic lineage was achieved by coculture with fully differentiated cardiomyocytes [161]. In the same study, it was also found that cell to cell contact within coculture was required for cardiomyogenic differentiation of mesenchymal stem cells, since culture medium conditioned by differentiated cardiomyocytes failed to elicit differentiation. Besides mesenchymal stem cells, adult endothelial progenitor cells were also reported to be capable of trans-differentiating into the cardiomyogenic lineage when subjected to coculture with fully differentiated cardiomyocytes [162]. A novel strategy to achieve stem cell differentiation through coculture would be to simulate tissue injury in vitro. In the study by Spees et al. [163], tissue injury was simulated by subjecting small airway epithelial cells to heat shock. When bone marrow derived mesenchymal stem cells were cocultured with the heatshocked cells, these trans-differentiated into epithelial cells.

The major advantage of coculture for stem cell differentiation is that it allows intimate contact between different cell types, which could in turn lead to a more efficient transduction of molecular signals that induce differentiation. The surface receptors of co-cultured cells come into direct physical contact, and the autocrine and paracrine factors secreted by one cell type readily interact with the other cell type. Hence, coculture provides a more physiological environment for stem cell differentiation in vitro that is more akin to in vivo conditions within tissues and organs. However, recently there is evidence that intimate physical contact during coculture may lead to fusion of different cell types, resulting in the formation of heterokaryons [164, 165]. In fact, cell-fusion phenomenon has been used to explain the ability of adult stem cells to trans-differentiate into cell types that are radically different from their tissue of origin, when transplanted in vivo [166]. At present, the usefulness of such fused heterokaryons in stem cell transplantation therapy is controversial. Perhaps, in tissues where polyploidy is commonplace, such as muscle, hepatocytes and Purkinje cells, the presence of such fused heterokaryons would be more acceptable. Coculture of two or more distinct cell populations also carries a strong risk of transmission of pathogens, in particular viruses. This would constitute a major obstacle to the clinical application of coculture for stem cell differentiation. Another major shortcoming of coculture is the difficulty of the separation of co-cultured cell populations. The highest degree of purity upon separation could be achieved by fluorescence activated cell sorting (FACS) [167]. However, FACS is labor intensive and requires expensive instrumentation. Even so, there is no guarantee of absolutely no contamination with another cell type after separation with FACS. This would pose a significant problem if the contaminating cell type is proliferating much more rapidly than the purified cell population of interest. Magnetic-affinity cell sorting (MACS) [168] is much cheaper compared to FACS, but the degree of purity upon separation is much lower. The problem of separating distinct cell populations, as well as the problem of cell fusion, may be overcome by keeping co-cultured cell populations physically separate. This could be achieved through the use of commercially available Transwell inserts [169] or artificially synthesized matrices such as microcapsules [170]. However, without intimate physical contact between different cell types during coculture, there could be less efficient transduction of molecular signals for differentiation. Additionally, the use of Transwell inserts and artificially synthesized matrices would also not alleviate the risk of viral transmission during coculture.

5.4. Conditioned Media

Because several studies have shown that coculture with differentiated somatic cells can be used to direct stem cell differentiation in vitro (see preceding section), it is possible that the culture media conditioned by differentiated somatic cells could contain various soluble factors that are capable of inducing stem cell differentiation. Provided that such putative differentiating factors are non-labile, the conditioned media could also be used to direct stem cell differentiation in vitro. The advantage of using cell-conditioned media over conventional coculture is that it alleviates the need to separate different cell populations. Additionally, filtering the conditioned media would negate the risk of contamination with a different cell type. However, it is important to note that filtration itself does not remove the risk of viral transmission.

At present, studies from various groups have successfully utilized conditioned media to direct stem cell differentiation in vitro. Culture media conditioned by bone marrow stromal cells [171] and embryonic retina [172] have been used to induce neuronal differentiation and proliferation of adult neural stem cells. In an interesting study, neural stem cells were trans-differentiated into retinal cells by medium that was conditioned by fully differentiated retinal cells [173]. Differentiation of ES cells into the neuronal lineage was achieved by culture in astrocyte-conditioned media [174], as well as in media conditioned by HepG2 cell line [175]. In another study [176], myeloma-conditioned medium directed the differentiation of bone marrow derived mesenchymal stem cells into the osteogenic lineage. However, the major limitation of conditioned media, as opposed to conventional coculture, is that there is no intimate physical contact and regulatory cross-talk between different cell types. This in turn limits the transduction of differentiating signals to the stem cells. Indeed, a number of studies have shown that, in some cases, conditioned media failed to elicit stem cell differentiation, when compared to coculture [161].

5.5. Induction of Intercellular Coupling through Gap Junction Formation

A common means of intercellular communication in vivo is through gap junctions, which form small connecting channels (pore size $\approx$ 1.5 nm) between the cytoplasm of adjacent cells [177, 178]. Gap junctions are primarily composed of transmembrane proteins that are generically referred to as connexins. Cells coupled by gap junctions can share small molecules (<1,500 Da) such as inorganic ions and metabolites, but not macromolecules (proteins, nucleic acid and polysaccharides). Even so, the electrochemical and metabolic coupling achieved through gap junctions, as a result of the passage of small molecules, plays an important role in regulating the physiological function of cells [177, 178].

Hence, a possible strategy that could be used to direct stem cell differentiation in vitro would be to induce intercellular coupling by gap junction formation between stem cells and differentiated somatic cells within coculture, or between stem cells themselves. This may be achieved by upregulating cellular expression of transmembrane connexin proteins, which are assembled into gap junctions. Previous studies have shown that increased connexin expression within in vitro culture is usually accompanied by enhanced intercellular coupling, as assessed by dye-coupling assays [179, 180].

Various techniques have been used to upregulate the expression of connexin proteins, with resultant enhancement of gap junction formation and intercellular coupling. With cardiomyocytes, Pimentel et al. [181] reported that stimulation with vascular endothelial growth factor (VEGF) could increase the expression of connexin protein subtype Cx43, while with neuronal cells cyclic cAMP signaling [180] and alltrans retinoic acid [182] have been implicated in upregulating Cx43 expression. Additionally, cAMP signaling has also been shown to enhance Cx43 expression and subsequent gap junction formation in osteoblasts, and the resultant increased intercellular coupling between osteoblasts was reported to enhance osteocalcin expression, which is a specific marker of osteoblast maturity [183]. With primary hepatocytes, chondroitin sulfate [184] and glucocorticoids [179] have both been shown to enhance intercellular coupling through gap junction formation. In an interesting study, forced expression of gap junction proteins Cx43 and Cx32, through DNA transfection within cell lines, led to an enhancement of intercellular coupling as assessed by dye-coupling assays [185]. Other molecular signals that have been reported to enhance gap junction formation and subsequent intercellular coupling include low-density lipoprotein [186] inhibitors of glycosylation [187], dopamine [188] and potassium conductance blockers, tetraethylammonium and 4-aminopyridine [189].

The major limitation of gap junction communication is that only inorganic ions and small molecular weight metabolites (<1,500 Da) can transverse through the cytoplasm of coupled cells, while much larger cytosolic signaling proteins and mRNA transcripts cannot pass through gap junctions [177, 178]. Hence, there might not be an effective transduction of differentiation signals through gap junctions alone.

5.6. Exposure to Free Radicals and Reactive Oxygen Species

Another interesting area of research that has received scant attention is the potential use of free radicals and reactive oxygen species (ROS) to direct stem cell differentiation in vitro. A free radical theory of development proposed by Allen and Balin [190] postulated that differential oxygen supplies to tissues in vivo result in the establishment of metabolic and oxidative gradients that in turn play an important role in differentiation and development. This concept is supported by the fact that cells and tissues at different stages of differentiation exhibit discrete changes in their antioxidant defenses and oxidation parameters. Generally, as cells become more differentiated, their intracellular environment becomes more pro-oxidizing or less reducing, when compared to undifferentiated or dedifferentiated cells [191]. This change in redox balance appears to be due to enhanced O_2 generation with increasing degree of cellular differentiation. Because ROS have been implicated as secondary messengers in many intracellular signalling cascades [192], changes to the oxidation parameter of the cell during differentiation would have profound effects on cellular function. This is best illustrated by the recent study of Noble et al. [193], which demonstrated that the intracellular redox state is a central modulator of the balance between self-renewal and differentiation in O-2A (oligodendrocyte-type-2 astrocyte) progenitor cells.

The molecular pathway by which intracellular redox state modulates cellular differentiation is at present not well characterized, and is just beginning to be understood. Recently, seminal studies by Fulco et al. [194] and Kyrylenko et al. [195] have implicated histone deacetylase Sir2 in the redox regulation of cellular differentiation.

Investigations by Sauer et al. [196] reported that cardiomyogenic differentiation of ES cells within embryoid bodies was enhanced in the presence of hydrogen peroxide. On the other hand, incubation with free radical scavengers and antioxidants had an inhibitory effect on cardiomyogenic differentiation. Further investigation implicated a NADPH oxidase-like enzyme in ROS generation during embryoid body formation. It was therefore concluded that ROS had an important role in early cardiac development. Besides cardiomyogenesis, oxidative stress induced by hydrogen peroxide has also been reported to stimulate osteogenic differentiation in vitro [197]. The free radical nitric oxide (NO) is known to have diverse physiological functions, and one of these is in neurogenesis. The differentiation of neural progenitors in vitro was reported to be enhanced in the presence of exogenous nitric oxide [198, 199]. In addition to nitric oxide, hydroxyl free radicals have also been reported to induce neuronal differentiation in vitro [200]. With cells of the immune system, the induction of oxidative stress in vitro has been reported to stimulate B lymphocyte activation and differentiation [201], as well as granulocytic maturation [202].

Hence, it is clear from this relatively small number of studies that exogenous free radicals and ROS could profoundly affect cellular differentiation in vitro. Further research needs to be carried out on the potential use of free radicals and ROS to direct stem cell differentiation in vitro.

6. Gene-Modulatory Approaches

6.1. Recombinant DNA

Directing stem cell differentiation and lineage commitment through genetic manipulation with recombinant DNA would avoid prolonged durations of ex vivo culture of autologous adult stem cells that can delay treatment to the patient and alter their immunogenicity [50, 51]. The advantage of this approach is that recombinant DNA constructs can also incorporate marker genes that would make it easier to purify and expand the transfected cells in vitro. Among the commonly used marker genes for the genetic manipulation of stem cells are: jellyfish green fluorescent protein [203], neomycin resistance [204] and lacZ [205].

The disadvantage of this approach is the potential risks associated with utilizing recombinant DNA technology in human clinical therapy. For example, the overexpression of any one particular protein or growth factor within transfected stem cells would certainly have unpredictable physiological effects upon transplantation in vivo. This problem may be overcome by placing the recombinant expression of the particular protein under the control of switchable promoters, several of which have been developed for expression in eukaryotic systems. Such switchable promoters could be responsive to exogenous chemicals, heat shock, or even light. Of particular interest are light-responsive promoters [206], because these would avoid the potentially toxic or pleiotropic effects of exogenous chemicals and heat treatment.

At present, there are as yet no reported studies on the coupling of lineage-specific genes to light-responsive promoters. Indeed, the creation of such recombinant constructs and their subsequent transfection within stem cells would certainly make an interesting study with potentially useful clinical applications. There is also a risk that such genetically manipulated stem cells may become malignant within the transplanted recipient. Moreover, there are safety concerns with regards to the use of recombinant viral-based vectors in the genetic manipulation of stem cells [207]. It remains uncertain as to whether legislation would ultimately permit the use of such vectors in human clinical therapy.

At present, a number of studies have successfully induced stem cell differentiation in vitro through genetic manipulation. Osteogenic differentiation of mesenchymal stem cells was enhanced through expression of bone morphogenetic protein-2, 4 and 6 [207, 208], and also with Notch protein [208]. Additionally, it was also reported that recombinant expression of bone morphogenetic protein-2 within adipogenic stem cells resulted in transdifferentiation into the osteogenic lineage [210]. With ES cells, expression of Noggin and Chordin proteins promoted differentiation into the neuronal lineage [211]. Similar results were reported with Id1 protein expression [212]. Gi alpha 2 expression in ES cells promoted terminal differentiation to adipocytes [213], while expression of the Wnt gene induced differentiation into the hematopoietic lineage [214]. At present, the potential detrimental effects of transplanting genetically modified stem cells in vivo are not well studied. More research needs to be carried out on animal models to address the safety aspects of such an approach.

6.2. Cybridization and Exposure to Cytoplasmic Extracts

Another possible strategy to directly influence gene regulatory networks controlling lineage commitment and differentiation in stem cells may be to fuse them with enucleated cytoplasts derived from differentiated somatic cells, to form cytoplasmic hybrids or cybrids. Techniques for generating enucleated cytoplasts from differentiated somatic cells have been developed and refined over the course of the past 3 decades [215-217]. Briefly, this usually involves treatment of the somatic cells with a microtubulin inhibitor (i.e. cytochalasin), followed by high-speed centrifugation within a layered density gradient (composed of either Ficoll or Percoll). The enucleated cytoplasts can then be fused with nucleated cells through a variety of different techniques utilizing electrical pulse [218], polyethylene glycol [219] or with Sendai virus [220].

Previous studies have shown that cybridization could be used to induce teratocarcinoma cells to express myoblast function [221, 222], as well as direct erythroid [223, 224] and myeloid [225] differentiation. One possible disadvantage of cytoplast fusion is the risk of viral transmission, although this would be clinically acceptable if cybrids were generated from autogenic cells. Moreover, the enucleation procedure by high speed centrifugation may lead to some cytoplasts retaining fragments of chromosomal DNA, which could result in aneuploidy upon fusion with nucleated cells. There could also be structural disruptions to the cytoskeleton and organelles within the enucleated cytoplasts, as a result of high-speed centrifugation. This could possibly lead to aberrant cellular function upon fusion with nucleated cells. Another major problem is the relatively low efficiency of cell fusion, and the relatively poor viability of the resulting cybrids [220]. This would pose a formidable challenge with relatively scarce adult stem cells [11].

Another variation on this theme may be to expose permeabilized stem cells cells (i.e. with streptolysin-O) to concentrated cytoplasmic extracts derived from differentiated somatic cells. Using such a technique, Hakelien and colleagues [226, 227] managed to reprogram 293T fibroblasts to express T-cell and neuronal function, by exposure to cytoplasmic extracts of T cells and neuronal progenitors respectively. Molecular analysis revealed that permeabilized fibroblasts exposed to T-cell cytoplasmic extracts displayed nuclear uptake and assembly of T-cell-specific transcription factors, activation of lymphoid specific genes, and expression of T-cell-specific surface antigens. Similarly, permeabilized fibroblasts exposed to the cytoplasmic extract of neuronal progenitor cells expressed neurofilament proteins, as well as displayed neurite-like outgrowths. These results therefore showed that reprogramming of differentiated somatic cells with cytoplasmic extracts could possibly be used for producing autogenic replacement cells for therapeutic applications. More recently, Qin et al. [228] demonstrated that cytoplasmic extracts can be used to direct the differentiation of murine ES cells to pneumocytes. It is important to note that a major limitation to the use of cytoplasmic extracts for cellular reprogramming is the risk of viral transmission. Nevertheless, this may be acceptable if cytoplasmic extracts were used to reprogram the differentiation of autogenic cells.

The advantage of cybridization is the intracellular exposure of the nuclei to cytosolic proteins and mRNA transcripts of another cell type, which could in turn lead to a more effective transduction of differentiation signals. Intracellular signaling molecules are likely to be extremely labile, and hence easily destroyed by the relatively harsh procedure used to make a cytoplasmic extract, whereas the stability of such labile molecules is likely to be preserved within intact cytoplasts.

6.3. Delivery of iRNA, Proteins or Their Synthetic Analogues into the Cell

To avoid permanent genetic modification with recombinant DNA transfection, an alternative strategy may be deliver RNA, proteins or their synthetic analogs, such as peptide nucleic acid [229], directly into the cell. Because such molecules have a limited active half-life in the cytosol and are obviously not incorporated into the genetic code of the cell, these would only exert a transient modulatory effect on gene expression. Even so, a transient effect may be preferable for clinical therapy, since this would ultimately avoid permanent genetic alteration to the cell. However, it remains an open question as to whether transient modulation of gene expression in stem cells could be effective for therapeutic strategies in the clinical setting. This approach could perhaps be useful for priming molecular signalling cascades within the cellular machinery, which are self-sustaining after the initial stimulus. Of particular interest in the context of stem cell therapy would be signaling molecules involved in making cell-fate decisions. For example, expression of Rho GTPase at a critical time point of mesenchymal precursor differentiation determines commitment to the myogenic rather than the adipogenic lineage [230], and there is no turning-back once the decision has been made. Nevertheless, transient modulation of gene expression in stem cells could very well bring about a state of malignancy, which would pose a significant safety risk. Hence, strategies for "weeding out" malignant cells have to be sought before transplantation or transfusion to patients. Of particular promise would be the new generation of chemotherapeutic drugs that are currently being developed and that possess a high degree of specificity toward transformed malignant cell types.

Of particular interest among the diverse array of intracellular proteins that can potentially modulate gene expression are transcription factors [231] that bind and activate specific promoter sequences on genomic DNA. These are in effect the most upstream modulators of gene expression and can be thought of as "master switches" that orchestrate the entire complexity of gene regulatory networks within stem cells [231]. In addition to transcription factors, other intracellular proteins of interest include those that have been implicated in the multitude of signaling cascades present within stem cells [232, 233]. These would exert more downstream control on gene regulation compared with transcription factors, but their effects on gene expression would be no less profound. Synthetic proteins may also be worthwhile looking at for modulating gene expression, in addition to naturally occurring intracellular proteins. For example, single chain variable fragment (scFv) antibodies [234] may be utilized to target intracellular signaling molecules and transcription factors, which could lead to a profound effect on gene expression.

RNA interference [235, 236] is another promising strategy for the transient modulation of gene expression in stem cells. To summarize briefly, the RNA interference pathway involves the processing of long double-stranded RNA (dsRNA) into 21–bp to 25-bp small interfering RNAs (siRNA) by an RNase-III-like enzyme called Dicer [237]. The siRNA is then incorporated within a multisubunit RNA-induced silencing complex, which specifically targets homologous cellular mRNA transcripts for degradation [238]. For exogenous administration, it is preferable to utilize short sequences of siRNA directly rather than long sequences of unprocessed dsRNA. Obviously, it is much more technically challenging and expensive to synthesize long dsRNA sequences. Furthermore, direct exposure of mammalian cells to long sequences of dsRNA can induce apoptosis through the activation of dsRNA-dependent protein kinase (PKR) and the type I interferon response [239]. This problem is much less acute or even nonexistent with short siRNA sequences (<30 bp long) [239]. Nevertheless, gene-repression by exogenously administered siRNA is a transient process that is largely dependent on cell cycling and on the half-life of the targeted protein, which on average lasts about three to five cell-doublings [240]. To overcome the relatively short active half-life of exogenously administered iRNA, it may be advantageous to look at synthetic analogs of RNA for antisense modulation of gene expression. Of particular interest would be peptide nucleic acids (PNA), in which the phosphate-sugar polynucleotide backbone of naturally occurring nucleic acids is replaced by a flexible pseudo-peptide polymer backbone, to which the nucleobases are linked [229]. This structure gives PNA the capacity to hybridize with high affinity and specificity to complementary sequences of RNA, in addition to conferring remarkable resistance to intracellular RNAses and proteinases [229]. Moreover, the pseudo-peptide backbone of PNA would enable it to be directly linked to peptide sequences, such as protein transduction domains [241], which could facilitate their direct delivery into the cell. This will be elaborated on below. The unique physio-chemical characteristics of PNA therefore make it a potentially powerful tool for modulating gene expression, as demonstrated by a number of studies [242, 243]. Attempting to deliver proteins, RNA, or its synthetic PNA analog directly into stem cells is undoubtedly technically challenging. This difficulty is further compounded by the relative scarcity of stem cells, which makes it imperative for the delivery technique not to compromise cellular viability. Hence, relatively "harsh" techniques such as electroporation [80] would be completely out of the question. Currently, two newly emerging delivery platforms appear particularly promising: (1) protein transduction domains [241], and (2) immunoliposomes [244]. Both of these have the potential for high efficiency delivery with little compromise to cellular viability and will each be critically examined in turn.

Protein transduction domains (PTD) are short peptide sequences that enable proteins to be translocated across the cell membrane and internalized within the cytosol, through atypical secretory and internalization pathways [241]. The exact mechanism by which this is achieved is not well understood, but it is likely to be independent of endocytotic mechanisms, transmembrane protein channels, and protein receptor binding. This is because in vitro studies have demonstrated that PTD-mediated translocation can occur even at low temperatures and does not have strong cellular specificity [245, 246]. A variety of naturally occurring and artificial PTD has been identified [241]. Among the first to be discovered were the TAT transactivator domain of the human immunodeficiency virus [247] and the homeodomain of Antennapedia transcription factor [248]. Additionally, a number of artificial peptide sequences have been demonstrated to possess the properties of naturally occurring PTD [249, 250]. A comprehensive review of "cell-permeable" transduction peptide domains is provided by Joliot and Prochiantz [240]. In addition to proteins, PTD may also be utilized for the delivery of both RNA and PNA, as has been demonstrated by a number of studies [251-253]. Because PNA possesses a pseudo-peptide backbone [229], it is relatively easy to covalently link PTD sequences to PNA via a peptide bond [251, 252]. By contrast, it is much more technically challenging to couple PTD directly to RNA sequences, because of their non-complementary molecular backbones (sugar-phosphate polynucleotide backbone in the case of RNA). Nevertheless, PTD can still facilitate the direct delivery of iRNA into the cell through incorporation within gene-delivery lipoplexes [253] or the use of PTD-poly-lysine conjugates [254].

Another promising delivery platform is immunoliposomes [244], which are lipid bilayer formulations impregnated with antibodies specific for surface molecules of the target cell. In addition to their high specificity to target cells, the other major advantage of immunoliposomes is the "protective coat" formed by the lipid bilayer around the molecules being delivered, shielding them from degradation by the digestive enzymes (i.e., RNAses, proteinases) that could be present in the extracellular environment [244]. This method would be advantageous for both in vitro and in vivo delivery of proteins, RNA, or PNA into stem cells.

Overwhelming safety and ethical concerns will preclude the clinical application of genetically modified stem cells in the foreseeable near future. Hence, there is a dire need to achieve the modulation of gene expression in stem cells without the direct application of recombinant DNA technology and the consequent genetic alteration that it entails. We hope that further rapid advances in the various emerging technologies that have been discussed will facilitate the attainment of this objective in the not too distant future.

7. Conclusions

Under natural physiological conditions, the differentiation and lineage commitment of stem cells probably involves multiple overlapping signalling pathways and gene regulatory networks that respond to a diverse array of extracellular cues. Perhaps, this may be best mimicked in vitro by using a combination of the various techniques discussed earlier, that are either milieu-based or directly involve gene modulation. However, it must be kept in mind that for clinical applications, protocol efficiency by itself is inadequate, and it is also imperative to develop xeno-free and pathogen-free conditions that comply to stringent cGMP requirements [255, 256].

References

[1] Weissman IL. Translating stem and progenitor cell biology to the clinic: barriers and opportunities. *Science*. 2000, 287:1442–1446.

[2] Burns CE, Zon LI. Portrait of a stem cell. *Dev. Cell*. 2002 Nov;3(5):612-3.

[3] Thomson JA, Itskovitz-Eldor J, Shapiro SS, Waknitz MA, Swiergiel JJ, Marshall VS, Jones JM. Embryonic stem cell lines derived from human blastocysts. *Science*. 1998 Nov 6;282(5391):1145-7.

[4] Reubinoff BE, Pera MF, Fong CY, Trounson A, Bongso A. Embryonic stem cell lines from human blastocysts: somatic differentiation in vitro. *Nat. Biotechnol.* 2000 Apr;18(4):399-404.

[5] Shamblott MJ, Axelman J, Wang S, Bugg EM, Littlefield JW, Donovan PJ, Blumenthal PD, Huggins GR, Gearhart JD. Derivation of pluripotent stem cells from cultured human primordial germ cells. *Proc. Natl. Acad. Sci. USA* 1998, 95:13726–13731.

[6] Andrews PW. pluripotent human EC cell lines. Review article. *APMIS* 1998, 106:158–167; discussion 167–168.

[7] Poulsom R, Alison MR, Forbes SJ, Wright NA. Adult stem cell plasticity. *J. Pathol.* 2002, 197:441–456.

[8] Erlandsson A, Morshead CM. Exploiting the properties of adult stem cells for the treatment of disease. *Curr. Opin. Mol. Ther*. 2006 Aug;8(4):331-7.

[9] Khosrotehrani K, Bianchi DW. Multi-lineage potential of fetal cells in maternal tissue: a legacy in reverse. *J. Cell Sci.* 2005 Apr 15;118(Pt 8):1559-63.

[10] Itskovitz Eldor J, Schuldiner M, Karsenti D, Eden A, Yanuka O, Amit M, Soreq H, Benvenisty N. Differentiation of human embryonic stem cells into embryoid bodies compromising the three embryonic germ layers. *Mol. Med.* 2000, 6:88–95.

[11] Verfaillie CM. Adult stem cells: assessing the case for pluripotency. *Trends Cell Biol.* 2001, 12:502–508.

[12] Orlic D, Kajstura J, Chimenti S, Jakoniuk I, Anderson SM, Li B, Pickel J, McKay R, Nadal Ginard B, Bodine DM, Leri A, Anversa P. Bone marrow cells regenerate infracted myocardium. *Nature* 2001, 410:701–705.

[13] Jackson KA, Majka SM, Wang H, Pocius J, Hartley CJ, Majesky MW, Entman ML, Michael LH, Hirschi KK, Goodell MA. Regeneration of ischemic cardiac muscle and vascular endothelium by adult stem cells. *J. Clin. Invest.* 2001, 107:1395–1402.

[14] Ferrari G, Cusella De Angelis G, Coletta M, Paolucci E, Stornaiuolo A, Cossu G, Mavilio F. Muscle regeneration by bone marrow-derived myogenic progenitors. *Science* 1998, 279:1528–1530.

[15] Gussoni E, Soneoka Y, Strickland CD, Buzney EA, Khan MK, Flint AF, Kunkel LM, Mulligan RC. Dystrophin expression in the mdx mouse restored by stem cell transplantation. *Nature* 1999, 401:390–394.

[16] Grant MB, May WS, Caballero S, Brown GA, Guthrie SM, Mames RN, Byrne BJ, Vaught T, Spoerri PE, Peck AB, Scott EW. Adult hematopoietic stem cells provide functional hemangioblast activity during retinal neovascularization. *Nat. Med.* 2002, 607–612.

[17] Brazelton TR, Rossi FM, Keshet GI, Blau HM. From marrow to brain: expression of neuronal phenotypes in adult mice. *Science* 2000, 290:1775–1779.

[18] Krause DS, Theise ND, Collector MI, Henegariu O, Hwang S, Gardner R, Neutzel S, Sharkis SJ. Multi-organ, multilineage engraftment by a single bone marrow-derived stem cell. *Cell* 2001, 105:369–377.

[19] Petersen BE, Bowen WC, Patrene KD, Mars WM, Sullivan AK, Murase N, Boggs SS, Greenberger JS, Goff JP. Bone marrow as a potential source of hepatic oval cells. *Science* 1999, 284:1168–1170.

[20] Udani VM. The continuum of stem cell transdifferentiation: possibility of hematopoietic stem cell plasticity with concurrent CD45 expression. *Stem Cells Dev.* 2006 Feb;15(1):1-3.

[21] Dominguez-Bendala J, Ricordi C. Stem cell plasticity and tissue replacement. *Cell Transplant.* 2005;14(7):423-5.

[22] Terada N, Hamazaki T, Oka M et al. Bone marrow cells adopt the phenotype of other cells by spontaneous cell fusion. *Nature* 2002;4:542–545.

[23] Ying QL, Nichols J, Evans EP et al. Changing potency by spontaneous fusion. *Nature* 2002;4:545–548.

[24] Goodell MA. Stem-cell "plasticity": befuddled by the muddle. *Curr. Opin. Hematol.* 2003;10:208–213.

[25] Hurlbut WB. Framing the future: embryonic stem cells, ethics and the emerging era of developmental biology. *Pediatr. Res.* 2006 Apr;59(4 Pt 2):4R-12R.

[26] Hug K. Sources of human embryos for stem cell research: ethical problems and their possible solutions. *Medicina* (Kaunas). 2005;41(12):1002-10.

[27] Ohara N. Ethical consideration of experimentation using living human embryos: the Catholic Church's position on human embryonic stem cell research and human cloning. *Clin. Exp. Obstet. Gynecol.* 2003;30(2-3):77-81.

[28] McCulloch CA, Strugurescu M, Hughes F et al. Osteogenic progenitor cells in rat bone marrow stromal populations exhibit self-renewal in culture. *Blood* 1991;77:1906–1911.

[29] Quarto R, Thomas D, Liang CT. Bone progenitor cell deficits and the age-associated decline in bone repair capacity. *Calcif. Tissue Int.* 1995;56:123–129.

[30] D'Ippolito G, Schiller PC, Ricordi C et al. Age-related osteogenic potential of mesenchymal stromal stem cells from human vertebral bone marrow. *J. Bone Miner Res.* 1999;14:1115–1122.

[31] Huttmann A, Li CL, Duhrsen U. Bone marrow-derived stem cells and "plasticity". *Ann. Hematol.* 2003 Oct;82(10):599-604.

[32] Yalniz M, Pour PM. Are there any stem cells in the pancreas? Pancreas. 2005 Aug;31(2):108-18.

[33] NIH Stem Cell Backgrounder, 2003. Available at: http://www.nih.gov/news/pr/mar2003/stemcellbackgrounder.htm. Accessed September 15, 2006.

[34] Wang L, Duan E, Sung LY, Jeong BS, Yang X, Tian XC. Generation and characterization of pluripotent stem cells from cloned bovine embryos. *Biol. Reprod.* 2005 Jul;73(1):149-55.

[35] Wakayama S, Ohta H, Kishigami S, Thuan NV, Hikichi T, Mizutani E, Miyake M, Wakayama T. Establishment of male and female nuclear transfer embryonic stem cell lines from different mouse strains and tissues. *Biol. Reprod.* 2005 Apr;72(4):932-6.

[36] Cobbe N. Why the apparent haste to clone humans? J Med Ethics. 2006 May;32(5):298-302.

[37] Taylor CJ, Bolton EM, Pocock S, Sharples LD, Pedersen RA, Bradley JA. Banking on human embryonic stem cells: estimating the number of donor cell lines needed for HLA matching. *Lancet.* 2005 Dec 10;366(9502):2019-25.

[38] Reddy S, Pedowitz DI, Parekh SG, Sennett BJ, Okereke E. The Morbidity Associated With Osteochondral Harvest From Asymptomatic Knees for the Treatment of Osteochondral Lesions of the Talus. *Am. J. Sports Med.* 2006 Sep 6; [Epub ahead of print]

[39] Cooke MJ, Stojkovic M, Przyborski SA. Growth of teratomas derived from human pluripotent stem cells is influenced by the graft site. *Stem Cells Dev.* 2006 Apr;15(2):254-9.

[40] Tzukerman M, Rosenberg T, Ravel Y, Reiter I, Coleman R, Skorecki K. (2003) An experimental platform for studying growth and invasiveness of tumor cells within teratomas derived from human embryonic stem cells. *Proc. Natl. Acad. Sci. USA.* 11;100(23):13507-12.

[41] Przyborski SA. Differentiation of human embryonic stem cells after transplantation in immune-deficient mice. *Stem Cells.* 2005 Oct;23(9):1242-50.

[42] Baier PC, Schindehutte J, Thinyane K, Flugge G, Fuchs E, Mansouri A, Paulus W, Gruss P, Trenkwalder C. (2004) Behavioral changes in unilaterally 6-hydroxy-dopamine lesioned rats after transplantation of differentiated mouse embryonic stem cells without morphological integration. *Stem Cells.* 22(3):396-404.

[43] Zhang SC, Wernig M, Duncan ID, Brustle O, Thomson JA. (2001) In vitro differentiation of transplantable neural precursors from human embryonic stem cells. *Nat. Biotechnol.* 19(12):1129-33.

[44] Fang B, Shi M, Liao L, Yang S, Liu Y, Zhao RC. Multiorgan engraftment and multilineage differentiation by human fetal bone marrow Flk1+/CD31-/CD34- Progenitors. *J. Hematother. Stem Cell Res.* 12(6):603-13.

[45] Mackenzie TC, Flake AW. (2001) Multilineage differentiation of human MSC after in utero transplantation. *Cytotherapy*. 3(5):403-5.

[46] Nieminen AL. (2003) Apoptosis and necrosis in health and disease: role of mitochondria. *Int. Rev. Cytol.* 224:29-55.

[47] Roth E, Manhart N, Wessner B. (2004) Assessing the antioxidative status in critically ill patients. *Curr. Opin. Clin. Nutr. Metab. Care*. 7(2):161-8.

[48] Rankin JA. (2004) Biological mediators of acute inflammation. *AACN Clin. Issues*. 15(1):3-17.

[49] Wang JS, Shum-Tim D, Chedrawy E, Chiu RC. (2001) The coronary delivery of marrow stromal cells for myocardial regeneration: pathophysiologic and therapeutic implications. *J. Thorac. Cardiovasc. Surg.* 2001 Oct;122(4):699-705.

[50] Johnson LF, S deSerres, SR Herzog, HD Peterson, Meyer AA. (1991). Antigenic cross-reactivity between media supplements for cultured keratinocyte grafts. *J. Burn Care Rehabil.* 12(4):306-12.

[51] Smythe GM, Grounds MD. (2000). Exposure to tissue culture conditions can adversely affect myoblast behavior in vivo in whole muscle grafts: implications for myoblast transfer therapy. *Cell Transplant*. 9(3):379-93.

[52] Tsuchiya H, Kitoh H, Sugiura F, Ishiguro N. (2003) Chondrogenesis enhanced by overexpression of sox9 gene in mouse bone marrow-derived mesenchymal stem cells. *Biochem. Biophys Res. Commun.* 301(2):338-43.

[53] Chen D, Zhang G. (2001) Enforced expression of the GATA-3 transcription factor affects cell fate decisions in hematopoiesis. *Exp. Hematol.* 29(8):971-80.

[54] Reid T, Warren R, Kirn D. (2002) Intravascular adenoviral agents in cancer patients: lessons from clinical trials. *Cancer Gene Ther.* 12 979-86.

[55] Wang JS, Shum-Tim D, Galipeau J, Chedrawy E, Eliopoulos N, Chiu RC. (2000) Marrow stromal cells for cellular cardiomyoplasty: feasibility and potential clinical advantages. *J. Thorac. Cardiovasc. Surg.* 120(5):999-1005.

[56] Davani S, Marandin A, Mersin N, Royer B, Kantelip B, Herve P, Etievent JP, Kantelip JP. (2003) Mesenchymal progenitor cells differentiate into an endothelial phenotype, enhance vascular density, and improve heart function in a rat cellular cardiomyoplasty model. *Circulation.* 108 Suppl 1:II253-8.

[57] Tang YL, Zhao Q, Zhang YC, Cheng L, Liu M, Shi J, Yang YZ, Pan C, Ge J, Phillips MI. (2004) Autologous mesenchymal stem cell transplantation induce VEGF and neovascularization in ischemic myocardium. *Regul. Pept.* 117(1):3-10.

[58] Zhang M, Methot D, Poppa V, Fujio Y, Walsh K, Murry CE. (2001) Cardiomyocyte grafting for cardiac repair: graft cell death and anti-death strategies. *J. Mol. Cell Cardiol.* 33(5):907-21.

[59] Zhang SJ, Zhu CJ, Zhao YF, Li J, Guo WZ. (2003) Different ischemic preconditioning for rat liver graft: protection and mechanism. *Hepatobiliary Pancreat Dis. Int.* 2(4):509-12.

[60] Drukker M, Katz G, Urbach A, Schuldiner M, Markel G, Itskovitz_Eldor J, Reubinoff, B, Mandelboim O, Benvenisty N. (2002) Characterization of the expression of MHC proteins in human embryonic stem cells. *Proc. Natl. Acad. Sci. USA.* 99:15 9864-9.

[61] Drukker M, Benvenisty N. (2004) The immunogenicity of human embryonic stem-derived cells. *Trends Biotechnol.* 22(3):136-41.

[62] Le Blanc K, Tammik C, Rosendahl K, Zetterberg E, Ringden O. (2003) HLA expression and immunologic properties of differentiated and undifferentiated mesenchymal stem cells. *Exp. Hematol.* 31(10):890-6.

[63] Ryan JM, Barry FP, Murphy JM, Mahon BP. Mesenchymal stem cells avoid allogeneic rejection. *J. Inflamm.* (Lond). 2005 Jul 26;2:8.

[64] Le Blanc K. Immunomodulatory effects of fetal and adult mesenchymal stem cells. *Cytotherapy.* 2003, 5(6):485-9.

[65] Inoue S, Popp FC, Koehl GE, Piso P, Schlitt HJ, Geissler EK, Dahlke MH. Immunomodulatory effects of mesenchymal stem cells in a rat organ transplant model. *Transplantation.* 2006 Jun 15;81(11):1589-95.

[66] Djouad F, Plence P, Bony C, Tropel P, Apparailly F, Sany J, Noel D, Jorgensen C. Immunosuppressive effect of mesenchymal stem cells favors tumor growth in allogeneic animals. *Blood.* 2003, 102(10):3837-44.

[67] Spradling A, Drummond-Barbosa D, Kai T. Stem cells find their niche. *Nature.* 2001 Nov 1;414(6859):98-104.

[68] Daley GQ, Goodell MA, Snyder EY. Realistic prospects for stem cell therapeutics. *Hematology Am. Soc. Hematol Educ. Program.* 2003;:398-418.

[69] Ehrhardt D. GFP technology for live cell imaging. *Curr. Opin. Plant Biol.* 2003 Dec;6(6):622-8.

[70] Cai Q, Rubin JT, Lotze MT. Genetically marking human cells--results of the first clinical gene transfer studies. *Cancer Gene Ther.* 1995 Jun;2(2):125-36.

[71] Akiyama H, Chaboissier MC, Martin JF, Schedl A, de Crombrugghe B. The transcription factor Sox9 has essential roles in successive steps of the chondrocyte differentiation pathway and is required for expression of Sox5 and Sox6. *Genes Dev.* 2002 Nov 1;16(21):2813-28.

[72] Lian JB, Javed A, Zaidi SK, Lengner C, Montecino M, van Wijnen AJ, Stein JL, Stein GS. Regulatory controls for osteoblast growth and differentiation: role of Runx/Cbfa/AML factors. *Crit. Rev. Eukaryot. Gene Expr.* 2004;14(1-2):1-41.

[73] Gersbach CA, Le Doux JM, Guldberg RE, Garcia AJ. Inducible regulation of Runx2-stimulated osteogenesis. *Gene Ther.* 2006 Jun;13(11):873-82.

[74] Gerhart J, Neely C, Stewart B, Perlman J, Beckmann D, Wallon M, Knudsen K, George-Weinstein M. Epiblast cells that express MyoD recruit pluripotent cells to the skeletal muscle lineage. *J. Cell Biol.* 2004 Mar 1;164(5):739-46.

[75] Berkes CA, Tapscott SJ. MyoD and the transcriptional control of myogenesis. *Semin. Cell Dev. Biol.* 2005 Aug-Oct;16(4-5):585-95.

[76] Hodgetts SI, Beilharz MW, Scalzo AA, Grounds MD. Why do cultured transplanted myoblasts die in vivo? DNA quantification shows enhanced survival of donor male myoblasts in host mice depleted of CD4+ and CD8+ cells or Nk1.1+ cells. *Cell Transplant.* 2000 Jul-Aug;9(4):489-502.

[77] Weil D, Garcon L, Harper M, Dumenil D, Dautry F, Kress M. Targeting the kinesin Eg5 to monitor siRNA transfection in mammalian cells. *Biotechniques.* 2002 Dec;33(6):1244-8.

[78] Wilson JA, Richardson CD. Induction of RNA interference using short interfering RNA expression vectors in cell culture and animal systems. *Curr. Opin. Mol. Ther.* 2003 Aug;5(4):389-96.

[79] Dass CR, Su T. Delivery of lipoplexes for genotherapy of solid tumours: role of vascular endothelial cells. *J. Pharm. Pharmacol.* 2000 Nov;52(11):1301-17.

[80] Chou TH, Biswas S, Lu S. Gene delivery using physical methods: an overview. *Methods Mol. Biol.* 2004;245:147-66.

[81] Lee CT, Park KH, Yanagisawa K, Adachi Y, Ohm JE, Nadaf S, Dikov MM, Curiel DT, Carbone DP. Combination therapy with conditionally replicating adenovirus and replication defective adenovirus. *Cancer Res.* 2004 Sep 15;64(18):6660-5.

[82] Hacein-Bey-Abina S, Von Kalle C, Schmidt M, McCormack MP, Wulffraat N, Leboulch P, Lim A, Osborne CS, Pawliuk R, Morillon E, Sorensen R, Forster A, Fraser P, Cohen JI, de Saint Basile G, Alexander I, Wintergerst U, Frebourg T, Aurias A, Stoppa-Lyonnet D, Romana S, Radford-Weiss I, Gross F, Valensi F, Delabesse E, Macintyre E, Sigaux F, Soulier J, Leiva LE, Wissler M, Prinz C, Rabbitts TH, Le Deist F, Fischer A, Cavazzana-Calvo M. LMO2-associated clonal T cell proliferation in two patients after gene therapy for SCID-X1. *Science.* 2003 Oct 17;302(5644):415-9.

[83] Wong M, Tuan RS. Nuserum, a synthetic serum replacement, supports chondrogenesis of embryonic chick limb bud mesenchymal cells in micromass culture. *In Vitro Cell Dev. Biol. Anim.* 1993;29A:917–922.

[84] Goldsborough MD, Tilkins ML, Price PJ et al. Serum-free culture of murine embryonic stem (ES) cells. *Focus* 1998;20: 8-12.

[85] Kallos MS, Sen A, Behie LA (2003) Large-scale expansion of mammalian neural stem cells: a review. *Med. Biol. Eng. Comput.* 41:271–282.

[86] Mobest D, Mertelsmann R, Henschler R (1998) Serum-free ex vivo expansion of CD34(+) hematopoietic progenitor cells. *Biotechnol. Bioeng.* 60:3 341–347.

[87] Almeida-Porada G, Brown RL, MacKintosh FR, Zanjani ED (2003) Evaluation of serum-free culture conditions able to support the ex vivo expansion and engraftment of human hematopoietic stem cells in the human-to-sheep xenograft model. *J. Hematother. Stem Cell Res.* 9:683–693.

[88] Lennon DP, Haynesworth SE, Young RG, Dennis JE, Caplan AI. A chemically defined medium supports in vitro proliferation and maintains the osteochondral potential of rat marrow-derived mesenchymal stem cells. *Exp. Cell Res.* 1995, 219:211–222.

[89] Yokoyama A, Sakamoto A, Kameda K, Imai Y, Tanaka J. NG2 proteoglycan-expressing microglia as multipotent neural progenitors in normal and pathologic brains. *Glia.* 2006 May;53(7):754-68.

[90] Quinn CM, Kagedal K, Terman A, Stroikin U, Brunk UT, Jessup W, Garner B. Induction of fibroblast apolipoprotein E expression during apoptosis, starvation-induced growth arrest and mitosis. *Biochem. J.* 2004 Mar 15;378(Pt 3):753-61.

[91] Mikkola M, Olsson C, Palgi J, Ustinov J, Palomaki T, Horelli-Kuitunen N, Knuutila S, Lundin K, Otonkoski T, Tuuri T. Distinct differentiation characteristics of individual human embryonic stem cell lines. *BMC Dev. Biol.* 2006 Aug 8;6:40.

[92] Hall FL, Han B, Kundu RK, Yee A, Nimni ME, Gordon EM. Phenotypic differentiation of TGF-beta1-responsive pluripotent premesenchymal prehematopoietic

progenitor (P4 stem) cells from murine bone marrow. *J. Hematother. Stem Cell Res.* 2001, 10:261–271.

[93] Bernhard H, Lohmann M, Batten WY, Metzger J, Lohr HF, Peschel C, zum Buschenfelde KM, Rose-John S. The gp130- stimulating designer cytokine hyper-IL-6 promotes the expansion of human hematopoietic progenitor cells capable to differentiate into functional dendritic cells. *Exp. Hematol.* 2000, 28:365–372.

[94] Flores-Guzm□n P, Guti□rrez-Rodriguez M, Mayani H (2002) In vitro proliferation, expansion, and differentiation of a CD34+ cell-enriched hematopoietic cell population from human umbilical cord blood in response to recombinant cytokines. *Arch. Med. Res.* 2002, 33:107–114.

[95] Yoshikawa Y, Hirayama F, Kanai M, Nakajo S, Ohkawara J, Fujihara M, Yamaguchi M, Sato N, Kasai M, Sekiguchi S, Ikebuchi K. tromal cell-independent differentiation of human cord blood CD34+CD38- lymphohematopoietic progenitors toward B cell lineage. *Leukemia* 2000, 14:727–734.

[96] Suzuki A, Iwama A, Miyashita H, Nakauchi H, Taniguchi H. Role for growth factors and extracellular matrix in controlling differentiation of prospectively isolated hepatic stem cells. *Development.* 2003 Jun;130(11):2513-24.

[97] Liu S, Qu Y, Stewart TJ, Howard MJ, Chakrabortty S, Holekamp TF, McDonald JW. Embryonic stem cells differentiate into oligodendrocytes and myelinate in culture and after spinal cord transplantation. *Proc. Natl. Acad. Sci. USA* 2000, 97:6126–6131.

[98] Guan K, Chang H, Rolletschek A, Wobus AM. Embryonic stem cell-derived neurogenesis. Retinoic acid induction and lineage selection of neuronal cells. *Cell Tissue Res.* 2001, 305:171–176.

[99] Kim BJ, Seo JH, Bubien JK, Oh YS (2002) Differentiation of adult bone marrow stem cells into neuroprogenitor cells in vitro. *Neuroreport* 13:1185–1188.

[100] Akita J, Takahashi M, Hojo M, Nishida A, Haruta M, Honda Y. Neuronal differentiation of adult rat hippocampusderived neural stem cells transplanted into embryonic rat explanted retinas with retinoic acid pretreatment. *Brain Res.* 2002, 954:286–293.

[101] Wobus AM, Kaomei G, Shan J, Wellner MC, Rohwedel J, Ji GJ, Fleischmann B, Katus HA, Hescheler J, Franz WM. Retinoic acid accelerates embryonic stem cell-derived cardiac differentiation and enhances development of ventricular cardiomyocytes. *J. Mol. Cell Cardiol.* 1997, 29:1525–1539.

[102] Phillips BW, Vernochet C, Dani C. Differentiation of embryonic stem cells for pharmacological studies on adipose cells. *Pharmacol. Res.* 2003, 47:263–268.

[103] Buttery LD, Bourne S, Xynos JD, Wood H, Hughes FJ, Hughes SP, Episkopou V, Polak JM. Differentiation of osteoblasts and in vitro bone formation from murine embryonic stem cells. *Tissue Eng.* 2001, 7:89–99.

[104] Rogers JJ, Young HE, Adkison LR, Lucas PA, Black AC. Differentiation factors induce expression of muscle, fat, cartilage, and bone in a clone of mouse pluripotent mesenchymal stem cells. *Am. Surg.* 1995, 61:231–236.

[105] Schinstine M, Iacovitti L. 5-Azacytidine and BDNF enhance the maturation of neurons derived from EGF-generated neural stem cells. *Exp. Neurol.* 1997, 144:315–325.

[106] Fukuda K. Molecular characterization of regenerated cardiomyocytes derived from adult mesenchymal stem cells. *Congenit. Anom. Kyoto* 2002, 42:1–9.

[107] Rangappa S, Fen C, Lee EH, Bongso A, Wei ES. Transformation of adult mesenchymal stem cells isolated from the fatty tissue into cardiomyocytes. *Ann. Thorac. Surg.* 2003, 75:775–779.

[108] Comper WP (1996) *Extracellular matrix*, vol 2, molecular components and interactions. Harwood Academic, Amsterdam.

[109] Volloch V, Kaplan D. Matrix-mediated cellular rejuvenation. *Matrix Biol.* 2002, 21:533–543.

[110] Chen SS, Revoltella RP, Papini S, Michelini M, Fitzgerald W, Zimmerberg J, Margolis L. Multilineage differentiation of rhesus monkey embryonic stem cells in three-dimensional culture systems. *Stem Cells* 2003, 21:281–295.

[111] Tarone G, Hirsch E, Brancaccio M, De Acetis M, Barberis L, Balzac F, Retta SF, Botta C, Altruda F, Silengo L, Retta F. Integrin function and regulation in development. *Int. J. Dev. Biol.* 2000, 44:725–731.

[112] Czyz J, Wobus A. Embryonic stem cell differentiation: the role of extracellular factors. *Differentiation* 2001, 68:167–174.

[113] Gronthos S, Simmons PJ, Graves SE, Robey PG (2001) Integrin mediated interactions between human bone marrow stromal precursor cells and the extracellular matrix. *Bone* 28:174–181.

[114] Roche P, Goldberg HA, Delmas PD, Malaval L. Selective attachment of osteoprogenitors to laminin. *Bone* 1999, 24:329–336.

[115] Madihally SV, Flake AW, Matthew HW (1999) Maintenance of CD34 expression during proliferation of CD34+ cord blood cells on glycosaminoglycan surfaces. *Stem Cells* 17:295–305.

[116] Chipperfield H, Bedi KS, Cool SM, Nurcombe V. Heparan sulfates isolated from adult neural progenitor cells can direct phenotypic maturation. *Int. J. Dev. Biol.* 2002, 46:661–670.

[117] Voermans C, Gerritsen WR, von dem Borne AE, van der Schoot CE. Increased migration of cord blood-derived CD34+ cells, as compared to bone marrow and mobilized peripheral blood CD34+ cells across uncoated or fibronectin-coated filters. *Exp. Hematol.* 1999, 27:1806–1814.

[118] Takano N, Kawakami T, Kawa Y, Asano M, Watabe H, Ito M, Soma Y, Kubota Y, Mizoguchi M. Fibronectin combined with stem cell factor plays an important role in melanocyte proliferation, differentiation and migration in cultured mouse neural crest cells. *Pigment Cell Res.* 2002, 15:192–200.

[119] Liu CY, Apuzzo ML, Tirrell DA. Engineering of the extracellular matrix: working toward neural stem cell programming and neurorestoration—concept and progress report. *Neurosurgery* 2003, 52:1154–1165; discussion 1165–1167.

[120] Griffith M, Hakim M, Shimmura S, Watsky MA, Li F, Carlsson D, Doillon CJ, Nakamura M, Suuronen E, Shinozaki N, Nakata K, Sheardown H (2002) Artificial human corneas: scaffolds for transplantation and host regeneration. *Cornea* 2002, 21:S54–S61.

[121] Poznansky MC, Evans RH, Foxall RB, Olszak IT, Piascik AH, Hartman KE, Brander C, Meyer TH, Pykett MJ, Chabner KT, Kalams SA, Rosenzweig M, Scadden DT. Efficient generation of human T cells from a tissue-engineered thymic organoid. *Nat. Biotechnol.* 2000, 18:729–734.

[122] Li Y, Ma T, Kniss DA, Yang ST, Lasky LC. Human cord cell hematopoiesis in three-dimensional nonwoven fibrous matrices: in vitro simulation of the marrow microenvironment. *J. Hematother. Stem Cell Res.* 2001, 10:355–368.

[123] Sasaki T, Takagi M, Soma T, Yoshida T. 3D culture of murine hematopoietic cells with spatial development of stromal cells in nonwoven fabrics. *Cytotherapy* 2002, 4:285–291.

[124] Tun T, Miyoshi H, Aung T, Takahashi S, Shimizu R, Kuroha T, Yamamoto M, Ohshima N. Effect of growth factors on ex vivo bone marrow cell expansion using three-dimensional matrix support. *Artif. Organs* 2002, 26:333–339.

[125] Tun T, Miyoshi H, Ema H, Nakauchi H, Ohshima N. New type of matrix support for bone marrow cell cultures: in vitro culture and in vivo transplantation experiments. *ASAIO J.* 2000, 46:522–526.

[126] Toquet J, Rohanizadeh R, Guicheux J, Couillaud S, Passuti N, Daculsi G, Heymann D (1999) Osteogenic potential in vitro of human bone marrow cells cultured on macroporous biphasic calcium phosphate ceramic. *J. Biomed. Mater. Res.* 44:98–108.

[127] Angele P, Kujat R, Nerlich M, Yoo J, Goldberg V, Johnstone B. Engineering of osteochondral tissue with bone marrow mesenchymal progenitor cells in a derivatized hyaluronangelatin composite sponge. *Tissue Eng.* 1999, 5:545–554.

[128] Gurevich O, Vexler A, Marx G, Prigozhina T, Levdansky L, Slavin S, Shimeliovich I, Gorodetsky R. Fibrin microbeads for isolating and growing bone marrow-derived progenitor cells capable of forming bone tissue. *Tissue Eng.* 2002, 8:661–672.

[129] Matsuda N, Morita N, Matsuda K, Watanabe M. Proliferation and differentiation of human osteoblastic cells associated with differential activation of MAP kinases in response to epidermal growth factor, hypoxia, and mechanical stress in vitro. *Biochem. Biophys. Res. Commun.* 1998, 249:350–354.

[130] Yoshikawa T, Peel SA, Gladstone JR, Davies JE. Biochemical analysis of the response in rat bone marrow cell cultures to mechanical stimulation. *Biomed. Mater. Eng.* 1997, 7:369–377.

[131] Altman GH, Horan RL, Martin I, Farhadi J, Stark PR, Volloch V, Richmond JC, Vunjak-Novakovic G, Kaplan DL. Cell differentiation by mechanical stress. *FASEB J.* 2002,16:270–272.

[132] Kapur S, Baylink DJ, William Lau KH. Fluid flow shear stress stimulates human osteoblast proliferation and differentiation through multiple interacting and competing signal transduction pathways. *Bone* 2003, 32:241–251.

[133] Maruyama T, Umezawa A, Kusakari S, Kikuchi H, Nozaki M, Hata J. Heat shock induces differentiation of human embryonal carcinoma cells into trophectoderm lineages. *Exp. Cell Res.* 1996, 224:123–127.

[134] Yamada T, Hashiguchi A, Fukushima S, Kakita Y, Umezawa A, Maruyama T, Hata J. Function of 90-kDa heat shock protein in cellular differentiation of human embryonal carcinoma cells. *In Vitro Cell Dev. Biol. Anim.* 32000, 6:139–146.

[135] Pirkkala L, Alastalo TP, Nykanen P, Seppa L, Sistonen L. Differentiation lineage-specific expression of human heat shock transcription factor 2. *FASEB J.* 1999, 13:1089–1098.
[136] Zhang WL, Tsuneishi S, Nakamura H. Induction of heat shock proteins and its effects on glial differentiation in rat C6 glioblastoma cells. *Kobe J. Med. Sci.* 2001, 47:77–95.
[137] Walsh D, Li Z, Wu Y, Nagata K. Heat shock and the role of the HSPs during neural plate induction in early mammalian CNS and brain development. *Cell Mol. Life Sci.* 1997, 53:198–211.
[138] Eremenko T, Esposito C, Pasquarelli A, Pasquali E, Volpe P. Cell-cycle kinetics of Friend erythroleukemia cells in a magnetically shielded room and in a low-frequency/low-intensity magnetic field. *Bioelectromagnetics* 1997, 18:58–66.
[139] McFarlane EH, Dawe GS, Marks M, Campbell IC. Changes in neurite outgrowth but not in cell division induced by low EMF exposure: influence of field strength and culture conditions on responses in rat PC12 pheochromocytoma cells. *Bioelectrochemistry* 2000, 52:23–28
[140] Sauer H, Rahimi G, Hescheler J, Wartenberg M. Effects of electrical fields on cardiomyocyte differentiation of embryonic stem cells. *J. Cell Biochem.* 1999 75:710–723.
[141] Xia Y, Buja LM, Scarpulla RC, McMillin JB. Electrical stimulation of neonatal cardiomyocytes results in the sequential activation of nuclear genes governing mitochondrial proliferation and differentiation. *Proc. Natl. Acad. Sci. USA* 1997, 94:11399–11404.
[142] Mie M, Endoh T, Yanagida Y, Kobatake E, Aizawa M. Induction of neural differentiation by electrically stimulated gene expression of NeuroD2. *J. Biotechnol.* 2003, 100:231–238.
[143] Hartig M, Joos U, Wiesmann HP. Capacitively coupled electric fields accelerate proliferation of osteoblast-like primary cells and increase bone extracellular matrix formation in vitro. *Eur. Biophys. J.* 29:499–506.
[144] Richards M, Fong CY, Chan WK, Wong PC, Bongso A. Human feeders support prolonged undifferentiated growth of human inner cell masses and embryonic stem cells. *Nat. Biotechnol.* 2002, 20:933–936.
[145] Yamaguchi M, Hirayama F, Murahashi H, Azuma H, Sato N, Miyazaki H, Fukazawa K, Sawada K, Koike T, Kuwabara M, Ikeda H, Ikebuchi K. Ex vivo expansion of human UC blood primitive hematopoietic progenitors and transplantable stem cells using human primary BM stromal cells and human AB serum. *Cytotherapy* 2002, 4:109–118.
[146] Fong CY, Bongso A. Derivation of human feeders for prolonged support of human embryonic stem cells. *Methods Mol. Biol.* 2006;331:129-35.
[147] Cheng L, Hammond H, Ye Z, Zhan X, Dravid G. Human adult marrow cells support prolonged expansion of human embryonic stem cells in culture. *Stem Cells* 2003, 21:131–142.
[148] Lanza F, Campioni D, Moretti S, Dominici M, Punturieri M, Focarile E, Pauli S, Dabusti M, Tieghi A, Bacilieri M, Scapoli C, De Angeli C, Galluccio L, Castoldi G. CD34(+) cell subsets and long-term culture colony-forming cells evaluated on both

autologous and normal bone marrow stroma predict long-term hematopoietic engraftment in patients undergoing autologous peripheral blood stem cell transplantation. *Exp. Hematol.* 2001 Dec;29(12):1484-93.

[149] Chute JP, Saini AA, Chute DJ, Wells MR, Clark WB, Harlan DM, Park J, Stull MK, Civin C, Davis TA. Ex vivo culture with human brain endothelial cells increases the SCID-repopulating capacity of adult human bone marrow. *Blood* 2002, 100:4433–4439.

[150] Palacios R, Golunski E, Samaridis J. In vitro generation of hematopoietic stem cells from an embryonic stem cell line. *Proc. Natl. Acad. Sci. USA* 1995, 92:7530–7534.

[151] Uzan G, Prandini MH, Rosa JP, Berthier R. Hematopoietic differentiation of embryonic stem cells: an in vitro model to study gene regulation during megakaryocytopoiesis. *Stem Cells* 1996, 14 Suppl 1:194–199.

[152] Li F, Lu S, Vida L, Thomson JA, Honig GR. Bone morphogenetic protein 4 induces efficient hematopoietic differentiation of rhesus monkey embryonic stem cells in vitro. *Blood* 2001, 98:335–342.

[153] Buttery LD, Bourne S, Xynos JD, Wood H, Hughes FJ, Hughes SP, Episkopou V, Polak JM. Differentiation of osteoblasts and in vitro bone formation from murine embryonic stem cells. *Tissue Eng.* 2001, 7:89–99.

[154] Mummery C, Ward-van Oostwaard D, Doevendans P, Spijker R, van den Brink S, Hassink R, van der Heyden M, Opthof T, Pera M, de la Riviere AB, Passier R, Tertoolen L. Differentiation of human embryonic stem cells to cardiomyocytes: role of coculture with visceral endoderm-like cells. *Circulation* 2003, 107:2733–2740.

[155] Schumm MA, Castellanos DA, Frydel BR, Sagen J. Enhanced viability and neuronal differentiation of neural progenitors by chromaffin cell co-culture. *Brain Res. Dev. Brain Res.* 2002, 137:115–125.

[156] Wan H, An Y, Zhang Z, Zhang Y, Wang Z. Differentiation of rat embryonic neural stem cells promoted by co-cultured Schwann cells. *Chin Med. J. (Engl)* 2003, 116:428–431.

[157] An YH, Wan H, Zhang ZS, Wang HY, Gao ZX, Sun MZ, Wang ZC. Effect of rat Schwann cell secretion on proliferation and differentiation of human neural stem cells. *Biomed. Environ Sci.* 2003, 16:90–94.

[158] Song HJ, Stevens CF, Gage FH. Neural stem cells from adult hippocampus develop essential properties of functional CNS neurons. *Nat. Neurosci.* 2002, 5:438–445.

[159] Liu Z, Martin LJ. Olfactory bulb core is a rich source of neural progenitor and stem cells in adult rodent and human. *J. Comp. Neurol.* 2003, 459:368–391.

[160] Galli R, Borello U, Gritti A, Minasi MG, Bjornson C, Coletta M, Mora M, De Angelis MG, Fiocco R, Cossu G, Vescovi AL. Skeletal myogenic potential of human and mouse neural stem cells. *Nat. Neurosci.* 2000, 3:986–991.

[161] Rangappa S, Entwistle JW, Wechsler AS, Kresh JY. Cardiomyocyte-mediated contact programs human mesenchymal stem cells to express cardiogenic phenotype. *J. Thorac. Cardiovasc. Surg.* 2003, 26:124–132.

[162] Condorelli G, Borello U, De Angelis L, Latronico M, Sirabella D, Coletta M, Galli R, Balconi G, Follenzi A, Frati G, Cusella De Angelis MG, Gioglio L, Amuchastegui S, Adorini L, Naldini L, Vescovi A, Dejana E, Cossu G. Cardiomyocytes induce

endothelial cells to trans-differentiate into cardiac muscle: implications for myocardium regeneration. *Proc. Natl. Acad. Sci. USA* 2001, 98:10733–10738.

[163] Spees JL, Olson SD, Ylostalo J, Lynch PJ, Smith J, Perry A, Peister A, Wang MY, Prockop DJ. Differentiation, cell fusion, and nuclear fusion during ex vivo repair of epithelium by human adult stem cells from bone marrow stroma. *Proc. Natl. Acad. Sci. USA* 2003, 100:2397–2402.

[164] Terada N, Hamazaki T, Oka M, Hoki M, Mastalerz DM, Nakano Y, Meyer EM, Morel L, Petersen BE, Scott EW. Bone marrow cells adopt the phenotype of other cells by spontaneous cell fusion. *Nature* 2002, 416:542–545.

[165] Ying QL, Nichols J, Evans EP, Smith AG. Changing potency by spontaneous fusion. *Nature* 2002, 416:545–548.

[166] Pauwelyn KA, Verfaillie CM. Transplantation of undifferentiated, bone marrow-derived stem cells. *Curr. Top Dev. Biol.* 2006;74:201-51.

[167] Herzenberg LA, Parks D, Sahaf B, Perez O, Roederer M. The history and future of the fluorescence activated cell sorter and flow cytometry: a view from Stanford. *Clin. Chem*. 2002, 48:1819–1827.

[168] Siegel DL. Selecting antibodies to cell-surface antigens using magnetic sorting techniques. *Methods Mol. Biol.* 2002, 178:219– 226.

[169] Giovino MA, Down JD, Jackson JD, Sykes M, Monroy RL, White-Scharf ME. Porcine hematopoiesis on primate stroma in long-term cultures: enhanced growth with neutralizing tumor necrosis factor-alpha and tumor growth factor-beta antibodies. *Transplantation* 2002, 73:723–731.

[170] Orive G, Gascon AR, Hernandez RM, Igartua M, Luis Pedraz J. Cell microencapsulation technology for biomedical purposes: novel insights and challenges. *Trends Pharmacol. Sci.* 2003, 24:207–210.

[171] Lou S, Gu P, Chen F, He C, Wang M, Lu C (2003) The effect of bone marrow stromal cells on neuronal differentiation of mesencephalic neural stem cells in Sprague-Dawley rats. *Brain Res.* 2003, 968:114–121.

[172] Kaneko Y, Ichikawa M, Kurimoto Y, Ohta K, Yoshimura N (2003) Neuronal differentiation of hippocampus-derived neural stem cells cultured in conditioned medium of embryonic rat retina. *Ophthalmic Res*. 35:268–275.

[173] Yuan HP, Ge J, Duan YH, Wang YM, Wang LN, Yang BB. *Induced differentiation of neural stem cells from subependymal zone into retinal cells in vitro*. Zhonghua Yan Ke Za Zhi 2003, 39:357–360.

[174] Nakayama T, Momoki Soga T, Inoue N. Astrocyte-derived factors instruct differentiation of embryonic stem cells into neurons. *Neurosci. Res*. 2003, :241–249.

[175] Rathjen J, Haines BP, Hudson KM, Nesci A, Dunn S, Rathjen PD. Directed differentiation of pluripotent cells to neural lineages: homogeneous formation and differentiation of a neurectoderm population. *Development* 2002, 129:2649–2661.

[176] Karadag A, Scutt AM, Croucher PI. Human myeloma cells promote the recruitment of osteoblast precursors: mediation by interleukin-6 and soluble interleukin-6 receptor. *J. Bone Miner Res.* 2000, 15:1935–1943.

[177] Evans WH, De Vuyst E, Leybaert L. The gap junction cellular internet: connexin hemichannels enter the signalling limelight. *Biochem. J.* 2006 Jul 1;397(1):1-14.

[178] Laird DW. Life cycle of connexins in health and disease. *Biochem. J.* 2006 Mar 15;394(Pt 3):527-43.

[179] Ren P, de Feijter AW, Paul DL, Ruch RJ. Enhancement of liver cell gap junction protein expression by glucocorticoids. *Carcinogenesis* 1994, 15:1807–1813.

[180] Dowling-Warriner CV, Trosko JE. Induction of gap junctional intercellular communication, connexin43 expression, and subsequent differentiation in human fetal neuronal cells by stimulation of the cyclic AMP pathway. *Neuroscience* 2000, 95:859–868.

[181] Pimentel RC, Yamada KA, Kl□ber AG, Saffitz JE. Autocrine regulation of myocyte Cx43 expression by VEGF. *Circ. Res.* 2002, 90:671–677.

[182] Carystinos GD, Alaoui-Jamali MA, Phipps J, Yen L, Batist G. Upregulation of gap junctional intercellular communication and connexin 43 expression by cyclic-AMP and alltrans-retinoic acid is associated with glutathione depletion and chemosensitivity in neuroblastoma cells. *Cancer Chemother. Pharmacol.* 2001, 47:126–132.

[183] Romanello M, Moro L, Pirulli D, Crovella S, D'Andrea P. Effects of cAMP on intercellular coupling and osteoblast differentiation. *Biochem. Biophys. Res. Commun.* 2001, 282:1138–1144.

[184] Kato S, Sugiura N, Kimata K, Kujiraoka T, Toyoda J, Akamatsu N. Chondroitin sulfate immobilized onto culture substrates modulates DNA synthesis, tyrosine aminotransferase induction, and intercellular communication in primary rat hepatocytes. *Cell Struct. Funct.* 1995, 20:199–209.

[185] Koffler L, Roshong S, Kyu Park I, Cesen-Cummings K, Thompson DC, Dwyer-Nield LD, Rice P, Mamay C, Malkinson AM, Ruch RJ. Growth inhibition in G(1) and altered expression of cyclin D1 and p27(kip-1) after forced connexin expression in lung and liver carcinoma cells. *J. Cell Biochem.* 2000, 79:347–354.

[186] Paulson AF, Lampe PD, Meyer RA, TenBroek E, Atkinson MM, Walseth TF, Johnson RG. Cyclic AMP and LDL trigger a rapid enhancement in gap junction assembly through a stimulation of connexin trafficking. *J. Cell Sci.* 2000, 113:3037–3049.

[187] Wang Y, Mehta PP. Facilitation of gap-junctional communication and gap-junction formation in mammalian cells by inhibition of glycosylation. *Eur. J. Cell Biol.* 1995, 67:285–296.

[188] Halliwell JV, Horne AL. Evidence for enhancement of gap junctional coupling between rat island of Calleja granule cells in vitro by the activation of dopamine D3 receptors. *J. Physiol.* 1998, 506:175–194.

[189] Kannan MS, Daniel EE. Formation of gap junctions by treatment in vitro with potassium conductance blockers. *J. Cell Biol.* 1978, 78:338–348.

[190] Allen RG, Balin AK. Oxidative influence on development and differentiation: an overview of a free radical theory of development. *Free Radic. Biol. Med.* 1989, 6:6 631–661.

[191] Sohal RS, Allen RG, Nations C. Oxygen free radicals play a role in cellular differentiation: an hypothesis. *J. Free Radic. Biol. Med.* 1986, 2:175–181.

[192] Sauer H, Wartenberg M, Hescheler J. Reactive oxygen species as intracellular messengers during cell growth and differentiation. *Cell Physiol. Biochem.* 2001, 11:173–186.

[193] Noble M, Smith J, Power J, Mayer Proschel M. Redox state as a central modulator of precursor cell function. *Ann. NY Acad. Sci.* 2003, 991:251–271

[194] Fulco M, Schiltz RL, Iezzi S, King MT, Zhao P, Kashiwaya Y, Hoffman E, Veech RL, Sartorelli V. Sir2 regulates skeletal muscle differentiation as a potential sensor of the redox state. *Mol. Cell* 2003, 12:51–62.

[195] Kyrylenko S, Kyrylenko O, Suuronen T, Salminen A. Differential regulation of the Sir2 histone deacetylase gene family by inhibitors of class I and II histone deacetylases. *Cell Mol. Life Sci.* 2003, 60:1990–1997.

[196] Sauer H, Rahimi G, Hescheler J, Wartenberg M. Role of reactive oxygen species and phosphatidylinositol 3-kinase in cardiomyocyte differentiation of embryonic stem cells. *FEBS Lett.* 2000, 7 476:218–223.

[197] Mody N, Parhami F, Sarafian TA, Demer LL. Oxidative stress modulates osteoblastic differentiation of vascular and bone cells. *Free Radic. Biol. Med.* 2001, 15 31:4 509–519.

[198] Moreno Lopez B, Noval JA, Gonzalez Bonet LG, Estrada C. Morphological bases for a role of nitric oxide in adult neurogenesis. *Brain Res.* 2000, 30 869:244–250.

[199] Gibbs SM. Regulation of neuronal proliferation and differentiation by nitric oxide. *Mol. Neurobiol.* 2002, 27:107–120.

[200] Oravecz K, Kalka D, Jeney F, Cantz M, Zs Nagy I. Hydroxyl free radicals induce cell differentiation in SK-N-MC neuroblastoma cells. *Tissue Cell* 2002, 34:33–38.

[201] Fedyk ER, Phipps RP. Reactive oxygen species and not lipoxygenase products are required for mouse B-lymphocyte activation and differentiation. *Int. J. Immunopharmacol.* 1994, 16:7 533–546.

[202] Nagy K, Pasti G, Bene L, Zs Nagy I. Induction of granulocytic maturation in HL-60 human leukemia cells by free radicals: a hypothesis of cell differentiation involving hydroxyl radicals. *Free Radic. Res. Commun.* 1993, 19:1–15.

[203] Stadtfeld M, Varas F, Graf T. Fluorescent protein-cell labeling and its application in time-lapse analysis of hematopoietic differentiation. *Methods Mol. Med.* 2005;105:395-412.

[204] Licht T, Herrmann F, Gottesman MM, Pastan I. In vivo drug-selectable genes: a new concept in gene therapy. *Stem Cells.* 1997;15(2):104-11.

[205] Bagnis C, Chabannon C, Mannoni P. Beta-galactosidase marker genes to tag and track human hematopoietic cells. *Cancer Gene Ther.* 1999 Jan-Feb;6(1):3-13.

[206] Shimizu-Sato S, Huq E, Tepperman JM et al. A lightswitchable gene promoter system. *Nat. Biotechnol.* 2002;20: 1041–1044.

[207] Ahrens M, Ankenbauer T, Schr□der D, Hollnagel A, Mayer H, Gross G. Expression of human bone morphogenetic proteins-2 or -4 in murine mesenchymal progenitor C3H10T1/2 cells induces differentiation into distinct mesenchymal cell lineages. *DNA Cell Biol.* 1993, 12:871–880.

[208] Gitelman SE, Kirk M, Ye JQ, Filvaroff EH, Kahn AJ, Derynck R. Vgr-1/BMP-6 induces osteoblastic differentiation of pluripotential mesenchymal cells. *Cell Growth Differ.* 1995, 6:827–836.

[209] Tezuka K, Yasuda M, Watanabe N, Morimura N, Kuroda K, Miyatani S, Hozumi N. Stimulation of osteoblastic cell differentiation by Notch. *J. Bone Miner Res.* 2002, 17:231–239.

[210] Dragoo JL, Choi JY, Lieberman JR, Huang J, Zuk PA, Zhang J, Hedrick MH, Benhaim P. Bone induction by BMP-2 transduced stem cells derived from human fat. *J. Orthop. Res.* 2003, 21:622–629.

[211] Gratsch TE, O'Shea KS. Noggin and chordin have distinct activities in promoting lineage commitment of mouse embryonic stem (ES) cells. *Dev. Biol.* 2002, 245:83–94.

[212] Jurga M, Buzan'ska L. The influence of Id1 protein on human stem cells differentiation into neuronal pathway. *J. Neurochem*. 2003, 85:Suppl 2:33.

[213] Su HL, Malbon CC, Wang HY. Increased expression of Gi alpha 2 in mouse embryo stem cells promotes terminal differentiation to adipocytes. Am J Physiol 265:C1729–1735 differentiation to adipocytes. *Am. J. Physiol.* 1993, 265:C1729–1735.

[214] Lako M, Lindsay S, Lincoln J, Cairns PM, Armstrong L, Hole N. Characterisation of Wnt gene expression during the differentiation of murine embryonic stem cells in vitro: role of Wnt3 in enhancing haematopoietic differentiation. *Mech. Dev*. 2001, 103:49–59.

[215] Goldman RD, Pollack R, Hopkins NH. Preservation of normal behavior by enucleated cells in culture. *Proc. Natl. Acad. Sci. USA* 193, 70:750–754.

[216] Wigler MH, Weinstein IB. A preparative method for obtaining enucleated mammalian cells. *Biochem. Biophys. Res. Commun*. 1975, 63:669–674.

[217] Stocco DM. Rapid, quantitative isolation of mitochondria from rat liver using Ficoll gradients in vertical rotors. *Anal. Biochem.* 1983, 131:453–457.

[218] Sligh JE, Levy SE, Waymire KG, Allard P, Dillehay DL, Nusinowitz S, Heckenlively JR, MacGregor GR, Wallace DC. Maternal germ-line transmission of mutant mtDNAs from embryonic stem cell-derived chimeric mice. *Proc. Natl. Acad. Sci. USA* 2000, 97:14461–14466.

[219] Davidson RL, Gerald PS. Induction of mammalian somatic cell hybridization by polyethylene glycol. *Methods Cell Biol*. 1977, 15:325–338.

[220] Lucas JJ, Kates JR. The construction of viable nuclearcytoplasmic hybrid cells by nuclear transplantation. *Cell* 1976, 7:397–405.

[221] Iwakura Y, Nozaki M, Asano M, Yoshida MC, Tsukada Y, Hibi N, Ochiai A, Tahara E, Tosu M, Sekiguchi T. Pleiotropic phenotypic expression in cybrids derived from mouse teratocarcinoma cells fused with rat myoblast cytoplasts. *Cell* 1985, 43:777–791.

[222] Tosu M, Terasaki T, Iwakura Y, Yoshida M, Sekiguchi T. Clonal isolation and characterization of myoblast-like reconstituted cells formed by fusion of karyoplasts from mouse teratocarcinoma cells with rat myoblast cytoplasts. *Cell Struct. Funct.* 1988, 13:249–266.

[223] Watanabe T, Nomura S, Oishi M. Induction of erythroid differentiation by cytoplast fusion in mouse erythroleukemia (Friend) cells. *Exp. Cell Res.* 1985, 159:224–234.

[224] Watanabe T, Nomura S, Kaneko T, Yamagoe S, Kamiya T, Oishi M. Cytoplasmic factors involved in erythroid differentiation in mouse erythroleukemia (MEL) cells. *Cell Differ Dev.* 1988, 25 Suppl:105–109.

[225] Okazaki T, Kato Y, Tashima M, Sawada H, Uchino H. Evidence of intracellular and trans-acting differentiation-inducing activity in human promyelocytic leukemia HL-60 cells: its possible involvement in process of cell differentiation from a commitment step to a phenotype-expression step. *J. Cell Physiol.* 1988, 134:261–268.

[226] Hakelien AM, Collas P. Novel approaches to transdifferentiation. *Cloning Stem Cells* 2002, 4:379–387.

[227] Hakelien AM, Landsverk HB, Robl JM, Sk□lhegg BS, Collas P. Reprogramming fibroblasts to express T-cell functions using cell extracts. *Nat. Biotechnol.* 2002, 20:460–466.

[228] Qin M, Tai G, Collas P, Polak JM, Bishop AE. Cell extract-derived differentiation of embryonic stem cells. *Stem Cells*. 2005 Jun-Jul;23(6):712-8.

[229] Marin VL, Roy S, Armitage BA. Recent advances in the development of peptide nucleic acid as a gene-targeted drug. *Expert Opin. Biol. Ther*. 2004, 4:337–348.

[230] Saltiel AR. Muscle or fat? Rho bridges the GAP. Cell 2003, 113:144–145.

[231] Urnov FD, Rebar EJ. Designed transcription factors as tools for therapeutics and functional genomics. *Biochem. Pharmacol*. 2002, 64:919–923.

[232] Loebel DA, Watson CM, De Young RA, Tam PP. Lineage choice and differentiation in mouse embryos and embryonic stem cells. *Dev. Biol.* 2003, 264:1–14.

[233] Moon RT, Kohn AD, De Ferrari GV, Kaykas A. WNT and beta-catenin signalling: diseases and therapies. *Nat. Rev. Genet*. 2004, 5:691–701.

[234] Cohen PA. Intrabodies. Targeting scFv expression to eukaryotic intracellular compartments. *Methods Mol. Biol.* 2002, 178:367–378.

[235] Shuey DJ, McCallus DE, Giordano T. RNAi: gene-silencing in therapeutic intervention. *Drug Discov. Today* 2002, 7:1040–1046.

[236] Milhavet O, Gary DS, Mattson MP (2003) RNA interference in biology and medicine. *Pharmacol. Rev*. 55:629–648.

[237] Hutvagner G, Zamore PD (2002) A microRNA in a multipleturnover RNAi enzyme complex. *Science* 2002, 297:2056–2060.

[238] Elbashir SM, Harborth J, Lendeckel W, Yalcin A, Weber K, Tuschl T (2001) Duplexes of 21-nucleotide RNAs mediate RNA interference in cultured mammalian cells. *Nature* 2001, 411:494–498.

[239] Gil J, Esteban M. Induction of apoptosis by the dsRNAdependent protein kinase (PKR): mechanism of action. *Apoptosis* 2000, 5:107–114.

[240] McManus MT, Sharp PA. Gene silencing in mammals by small interfering RNAs. *Nat. Rev. Genet* 2002, 10:737–747.

[241] Joliot A, Prochiantz A. Transduction peptides: from technology to physiology. *Nat. Cell Biol.* 2004, 6:189–196.

[242] Wickstrom E, Urtishak KA, Choob M, Tian X, Sternheim N, Cross LM, Rubinstein A, Farber SA. Downregulation of gene expression with negatively charged peptide nucleic acids (PNAs) in zebrafish embryos. *Methods Cell Biol.* 2004, 77:137–158.

[243] Shiraishi T, Nielsen PE. Down-regulation of MDM2 and activation of p53 in human cancer cells by antisense 9-aminoacridine-PNA (peptide nucleic acid) conjugates. *Nucleic. Acids Res*. 2004, 32:4893–4902.

[244] Bendas G. Immunoliposomes: a promising approach to targeting cancer therapy. *BioDrugs* 2004, 15:215–224.

[245] Derossi D, Calvet S, Trembleau A, Brunissen A, Chassaing G, Prochiantz A. Cell internalization of the third helix of the Antennapedia homeodomain is receptor-independent. *J. Biol. Chem.* 1996, 271:18188–18193.

[246] Vives E, Brodin P, Lebleu B. A truncated HIV-1 Tat protein basic domain rapidly translocates through the plasma membrane and accumulates in the cell nucleus. *J. Biol. Chem.* 1997, 272:16010–16017.

[247] Frankel AD, Pabo CO. Cellular uptake of the TAT protein from human immunodeficiency virus. Cell penetrating transportan and penetratin analogues. *Bioconjug. Chem.* 2000, 11:619–626.

[248] Joliot A, Pernelle C, Deagostini-Bazin H, Prochiantz A. Antennapedia homeobox peptide regulates neural morphogenesis. *Proc. Natl. Acad. Sci. USA* 1991, 88:1864–1868.

[249] Lindgren M, Gallet X, Soomets U, Hallbrink M, Brakenhielm M, Pooga M et al. Translocation properties of novel cell penetrating transportan and penetratin analogues. *Bioconjug. Chem.* 2000, 11:619–626.

[250] Rousselle C, Smirnova M, Clair P, Lefauconnier JM, Chavanieu A, Calas B et al. Enhanced delivery of doxorubicin into the brain via a peptide-vector-mediated strategy: saturation kinetics and specificity. *J. Pharmacol. Exp. Ther.* 2001, 296:124–131.

[251] Pooga M, Soomets U, Hallbrink M, Valkna A, Saar K, Rezaei K, Kahl U, Hao JX, Xu XJ, Wiesenfeld-Hallin, Z, Hokfelt T, Bartfai T, Langel U. Cell penetrating PNA constructs regulate galanin receptor levels and modify pain transmission in vivo. *Nat. Biotechnol.* 1998, 16:857–861.

[252] Gallazzi F, Wang Y, Jia F, Shenoy N, Landon LA, Hannink M, Lever SZ, Lewis MR. Synthesis of radiometal-labeled and fluorescent cell-permeating peptide-PNA conjugates for targeting the bcl-2 proto-oncogene. *Bioconjug. Chem.* 2003, 14:1083–1095.

[253] Hyndman L, Lemoine JL, Huang L, Porteous DJ, Boyd AC, Nan X. HIV-1 Tat protein transduction domain peptide facilitates gene transfer in combination with cationic liposomes. *J. Control Release* 2004, 99:435–444.

[254] Hashida H, Miyamoto M, Cho Y, Hida Y, Kato K, Kurokawa T, Okushiba S, Kondo S, Dosaka-Akita H, Katoh H. Fusion of HIV-1 Tat protein transduction domain to poly-lysine as a new DNA delivery tool. *Br. J. Cancer* 2004, 90:1252–1258.

[255] Bosse R, Kulmburg P, von Kalle C, Engelhardt M, Dwenger A, Rosenthal F, Schulz G. Production of stem-cell transplants according to good manufacturing practice. *Ann. Hematol.* 2000 Sep;79(9):469-76.

[256] Bosse R, Singhofer-Wowra M, Rosenthal F, Schulz G. Good manufacturing practice production of human stem cells for somatic cell and gene therapy. *Stem Cells.* 1997;15 Suppl 1:275-80.

In: Stem Cell Research Trends
Editor: Josse R. Braggina, pp. 155-186

ISBN: 978-1-60021-622-0

Chapter V

Effect of Parathyroid Hormone (PTH (1-34)) on Hematopoietic and Mesenchymal Stem Cells and Their Progeny

I. N. Nifontova, D. A. Svinareva, N. V. Saz, V. L. Surin, Yu. V. Olshanskaya, J.L. Chertkov and N.J. Drize
National Hematology Research Centre, Moscow, Russian Federation

Abstract

Parathyroid hormone (PTH) is known as a regulator of calcium exchange; it also stimulates and activates proliferation of spindle-shaped N-cadherin positive osteoblasts which are able to control hematopoietic stem cells (HSC) in the niche. Because PTH could be an attractive drug for expansion of long-term repopulating HSC *in vivo*, the influence of this hormone on hematopoietic and stromal stem cells was studied *in vivo* and *in vitro*. Limiting dilution analysis *in vivo* indicated a significant increase in the frequency of competitive repopulation units (CRU) in the bone marrow of PTH treated mice. The calculated frequency of CRU was dependent upon the time of the recipient analysis; it increased from 2.5- to 4- fold between 10 and 16 months since reconstitution. The LTC-IC frequency increased 5-fold and CAFC 28-35 2.5-fold, whereas the number of differentiated progenitors CFU-S and CFU-GM, as well as terminally differentiated mature hematopoietic cells, did not change after PTH treatment. Possible explanation of such disproportion between early and late hematopoietic progenitors lies in discriminate PTH influences on osteoblastic and vascular niches specialized for quiescent and differentiating hematopoietic progenitors, respectively. Seeding efficiency (F-24) in the bone marrow stromal microenvironment differed significantly between long-term repopulating cells (CAFC 28-35) that did not change after PTH treatment and short-term repopulating cells (CFU-S) that decreased dramatically in PTH treated animals. On the contrary, the spleen microenvironment was not sensitive to long-term PTH exposure. Such functional discrepancies in stromal microenvironments could be explained by different impacts of PTH on cell components of niches specialized for long-term and short-term hematopoietic precursor cells. Moreover, it is not clear whether PTH also affects mesenchymal stem cells (MSC). MSCs from mice bone marrow are able to rebuild the hematopoietic microenvironment de novo after transplantation of a bone marrow plug under the renal capsule of a syngeneic animal. In ectopic hematopoietic foci formed 6 weeks later, the stromal cells present belong to the donor of the bone marrow; the hematopoietic cells are of recipient origin. Using this method, it was shown that 4 weeks of PTH treatment has no influence on MSC either during de novo foci formation or in the case of transplantation of pretreated with PTH bone marrow. Therefore PTH does not affect MSC directly, but affects its more differentiated progeny – osteoblasts and probably cells of vascular niches. PTH stimulates osteoblasts in completely formed niches, as was shown by *in vivo* ectopic foci formation de novo, as well as the experiments *in vitro*. In long-term bone marrow culture, the most pronounced stimulating effect of PTH addition on hematopoietic precursors was revealed after 10 weeks of cultivation. Simultaneously, the expression level of genes specific for osteogenic differentiation increased. The expression level of Notch-1 ligand Jagged-1 also increased steadily and significantly after 7 weeks in culture. The obtained data suggest PTH is a possible tool for specific expansion of early hematopoietic precursors only.

INTRODUCTION

The expansion and activation of hematopoietic stem cells (HSC) *ex vivo* can be used as a tool for achieving clinically applicable amounts of HSC. The capacity for sustained self-renewal – generation of daughter cells with the same regenerative properties as the parent cell and long-term repopulating ability – is the main feature of early HSC. The necessity of HSC expansion, without loss of their self-maintaining ability, is one of the problems attendant stem cell transplantation. The lack of HSC is a common barrier to stable reconstitution of hematopoiesis in patients after chemo- and radio- therapies. The destruction of HSC, and the depletion or even "empting" of cancer patient's bone marrow, demands either stimulation of self HSC or transplantation of exogeneous stem cells. Regulation of the self-renewal process in HSC is only dimly understood. Nevertheless, there is strong evidence that *in vivo* self-renewal of HSC is controlled by extrinsic factors, moreover intrinsic and extrinsic cellular mechanisms, which regulate the balance of self-renewal and differentiation (Sauvageau et al., 2004), (Moore and Lemischka, 2006).

The stromal microenvironment and hematopoietic stem cells make up a structure that coordinates normal hematopoiesis, a balance of stem cell quiescence and activity. The "niche" concept was introduced by Schofield about 30 years ago (Schofield, 1978) and was defined only by the functional unit, until studies on Drosophila supported the idea by anatomical structure (Lin, 2002). Bone marrow stromal cells represent several different populations, including fibroblasts, macrophages, endothelial cells, adipocytes and ostoblasts. Osteoblasts and hematopoietic cells are closely associated in the adult bone marrow. Two recent publications highlight the role of spindle-shaped N-cadherin+CD45- osteoblasts in controlling the development and the number of HSC (Zhang et al., 2003), (Calvi et al., 2003). These cells directly contact HSC and maintain HSC in the "stemness" state due to activating and modulating signals. The selective loss of osteoblasts caused a progressive loss of bone, bone marrow cellularity, and early hematopoietic progenitors. The former fetal hematopoietic territories like liver and spleen became the main hematopoietic organs. The reappearance of osteoblasts induces hematopoiesis at sites of new bone formation (Visnjic et al., 2004). In another study, osteoblasts were shown to facilitate engraftment of HSC in an allogeneic environment. When purified osteoblasts were cotransplanted with HSC into allogeneic mouse strains, the transplanted recipient mice demonstrated long-term survival and complete engraftment by the donor cells (El Badri et al., 1998). It appears that the ability of the osteoblastic niche to retain stem cells in a quiescent state is an important mechanism in maintaining hematopoietic tissue homeostasis (Arai et al., 2004).

HSC stick to osteoblasts with receptor-ligand complexes, signaling wires, and matrix components such as collagen and osteopontin (Haylock and Nilsson, 2005), (Stier et al., 2005), (Zhu and Emerson, 2004), (Moore and Lemischka, 2004). Osteoblasts present Ang-1(Angiopoietin-1), which in turn stimulates the HSC to express Tie-2 – a tyrosine kinase receptor which "ties" the HSC to the trabecular bone- lining cells (Moore and Lemischka, 2004). The signals sent into the HSC from the Ang-1-Tie-2 complex play an extremely important role in the maintenance of HSC-"stemness". The signals from Ang-1-Tie-2 complexes prevent the HSC from cycling, keep them sticking to the bone surface, keep their β1-integrin signaling from fading, and lastly keep them able to generate proliferating-amplifying progenitors (Arai et al., 2004), (Arai et al., 2005). There are several other important players in HSC control in niche; for example, Wnt that signals to β-catenine promotes HSC self-renewal rather than differentiation (Reya et al., 2003), (Staal and Clevers, 2005), (Van Den Berg et al., 1998), (Willert et al., 2003). The accumulating cytoplasmic β-catenine contributes to the mooring of HSC to the osteoblasts by binding to N-cadherins (Zhang et al., 2003). Another crucial pair of molecules is Notch-1 and its activator Jagged-1 which collaborates with the Wnt receptor in maintaining the self-renewing "stemness" of the HSC (Duncan et al., 2005). The proto-oncogene Bmi-1 is also important in HSC regulation; expression of Bmi-1 was shown to be a key factor in normal HSC maintenance (Park et al., 2003).

When a niche-bound stem cell receives a signal to initiate a growth-division cycle, one daughter cell stays in the niche and returns into a Go state, while the other loses contact with the osteoblast. If it cannot find another niche, it proliferates and moves along trails marked with appropriate cytokines and adhesives in order to find a vascular endothelial cell niche where it could terminally differentiate and it or its progeny can enter the blood (Whitfield, 2006). The interaction of HSC with sinusoidal endothelium suggests that endothelial cells create an alternative niche in bone marrow (Li et al., 2004) (Cardier and Barbera-Guillem, 1997), (Yin and Li, 2006b). It was shown that *in vivo* ablation of endothelial cells by anti-vascular endothelial cadherin antibody leads to hematopoietic failure (Avecilla et al., 2004). HSC expressing a member of the signaling lymphocyte activation molecule (SLAM) family were found adjacent to the bone surface as well as to endothelial cells in both bone marrow and spleen (Kiel et al., 2005). Therefore, at least two distinct niches supporting HSC function have been identified in bone marrow: the osteoblastic niche and the vascular niche (Kopp et al., 2005), (Wagers, 2005), and HSC fate and niche choice are interconnected.

Stimulation and activation of spindle-shaped osteoblasts by parathyroid hormone (PTH) induced an increase in HSC pull, engraftment of bone marrow after transplantation, and the survival rate of lethally irradiated mice (Calvi et al., 2003), (Whitfield, 2006).

Parathyroid hormone is an 84-amino-acid polypeptide, a major mediator of bone remodeling, and an essential regulator of calcium homeostasis (Strewler et al., 1987). Both *in vivo* and *in vitro* studies have shown that the N-terminal 1-34 synthetic fragment of PTH mediates full PTH activity (Habener et al., 1984). During the last 75 years it was learned that bovine parathyroid extract could stimulate osteogenesis (Selye H., 1932) and increase survival of irradiated rats (Rixon et al., 1958). PTH directly stimulates the initiation of DNA replication in murine spleen colony-forming units (CFU-S) via cycling AMP, stimulates the proliferation of normal and irradiated rodent bone marrow cells, controls hematopoiesis, and increases the survival of lethally irradiated mice when injected any time between 18 h before and 3 h after irradiation (Rixon and Whitfield, 1961), (Whitfield, 2006). PTH and PTH-related protein indirectly activate osteoclasts in paracrine ways by the factors produced by osteoblasts (Swarthout et al., 2002).

Three of its N-terminal fragments (rhPTH-(1-34)/Forteo^{TM}/, rhPTH-(1-31) /Ostabolin^{TM}/ and rhPTH /Ostabin-C^{TM}) have become very important clinically because of their ability to stimulate bone formation and consequently treat osteoporosis, accelerate fracture healing (Whitfield, 2005), and strengthen bone microarchitecture in humans, monkeys and rodents--practically without side effects (Whitfield et al., 2003). However, PTH alone does not cure osteoporosis, but it is able to greatly restore bone mass (especially trabecular bone), increase bone strength, and quickly reduce fracture incidence (Dempster et al., 2001), (Lane et al., 1998), (Neer et al., 2001). These facts at the first glance could be quite surprising and paradoxical because hyperparathyroidism is invariably associated with bone loss and patients with multiple myeloma with increased concentration of PTH in sera suffer from fractures and bone lesions (Potts, 2005), (Patel et al., 2005). Later it was shown that PTH is an anabolic agent for osteoporosis and directly stimulates bone formation. However, only after intermittent, short elevations given with continuous elevation does it lead to bone loss. In humans, intermittent high doses of PTH are anabolic for bone, while continuous elevation is catabolic (Rodan and Martin, 2000), (Harada and Rodan, 2003). Therefore, timing is absolutely critical in PTH administration. In bone marrow, the high-affinity PTH receptor is expressed only on osteoblasts; its activation by hormone prolongs osteoblast life and increases its activity, leading to bone formation (Calvi et al., 2001). However, osteoblast activation leads to stimulation of osteoclasts. Fine dose and treatment duration adjustment can achieve the desired balance of bone resorption by osteoclasts occuring less rapidly than the initial direct stimulation of osteoblasts (Schiller et al., 1999), (Ejersted et al., 1994). Hence, transient elevations may stimulate osteoblast anabolic activity, with or whithout triggering the coupled catabolic response through osteoclasts.

In addition to timing, the dose of PTH given is also very important. In low doses (10-20 mkg/day) it stimulates osteoblasts; augmentation of the dose activates osteoclasts and causes bone fractures. This statement was confirmed in human clinical trails with prior vertebral fractures, where PTH strikingly reduced fracture incidence by 70% when compared to the placebo group (Neer et al., 2001a). In mice constitutively expressing the PTH receptor PTH-R1, increased bone formation was observed. (Zhang et al., 2003). However, there was a delay in the formation of bone marrow cavities and several types of stromal cells, for instance adipocytes, so there were a reduced number of stromal cells in these animals (Kuznetsov et al., 2004). Activation of osteoblasts by PTH demands endogenous FGF-2 expression. In

FGF-2 deficient mice, PTH does not influence the bone formation; this is in contrast to wild type animals where bone formation is dramatically increased (Hurley et al., 1999), (Hurley et al., 2006). Wnt - β-catenin signaling activated by PTH promotes the survival of osteoblasts (Tobimatsu et al., 2006), (Kulkarni et al., 2005). PTH could also suppress the maturation of osteoblasts by inhibiting expression of the hedgehog gene family and the gene RUNX2, which are associated with differentiation of osteoblasts and mineralization of extracellular matrices (van der Horst et al., 2005).

Low doses of PTH stimulate the proliferation of osteoblasts *in vitro*; in addition, activation of MAPK (Swarthout et al., 2002) and subsequent expression of different transcription factors (phospholypase A2, c-myc, c-jun) involved in regulation of cell division can stimulate osteoblast proliferation. PTH inhibits apoptosis by protein kinase A-dependent phosphorilation, inactivation of pro-apoptotic protein Bad, and inactivation of bcl-2 transcription. The longevity of the effect depends on the level of RUNX2 expression, which in turn is decreased by PTH (Bellido et al., 2003).

Daily injections of PTH decrease the intensity of osteoblast apoptosis thus increasing their number and subsequently enhancing bone formation; the osteoclast number remains unchanged. At the same time, an increased level of PTH due to a Ca2+ deficient diet did not influence the apoptosis of osteoblasts; instead it caused elevation in the number of osteoclasts. The administration of high doses of PTH (up to 40 mkg) increased bone mass more efficiently, but previously present osteoclast expansion augmented the risk of fractures (Teitelbaum, 2004), (Finkelstein et al., 2003) and as a result lead to bone destruction accompanied with the release of Ca, P, and elements of organic matrices. The dual effect of PTH on the process of bone remodeling regulates the balance between bone resorption and formation.

The suggestion concerning the role of osteoblasts in hematopoiesis was first proposed by Taichman and Emerson; the data of Calvi (Calvi et al., 2003) and Zhang (Zhang et al., 2003) provided direct *in vivo* support. As the size of the HSC pool depends on available trabecular bone niche space (Zhang et al., 2003), (Zhu and Emerson, 2004), PTH could increase the HSC pool size (Calvi et al., 2003) and subsequently increase the effectiveness of the response to bone marrow injury (Calvi, 2006). Regulation of the HSC pool size by PTH can be controlled in several ways. First, PTH stimulates the growth of trabecular bone and directly increases the number of osteoblast niches (Calvi et al., 2003). Second, PTH promotes the attachment of HSC cells to the expanding niches by stimulating bone-lining cells to make more N-cadherin for binding to the HSC cells' β-catenin (Marie, 2002), (Zhang et al., 2003c). Third, PTH stimulates and activates osteoblasts and as a result the expression of Jagged-1 increases. This, in turn, activates Notch-1 on primitive HSC, resulting in expansion of the stem cell compartment (Weber et al., 2006), (Calvi et al., 2003). Fourth, SDF-1 synthesis could be increased by PTH and could attract more HSC to the niche. SDF-1 produced by marrow stroma and bone tissue forms a decreasing gradient from the marrow extravascular compartment toward the lumen of infiltrating vessels (Ponomaryov et al., 2000). This SDF-1 gradient has been shown to be essential in the process of HSC migration to bone marrow (Nilsson et al., 2006).

PTH could also modulate the number of HSC in bone marrow by induction of osteopontine expression in osteoblasts (Haylock and Nilsson, 2005). Osteopontin inhibits proliferation of HSC by suppressing Notch expression (Stier et al., 2005), (Iwata et al., 2004).

Now, strategies for affecting stem cells in order to achieve therapeutic outcomes have become feasible. The identification of osteoblasts as participants in the hematopoietic stem cell niche has created ground for pharmacological manipulation of expanding primitive HSC. It was clearly shown that through changing the niche size, one could affect the number of stem cells. Obviously, PTH may be useful in stem cell expansion *ex vivo* or *in vivo*, and has potential implications for stem cell harvesting and recovery after transplantation. The practical method of manipulating stem cells by impacting the microenvironment would be a valuable addition for the treatment of cancer patients. Detailed studies addressing rigorous HSC self-renewal by long-term reconstitution and serial transplantation have not yet been performed (Moore and Lemischka, 2006). Taking into consideration the usage of PTH in clinical practice for treatment of patients with osteoporosis, and the idea of its administration for cancer patients, we studied the long-term repopulating ability of HSC from mice treated with PTH for more than a year using a competitive repopulation assay. The alterations of various types of hematopoietic precursors and functional changes in the stromal microenvironment and mesenchymal stem cells after PTH treatment were studied both *in vivo* and *in vitro*.

MATERIALS AND METHODS

Animals

(CBA x C57Bl6)F1 9-26 weeks-old male (for the competitive repopulation assay) and female mice were obtained from Stolbovaya Animal Centre, Moscow Region, Russian Federation. All animals were kept in the department for animal care of National Hematology Research Centre at 20-25°C, and received acidulous water (ph 2.4) with standard fodder.

Irradiation Conditions

Mice were irradiated by ^{137}Cs exciter IPK with 16.6 sGy/minute; cell cultures were irradiated with 435 sGy/minute. In the CFU-S assay, mice were lethally irradiated with 10.5 Gy divided into two equal fractions given 3 hours apart.

PTH Treatment

Rat synthetic PTH (1-34) was injected intraperitoneally in doses known to increase the number of osteoblasts (10, 30, and 80 mkg/kg, 5 days a week for 4 weeks). Two weeks after the last injection the bone marrow from the femurs of the PTH treated and control groups was used for the experiments described later.

Long-Term Bone Marrow Cultures

In order to establish murine long-term bone marrow culture (LTBMC) Dexter's type (Dexter et al., 1977) the content of each femur was flashed into a 25 cm^2 cultivation flask under sterile conditions using 5 ml of Fisher's medium (ICN). Another 5 ml were added, bringing the final volume to 10 ml. Culture medium consisted of 80% of Fisher's medium, 2 mM gluthamin (ICN), antibiotic mixture (100 U/ml of penicillin (Ferein) and 50 mkg/ml of srepomycine (Ferein)), 10^{-6} M hydrocortisone (Sigma), 20% of serum mixture (1/3 of fetal calf serum, FCS (HyClone) and 2/3 of horse serum (GibcoBRL)). Rat synthetic parathyroid hormone, PTH (1-34) (Bachem, USA) was added at the moment of culture initiation and weekly during the medium change thereafter. PTH concentrations used in the experiments were 10^{-8} M, $5x10^{-8}$ M and 10^{-7} M. Cultures were kept at 33°C and 5% CO_2 in a humidified atmosphere.

Cobblestone Area Forming Cells (CAFC) Assay

CAFC frequency was counted using the method of Ploemacher (Ploemacher et al., 1989a). Sixty out of 96 wells of the cultivation plates were previously preincubated with 0.1% gelatine for 30 min. Next, $2\text{-}3x10^3$ cells of the MS-5 (Kobari et al., 1995) cell line were plated per well in the α-MEM with 10% of FCS. is a murine cell line which produces SDF-1 and is known to maintain murine as well as human CAFC formation. All the margin wells were filled with 0.1 M NaOH (ICN) to prevent the wells from drying. All plates were kept in 37°C and 5% CO_2 in a humidified atmosphere. On the next day simultaneously with the media exchange, tested cells from the PTH treated and control groups were implanted in 4 subsequent dilutions. Usually $3x10^4$, $1.5x10^4$, $0.75x10^4$ and $0.375x10^4$ cells per well were plated in each dilution; 15 wells were used for one dilution. Wells were analyzed weekly using an inverted microscope. Plates were cultivated with 33°C and 5% CO_2 in a humidified atmosphere with weekly exchange of the half of the medium. In order to calculate the CAFC frequency, the number of negative (not containing CAFC) wells was counted so the Poisson's statistic could be applied.

CFU-S Assay

To estimate CFU-S frequency (Till, 1961) mice were lethally irradiated, with the number of endogenous colonies generated being less than 0.2 colonies per spleen. One to three hours after irradiation, $2\text{-}3x10^5$ tested cells (either from the control or PTH treated bone marrow, or from the suspension fraction of LTBMC cultivated with or without of PTH) were injected intravenously into each recipient mouse. Ten days later, the number of colonies was calculated in the spleens that were fixated in Buen's solution.

CFU-C Assay

CFU-C frequency was estimated in 0.8% methylcellulose with 25% FCS, 5% concentrated nutrition medium (0.06M glutamine, 1.1% BSA (GibcoBRL, V fraction), 1.4×10^{-4} β-mercaptoethanol (Sigma), 28 mM HEPES and antibiotic mixture), 2 U/ml erythropoietin and 8% condition mediums from L929 (Mayer, 1983) and WEHI 3B (Prestidge et al., 1984) (1:3 mixture) as a source of growth factors. A hundred thousand cells per well were cultivated in 24-well plates at 37° and 5% CO_2 in a humidified atmosphere. The number of colonies was counted on the 7th day of cultivation using an inverted microscope.

Estimation of the Proliferation Rate of the Various Hematopoietic Precursors

One half of the tested cells fraction (10^6/ml) was incubated with hydroxyurea (HU) (Serva, 1 mg/ml) for 2 hours in Fisher's medium with 2% FCS at 37° and 5% CO_2. The remaining half of the cell fraction was incubated in the same conditions without the cytostatic agent. Next, cells were washed with Fisher's medium to get rid of the hydroxiurea and the frequency of different hematopoietic precursors was estimated using the methods described above. The proportion of proliferating cells was calculated as

$$N=[(N_{without\ HU}-N_{with\ HU})/N_{wthout\ HU}]x100\% ,$$

where N is the proportion of proliferating cells, $N_{without\ HU}$ is the number of hematopoietic precursors among non HU-treated cells, and $N_{with\ HU}$ is the number of hematopoietic precursors subjected to HU treatment.

Competitive Repopulation Assay

In order to functionally characterize the most immature HSC capable of long-term reconstitution of hematopoiesis, a competitive repopulation assay was used. A stem cell defined by this assay is termed a competitive repopulating unit (CRU) (Miller and Eaves, 1997b). The method used is based on the fundamental ability of HSC to reconstitute hematopoiesis in a lethally irradiated recipient. Histocompatible but genetically distinguishable "test" stem cells are injected into lethally irradiated mice, together with a large excess of marrow cells containing a normal frequency of long-term repopulating HSC. Natural sexual differences in genotypes were used as genetic markers.

Male CBF1 mice were pretreated with PTH as described above. At the end of the PTH course, the bone marrow cells of treated males were mixed with female bone marrow cells in different relations (1:1, taking 250×10^3 cells of each; 1:3, taking 125×10^3 male and 375×10^3 female cells; and 1:19, taking 25×10^3 male and 475×10^3 female bone marrow cells) and were injected i.v. into lethally irradiated female recipients. The same procedures were performed using untreated male bone marrow; such groups were considered as controls.

To determine the genotype of the colony forming units-spleen (CFU-S) among the marrow cells of the reconstituted mice 3, 10 and 16 months after the transplantation, bone marrow from 5 control and 5 experimental mice of all dilution groups was aspirated through a knee joint under light diethyl ether anesthesia. Cells from each mouse were injected intravenously into 7 lethally irradiated secondary female recipients. Individual macroscopic spleen colonies were isolated using a dissection microscope 10 days later and analyzed with PCR for the gender attribute (see below).

The genotype of the bone marrow and peripheral blood cells was analyzed with FISH. Direct DNA-probes to the centromere of the Y-chromosome, conjugated with fluorochrome 1200-*MCy3 (Cambio Ltd) were used. 400 to 1000 nuclei were analyzed on each slide--nuclei with red signals were designated XY; nuclei without any signal, XX.

CRU frequency in the test cell suspension was determined by limiting dilution analysis (Miller and Eaves, 1997a). Three, 10 and 16 months following repopulation, the proportion of Y-negative CFU-S was determined by PCR and plotted against the number of PTH treated male marrow cells injected. A best-fit line was generated using the maximum-likelihood method without forcing the data through the origin; the frequency of CRU was then determined using standard statistical methods by interpolation of the number of test cells required, obtaining a 37% negative response.

Other statistical analysis was done using the Student's t-test.

Seeding Efficiency of HSC in Bone Marrow and Spleen

In order to analyze the possible influence of PTH treated hematopoietic microenvironments on the homing of different types of HSC, seeding efficiency for 24 hours (F24) was determined.

F24 for CFU-S was estimated using the following design: mice were first subjected to a 4-week long PTH course; the control group was injected with placebo. After the end of the course both groups were lethally irradiated (intermediate recipients) and injected with 16×10^6 pooled bone marrow cells from 8 intact donor mice. These cells were previously incubated in plastic culture flasks in αMEM with 10% FCS for 2 hours at 37°C and 5% CO_2. Then, non-adherent cells were collected and used for the following manipulations: twenty four hours after injection, the spleen and bone marrow of the intermediate recipients were taken out and dispersed. These cells were injected into the lethally irradiated mice (final recipients). The number of cells injected was equivalent to 1/30 of the spleen or 1/5 of one femur. Ten days later the CFU-S number was counted. To estimate the initial number of CFU-S in the pooled bone marrow mixture injected into the intermediate recipients, 2×10^4 cells per mouse were injected into lethally irradiated mice and 10 days later the number of spleen colonies was determined. F24 was calculated as

$$F24 = (a/N) \times 100,$$

where a is the mean CFU-S number in the final recipients and N is the initial number of CFU-S in the used pooled bone marrow.

F24 for the CAFC 28 was determined using a similar design. The initial CAFC 28 number in the pooled bone marrow mixture was estimated using the standard method (Ploemacher et al., 1989b). When analyzing the intermediate recipients, 1/24 of the spleen and ¼ of the bone marrow from the femur and tibia were plated per well in the first dilution. Three additional two-fold dilutions were performed. Fifteen wells were plated in each dilution; cells were cultivated on long-term bone marrow culture medium (see above). Half of the media was replaced weekly with fresh media. After 28 days of cultivation, the wells were analyzed in order to count the number of wells not containing CAFC. CAFC frequency was calculated with Poisson's statistics. It must be admitted that because it was necessary to plate a large number of dying cells from lethally irradiated intermediate recipients, some amount of data corruption is possible. Therefore, errors of measurement in these experiments can be quite considerable.

Analysis of the DNA Gender Attribute by PCR

DNA from 25 spleen colonies generated from bone marrow of each initial mouse was analyzed by PCR for its gender attribute. Primers for Smc gene (5'-CTGAACTATTTGGATCAGATTGC-3'(exon 3) and 5'-CACCGACGGTCCTTGCAGAT-3' (exon 4)) permit the simultaneous amplification of the 391 bp fragment from the Y-chromosome localized copy of the gene and the 429 bp fragment from the X-chromosome localized copy. Differences in fragment length are conditioned on different intron length on the sex chromosomes. There were 32 cycles in the PCR, with the cycling conditions being 94°C for 1 minute, 62°C for 1 minute, and 72°C for two minutes. The PCR products were analyzed by electrophoresis using a 2% agarose gel (Sigma).

Ectopic Foci Formation

Mice were subjected to a 4-week PTH course; the control group received placebo injections. Two weeks later, bone marrow plugs of mice from both groups were implanted under the renal capsule of syngeneic animals. Half of the recipients were treated with PTH in the same manner during foci formation. Six weeks later, the size of the foci was estimated by the number of nucleated cells it contained; osteogenic activity of the stromal progenitors was estimated by the weight of the formed bone shell. Frequency of the different hematopoietic precursors (CAFC 7, CAFC 28-35, CFU-S and CFU-C) in the foci was determined using the corresponding methods described above. The self-renewal ability of the stromal cells capable to transfer the hematopoietic microenvironment from the bone marrow of PTH treated and control mice was determined by retransplantation of the formed ectopic foci under the renal capsule of the secondary recipients. Six weeks later the size of the foci formed was estimated by its nucleated cell number.

Analysis of Gene Expression by Reverse Transcription-Polymerase Chain Reaction (RT-PCR)

RNA was obtained from various cell samples by standard protocol (Chomczynski and Sacchi, 1987). cDNA was synthesized using M-MLV reverse transcriptase (Promega) by manufacturer protocol. For identification of the genes of interest among total cDNA, PCR with specific primers (Table 1) was applied.

Table 1. Characteristics of specific primers used in the study

Gene	Primers	Fragment size
β-actin	sense-ACCGTGAAAAGATGACCCAG	416
	antisense-CGTTGCCAATAGTGATGACC	
Jagged-1	sense -ACGTTGTTGGTGGTGTTG	280
	antisense -AGTAAACGTGATGGAAACAG	
Notch-1	sense -GTGGTGCCTCCTAGAGAAAA	280
	antisense -TGAGCACAGCCCTGAACC	
BMI-1	sense -CAC AAA ACC AGA CCA CTC CT	569
	antisense -TCA CTT TCC AGC TCT CCA GC	
Osteopontin	sense -CACTTTCACTCCAATCGTCCC	495
	antisense -GCCTCTTCTTTAGTTGACCTC	
COMP	sense -CAGAGTGACAGTGATGGTGA	413
	antisense -CTGAAGTCGGTGAGGGTGAC	

The cycling conditions were: 94°C for 30 seconds for denaturation and 72°C for 1 minute for synthesis. Annealing always lasted for 30 seconds, with the temperature used depending on the primer sequence. For quantative analysis of the PCR products, the reaction lasted from 10 to 28 cycles. PCR-products were analyzed by electrophoresis in a 1.5% agarose gel (Sigma) and were then subjected to Southern-blot hybridization according to standard protocol (Maniatis et al., 1982). The membranes were analyzed with Phosphoimager Cyclone (Packard, USA).

RESULTS

Estimation of Competitive Repopulation Units (CRU) Number in the Bone Marrow of Mice Treated with PTH

The number of competitive repopulation units (CRU) in chimeras repopulated with the mixture of intact and PTH treated bone marrow cells was estimated by its clonogenic polypotent progeny CFU-S in the bone marrow, as well as by the mature hematopoietic cells in the bone marrow and peripheral blood. Concentrations of CFU-S in the bone marrow of mice reconstituted with the mixture of intact and PTH treated bone marrow cells did not differ from the control group for one year following reconstitution (Table 2).

Table 2. CFU-S concentration per 10^5 cells in the bone marrow of reconstituted mice

Injected bone marrow	Time after reconstitution, months		
	3	10	16
$250x10^3$ female + $250x10^3$ male	3,2 ± 0,3	3.6 ± 0.8	4.3 ± 0.3
$375x10^3$ female + $125x10^3$ male	2,7 ± 0.2	2.9 ± 0.8	4.0 ± 0.9
$475x10^3$ female + $25x10^3$ male	4,3 ± 0,7	3,5 ± 0,8	3.6 ± 1.0
$250x10^3$ female + $250x10^3$ male PTH treated	7,5 ± 0.5	2.8 ± 0.4	6.6 ± 0.6
$375x10^3$ female + $125x10^3$ male PTH treated	3,1 ± 0.4	2.7 ± 1.0	5.8 ± 0.7
$475x10^3$ female + 25 $x10^3$ male PTH treated	2,7 ± 0,3	3,5 ± 1,1	5.3 ± 1.9

However 4 months later (16 months since the repopulation), CFU-S concentration became significantly higher in the mice which received PTH treated bone marrow cells ($p < 0.01$ for the group received $250x10^3$ treated cells and $p < 0.02$ for the group received $125x10^3$ cells). It is known that CFU-S concentration increases with age for some strains of mice (Harrison, 1980), (Harrison, 1983) and culminates in the oldest animals. In our study only a tendency with no significant differences was observed. It is probable that the mice were not old enough at the moment of analysis.

In the control groups reconstituted with the mixture of PTH untreated male and female bone marrow cells, the proportion of Y+ CFU-S gradually decreased with time (Figure 1). This could be a result of reversion of hematopoiesis to the recipient type and its own hematopoietic stem cells beginning to function. When female mice were repopulated with male bone marrow cells, only 66.7 ± 12.8 % of CFU-S were male 3 months after reconstitution and 41.3 ± 11.7 % were male after another 7 months (data from 4 independent experiments, 31 animal analyzed). However, in mice reconstituted with PTH treated bone marrow, the proportion of Y-positive CFU-S steadily rose during the entire time of observation. The proportion of Y-positive CFU-S had significantly exceeded the control level at the 16 months since transplantation; that clearly demonstrated an impact of PTH treated CFU-S on hematopoiesis. The ratio of Y-positive CFU-S and mature cells was changing during the life-time of the mouse (Table 3).

The frequency of Y-positive CRU was changing dramatically during the lifetime of the mouse. In the control group 3 months after reconstitution, 1 out of 38 $x10^4$ transplanted male cells was Y-positive CRU, while in the PTH treated group, it was 1 out of 68 $x10^4$ transplanted male cells (Figure 2a). Ten months after reconstitution the frequency of Y-positive CRU in PTH treated group (1 out of 20 $x10^4$ cells) exceeded the control one (1 out of 54 $x10^4$ cells) by about 2.5-fold (Figure 2b). Six months later, the frequency of PTH treated Y-positive CRU became 4-fold higher as compared to controls (1 out of 15 $x10^4$ versus 1 out of 56 $x10^4$ in the control group) (Figure 2c).

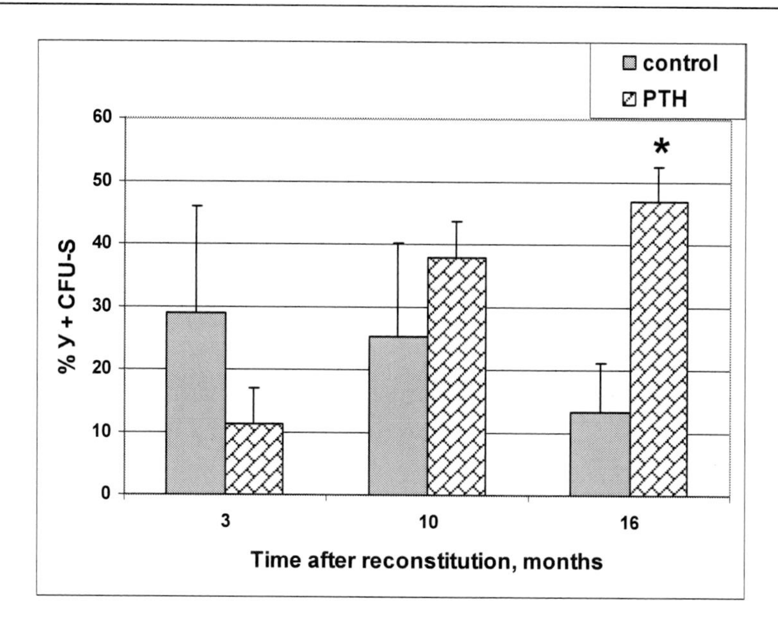

Figure 1. Proportion of Y- positive CFU-S in bone marrow of reconstituted mice. Bone marrow cells from each mouse reconstituted with male and female bone marrow mixtures were injected to 7 irradiated recipients and the number of spleen colonies was estimated 10 days later. DNA from each isolated spleen colony was analyzed by PCR for male sequences. Eleven to 15 reconstituted mice were analyzed per group, and 25 spleen colonies per donor mouse were tested. Each column with error bar represents the mean ± s.e.m., * P<0.01, compared to the control group.

Table 3. Proportion of Y-positive CFU-S and nuclear hematopoietic cells in reconstituted mice

Injected bone marrow cells/ mouse number	% Y+ cells				
	Method of detection				
	PCR			FISH	
	CFU-S in bone marrow			Bone marrow	Periphal blood
	Time after reconstitution, months				
	3	10	16	16	
$250x10^3$ female + $250x10^3$ male / 1	20	0	0	nd *	12,1
$250x10^3$ female + $250x10^3$ male / 2	62,5	48	32	nd	11.8
$250x10^3$ female + $250x10^3$ male / 3	4,3	28	8	nd	2.8
$250x10^3$ female + $250x10^3$ male +PTH/ 4	20	20,8	56,7	48,7	52,8
$250x10^3$ female + $250x10^3$ male +PTH/ 5	32	26,9	38,7	43,8	13,0
$250x10^3$ female + $250x10^3$ male +PTH/ 6	0	52,0	33,3	22,5	23,1
$250x10^3$ female + $250x10^3$ male +PTH/ 7	4	48,0	56,7	32,6	37,6
$250x10^3$ female + $250x10^3$ male +PTH/ 8	0	42,3	26,9	25.7	17.3
$375x10^3$ female + $125x10^3$male +PTH/ 9	12.5	0	4.5	32	1.4
$375x10^3$ female + $125x10^3$male +PTH/ 10	0	8.3	4	1.3	2.3
$375x10^3$ female + $125x10^3$male +PTH/ 11	24	24	40	33.7	16.8

* Not done.

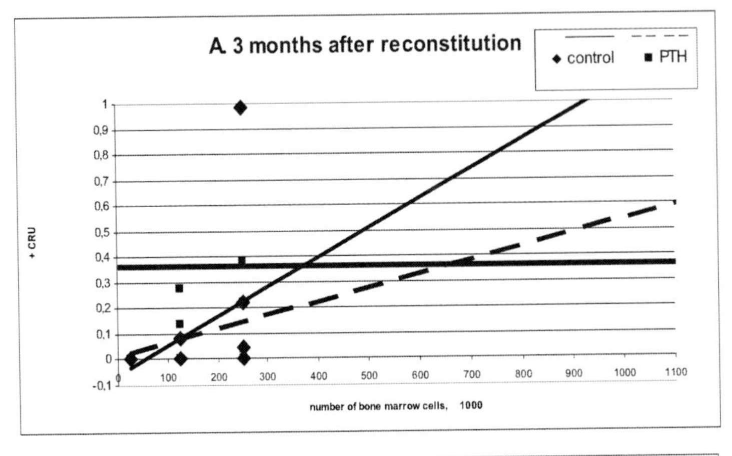

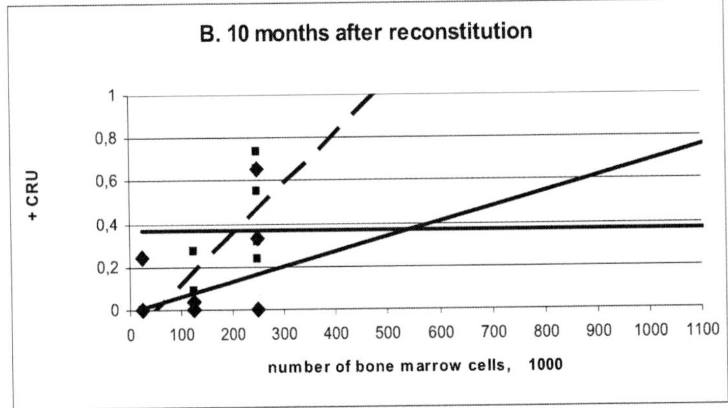

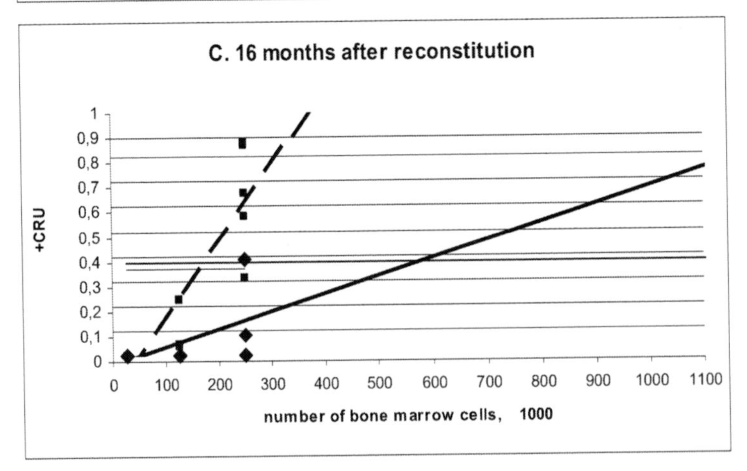

Figure 2. Limiting dilution analysis of the frequency of CRU in reconstituted mice (A) 3 months after reconstitution, (B) 10 months after reconstitution and (C) 16 months after reconstitution. Limiting number of male PTH pretreated or not treated (control group) bone marrow cells were injected together with female marrow cells into groups of 4-10 lethally irradiated female mice. Each line was performed as best fit and was generated using the maximum-likelihood method without forcing the data through the origin. Twelve to 15 mice of the PTH treated and control groups were analyzed by PCR for CFU-S gender attribution. The frequency of CRU was determined using standard statistical methods by interpolation of the number of test cells required to obtain a 37% negative response (intersection with horizontal line).

CAFC Frequency

The frequency of HSC less primitive than CRU increased 2- to 3-fold in the bone marrow of mice treated with 80 μg/kg of PTH when compared with the control mice (Table 4). Unlike the control group, almost half of the PTH treated progenitors were proliferating. PTH also stimulated CAFC 28-35 to proliferate in ectopic foci which were formed during PTH injections. However, the frequency of hematopoietic progenitor cells in the foci did not increase. One possible explanation could be that for a large part of PTH treatment, the hematopoietic microenvironment is in the process of rebuilding itself and is only later repopulated with recipient hematopoietic cells. Therefore, the targets for the hormone are differentiating along with the treatment and are probably not able to be activated completely. In addition, the impact of PTH on CAFC 28-35 frequency is dose dependent (Figure 3). If 10 μg/kg of PTH was used, CAFC 28-35 frequency in the bone marrow was 2 -fold higher than in the group treated with 80 μg/kg of PTH; the intermediate dose (30 μg/kg) had no significant effect on the frequency of these hematopoietic precursors. More mature precursors such as CAFC 7 that frequency correlate with CFU-S concentration respond to PTH in a different manner. Their frequency in the bone marrow decreased after PTH treatment (Table 4) and unlike the controls, they did not proliferate. In ectopic hematopoietic foci, the concentration of CAFC 7 precursors was comparable to the controls, while the proliferation index decreased.

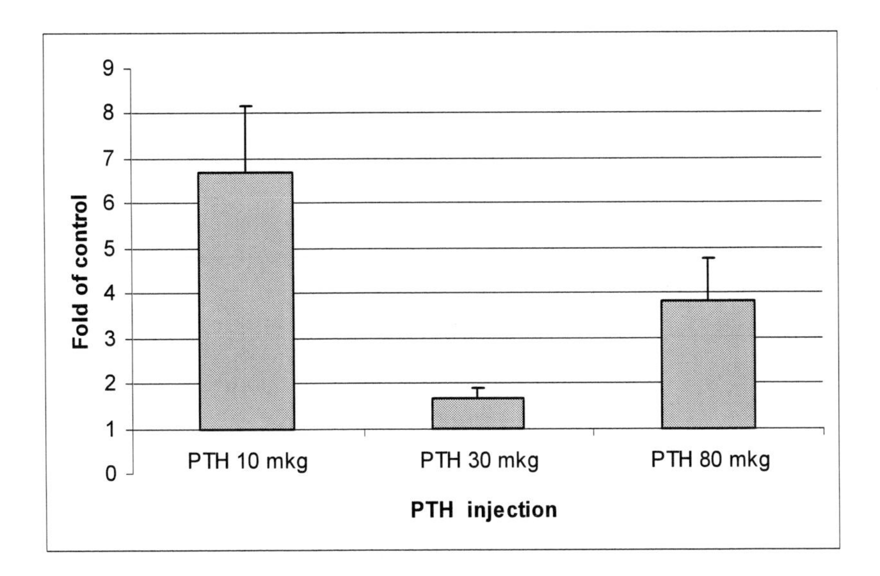

Figure 3. Alteration of CAFC 28-35 frequencies in bone marrow of PTH treated mice. Bone marrow cells from mice injected with PTH for 4 weeks in different concentrations were plated in limiting dilutions on a MS5 cell line. The frequency of CAFC 28-35 was determined using a standard Poisson's statistical method. Each column with error bar represents the mean ± s.e.m from 5 animals compared with the frequency of CAFC 28-35 in control group, accepted for unity. (Fold of control).

Table 4. Frequencies of CAFC per 10^5 cells and their proliferation rate in bone marrow and ectopic foci after PTH treatment

PTH treatment	CAFC 28-35 days		CAFC 7 days	
	Bone marrow	Suicide %	Bone marrow	Suicide %
control	2.5 ± 0.3	0	43.3	40.5
+ PTH	7.1 ± 2.2	43	10.8	0
	Ectopic foci		Ectopic foci	
control	0.75	0	22.5	67.3
+ PTH	0.85	62	17.6	12.1

CFU-S and CFU-C Frequency

Unlike the data observed by Calvi (Calvi et al., 2003d), our experiments demonstrated a CFU-S concentration which decreased almost 2 -fold in the bone marrow of mice treated with PTH (Table 5). The concentration of the more differentiated oligopotent hematopoietic precursors, CFU-C, was not affected by PTH injections in either bone marrow or ectopic foci. However the precursors of both cell types proliferated more intensely in the PTH treated group.

Table 5. Concentration of CFU-S and CFU-C per 10^5 cells and their proliferation rate in bone marrow and ectopic foci after PTH treatment

	CFU-S per 10^5 cells Bone marrow	Suicide %
control	45.5 ± 2.2	nd*
+ PTH	24.5 ± 2.8	nd
	CFU-C per 10^5 cells	Suicide %
	Bone marrow	
control	64.5 ± 1.89	11.95
+ PTH	62.0 ± 3.4	50
	Ectopic foci	
control	33.3 ±12.1	32
+ PTH	21.3 ± 4.6	50

* not done

Seeding Efficiency (F24)

The increase in the frequency of CRU and CAFC 28-35 was accompanied by a slight decrease of more mature progenitors; this suggests that some alterations in interactions between stromal and hematopoietic cells had occurred under the PTH treatment. In connection with that altered interaction, the seeding efficiency of CAFC 28-35 and CFU-S to the microenvironment of the PTH treated mice 24 hours after transplantation was determined.

It was found that F24 of CFU-S from the spleen was not affected by PTH. Seeding efficiency to the bone marrow, on the contrary, decreased almost 4-fold following any dose (Figure 4a). There was no effect on the CAFC 28-35 if 80 μg/kg of PTH was used and only a slight decrease was noted if 10 μg/kg of PTH was administrated (Figure 4b). Obviously some features of the stromal microenvironment regulating CFU-S are affected by PTH.

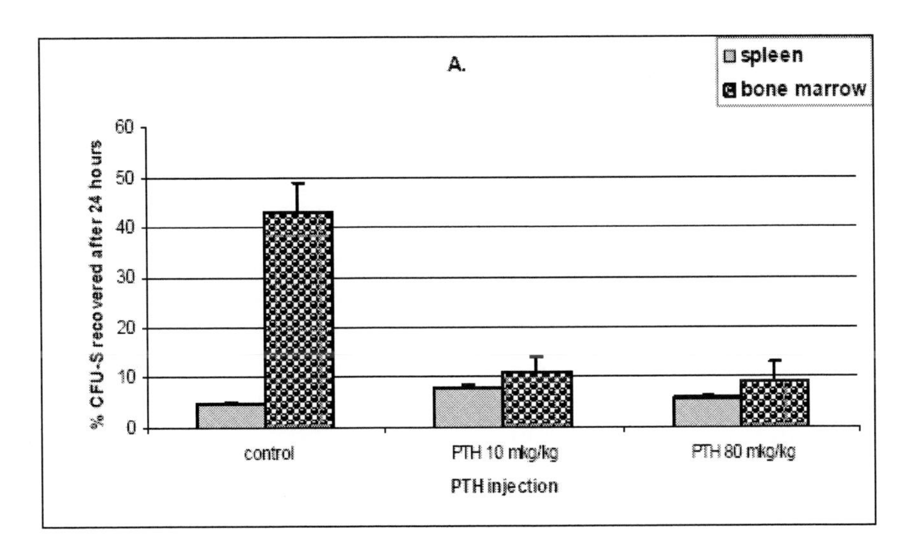

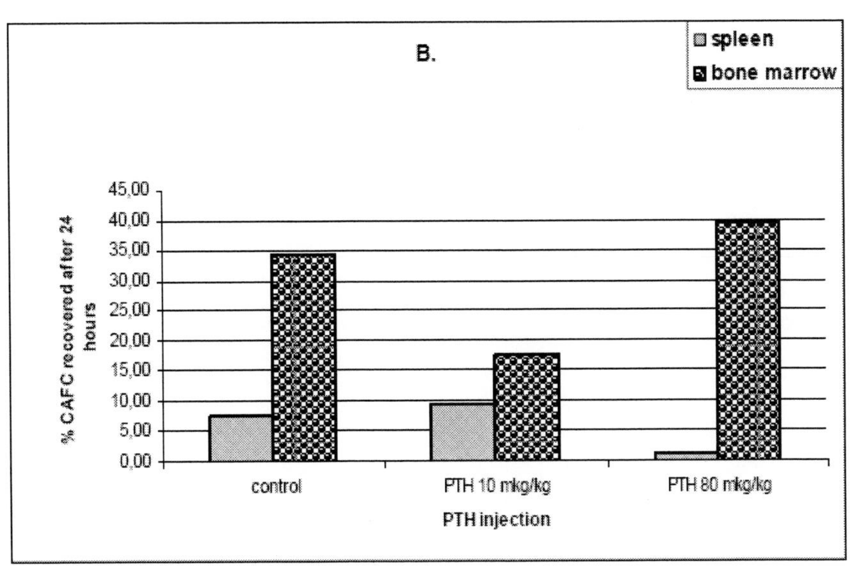

Figure 4. Seeding efficiency of hematopoietic precursor cells to bone marrow and spleen of PTH treated mice 24 hours after transplantation (A) Proportion of CFU-S seeded, (B) Proportion of CAFC 28-35 seeded. Bone marrow cells were injected into lethally irradiated PTH treated mice; 24 hours later the number of CFU-S and CAFC 28-35 were estimated in bone marrow and spleen of the recipients. Each column represents the mean from 3 animals.

Mesenchymal Stem Cells

The observed alterations of the stromal microenvironment had raised the question of the effect of PTH on mesenchymal stem cells (MSC) capable of transferring the hematopoietic microenvironment. The method of ectopic foci formation offers possibilities for MSC studies. In mice bone marrow there are cells which are able to transfer and rebuild the hematopoietic microenvironment de novo after transplantation of the bone marrow plug under the renal capsule of a syngeneic animal. In hematopoietic ectopic foci formed for 6 weeks, the stromal cells belong to the donor of the bone marrow while the hematopoietic cells are of recipient origin. The size of the foci formed (estimated by the nuclear cell number) is proportional to the femur equivalent transplanted; this can be used for semi-quantitative determination of such cell numbers. The capability of these cells to differentiate into all stromal cell types which form functional hematopoietic microenvironments completely fits the criteria of stem cell differentiation. Self-renewal ability of these cells was also proven previously (Chertkov and Gurevitch, 1979), (Chertkov et al., 1980). So the cells capable to transfer hematopoietic microenvironment could definitely be concerned as mesenchymal stem cells (MSC).

PTH did not influence MSC in the administered doses neither when given during the process of foci formation, nor in case of pretreated MSC (Table 6). The hormone also did not affect the MSC self-renewal abilities (Svinareva et al., 2004b). Thus *in vivo*, PTH has an influence not on MSC, but on its more differentiated progeny.

Table 6. The size of ectopic foci after PTH treatment

PTH concentration	Foci size in mice, treated with PTH (M ± m)	Foci size after transplantation of PTH treated bone marrow (M ± m)
control	20.7 ± 3.69	10.02 ± 1.84
PTH 10 mkg/kg	21.3 ± 5.66	11.4 ± 2.87
PTH 30 mkg/kg	13.7 ± 3.66	7.3 ± 1.9
PTH 80 mkg/kg	11.3 ± 2.04	11.6 ± 2.85

Effects of PTH In Vitro

Addition of PTH to the media for long-term bone marrow cultures (LTBMC) for 10 weeks did not affect total cell production (Svinareva et al., 2005b). Treatment of adherent cell layers with 5×10^{-8} M of PTH led to a 5.7-fold increase in CAFC 28-35 frequency after 3 weeks of cultivation, a 7.4-fold increase after 6 weeks, and up to 9.6-fold increase after 10 weeks in culture (Figure 5a). These precursors were actively proliferating both in the treated and untreated LTBMC after first 3 weeks of cultivation, and the proliferation index in the PTH treated cultures was twice as high as controls (Figure 5b). After the 6th week in culture, proliferation of CAFC 28-35 was detected only in the PTH treated LTBMC; after 10 weeks all proliferation stopped. The frequency of more mature CAFC 7 did not differ significantly during 10 weeks of cultivation with PTH (Figure 6a).

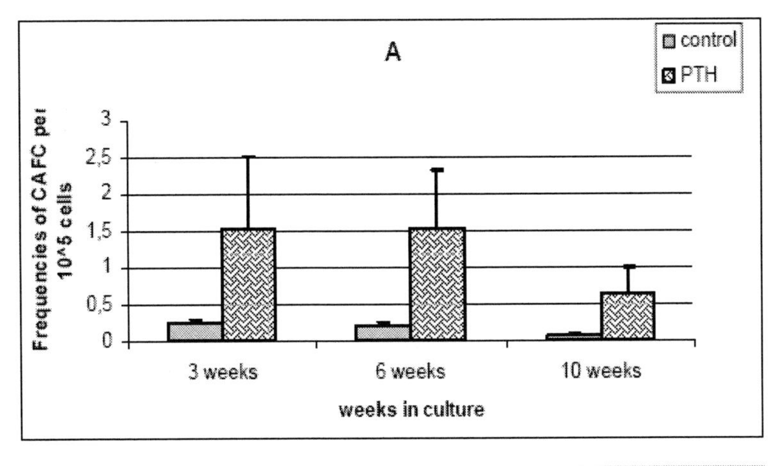

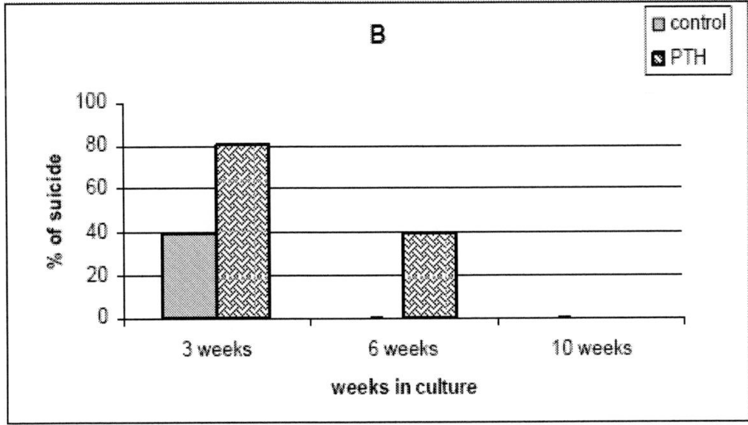

Figure 5. CAFC 28-35 in PTH treated long-term bone marrow cultures (A) the frequencies of CAFC in suspension fraction of cultures, (B) the proliferation rate of CAFC 28-35. The cells from the suspension fraction from 4 independent cultures per group were tested at 3, 6 and 10 weeks in culture. The proliferation rate was measured by the proportion of precursor cells suicide after 2 hours incubation with hydroxyurea.

After 3 weeks in culture about half of these progenitors were proliferating in both types of cultures. After 6 weeks no proliferation was detected among PTH treated CAFC 7, while 4 weeks later, unlike controls, they began to proliferate once again (Figure 6b).

Production of more mature hematopoietic progenitors able to generate colonies in semisolid medium (CFU-C) did not change after PTH treatment for 6 weeks. However up to the 10th week in culture, their concentration had increased 6-fold when compared to control cultures (Figure 7a). This could be a consequence of active proliferation of CAFC 7 detected at that moment. CFU-C proliferation rate did not differ between PTH treated and control LTBMC through the entire observation period (Figure 7b). On the whole, the results describing the effects of PTH on the various types of hematopoietic precursors in vitro verified the data concerning the same cells *in vivo*.

In order to clear up the possible effect of PTH on MSC in culture, adherent layers of stromal cells from the control and PTH treated long-term bone marrow cultures were implanted under the renal capsule of control mice or mice treated with PTH. In the untreated recipients, the cellularity of foci formed from PTH treated LTBMC were approximately 2.5-fold higher than foci formed from the untreated cultures (Table 7).

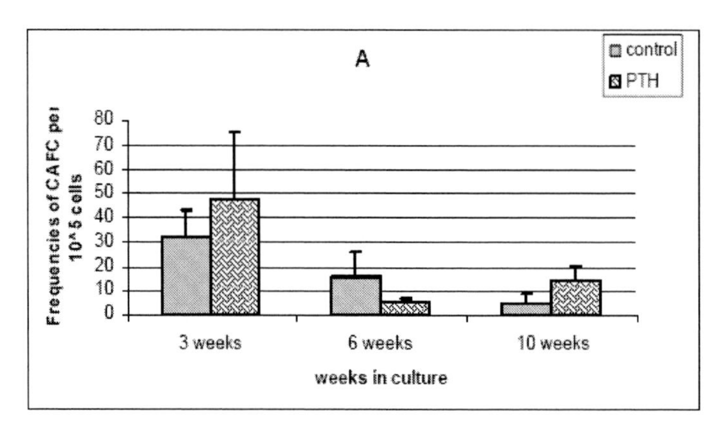

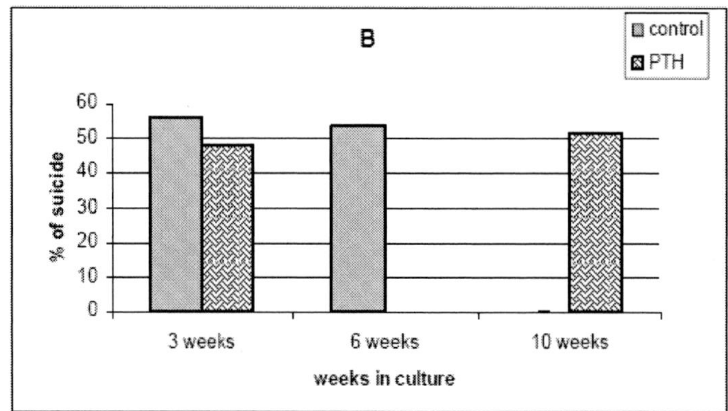

Figure 6. CAFC 7 in PTH treated long-term bone marrow cultures (A) the frequencies of CAFC in suspension fraction of cultures, (B) the proliferation rate of CAFC 7. The cells from the suspension fraction from 4 independent cultures per group were tested at 3, 6 and 10 weeks in culture. The proliferation rate was measured by the proportion of precursor cells suicide after 2 hours incubation with hydroxyurea.

Bone shell weight of the foci formed from the PTH treated cultures also increased 4-fold. PTH injections (80 μg/kg) to the recipients during the foci formation led to an increase in the size of the foci formed when compared to the foci formed in the untreated recipients. Nevertheless, there were no differences between foci formed from the PTH treated LTBMC and control LTBMC. Therefore the data obtained *in vitro* confirmed the *in vivo* results and demonstrated that PTH had an influence on the more differentiated cells rather than MSC stromal precursor cells.

The expression level of genes responsible for self-maintenance and differentiation to stromal lineages changes in the adherent cells form PTH treated LTBMC. Significant increases were detected in the expression level of Bmi-1 and Jagged-1, while Notch expression was less affected. The expression levels of genes from the osteogenic (osteopontin) and cartilage (COMP) differentiation pathways were also increased in PTH treated cultures (Figure 8).

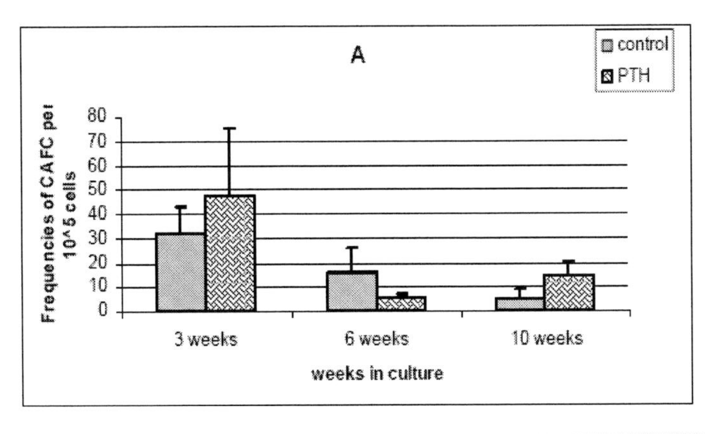

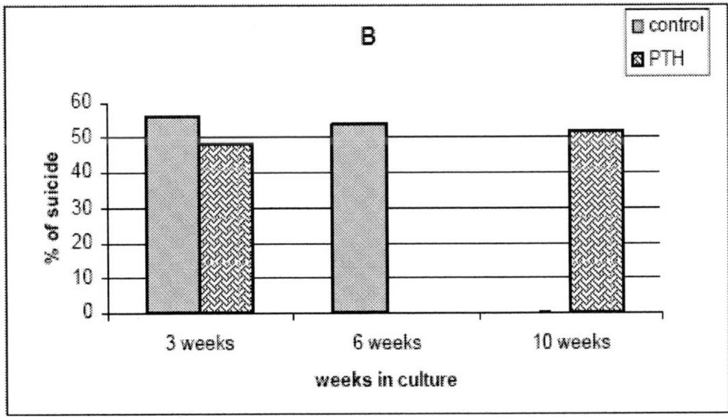

Figure 7. CFU-C in PTH treated long-term bone marrow cultures (A) concentration of CFU-C in suspension fraction of cultures, (B) the proliferation rate of CFU-C. The cells from the suspension fraction from 4 independent cultures per group were tested at 3, 6 and 10 weeks in culture. The proliferation rate was measured by the proportion of precursor cells suicide after 2 hours incubation with hydroxyurea.

Table 7. The size of ectopic foci formed from ACLs of LTBMC after 6 weeks of PTH treatment

Culture type	Foci size, x 10^6 (M ± m)	Weight of bone shell, mg
Untreated recipients		
control	2.6 ± 0.69	0.75
PTH $5x10^{-7}$M	6.8 ± 3.08	3
PTH treated recipients		
control	9.23 ± 2.82	4.5
PTH $5x10^{-7}$M	10.13 ± 1.0	3.75

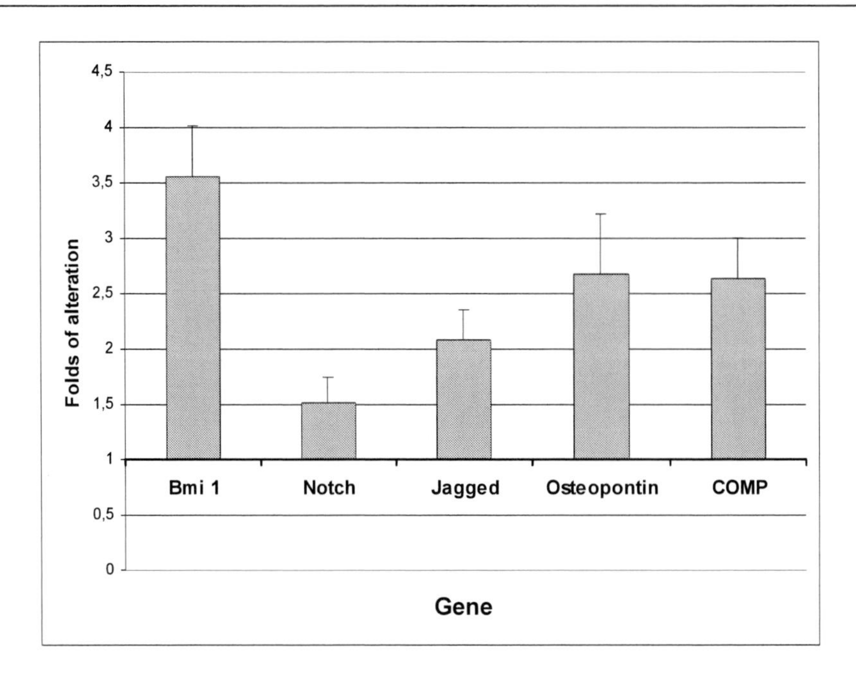

Figure 8. Relative expression level of several genes in adherent cell layers of PTH treated long-term bone marrow cultures. RNA was isolated from adherent cell layers of 6 week-old cultures. Gene expression levels were measured semi-quantitatively by RT-PCR with subsequent analysis of PCR products by a Phosphoimager after Southern-blot hybridization with the appropriate DNA-zond. Each column with error bar represents the mean ± s.e.m., compared to the relative expression level of this gene in untreated adherent cell layers, accepted for unity.

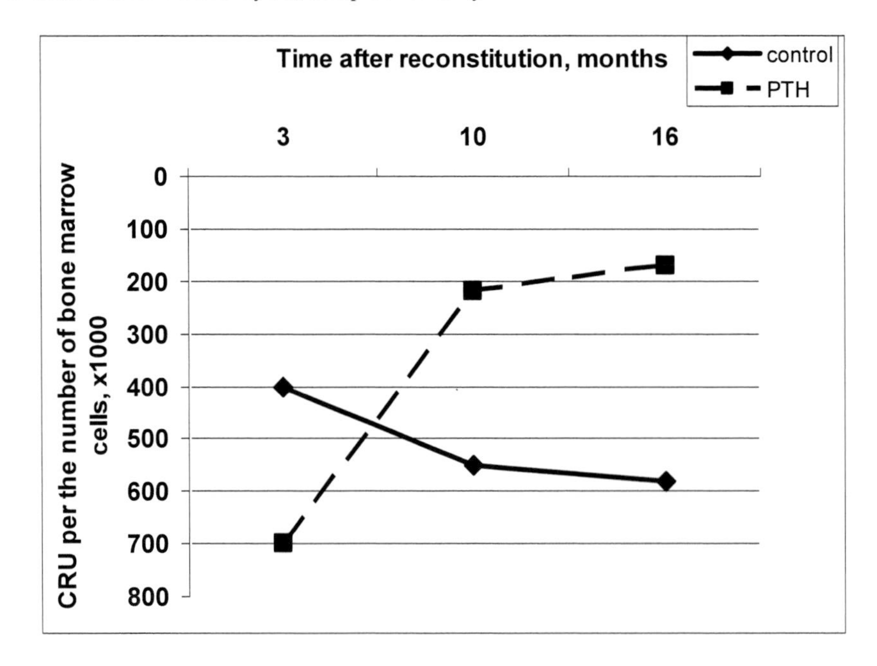

Figure 9. Dynamic of competition between PTH treated and control HSC in bone marrow of reconstituted mice.

Conclusion

It was previously shown that prolonged treatment of mice with PTH leads to doubling of the Lin-c-kit+Sca1+ HSC number (Calvi et al., 2003c). However, it is still not known whether these cells are true primitive HSC, able to produce long-term repopulation of the bone marrow. This study, for the first time, demonstrated that early hematopoietic stem cells (defined by the competitive repopulation method) had increased 4-fold following PTH treatment. Analyzing all the data obtained, it is possible to track the dynamic of HSC competition for an extended period following reconstitution (Figure 9).

Three months after transplantation of the HSC mixture, the frequency of competitive repopulation units (CRU) in the group reconstituted with PTH treated bone marrow was lower than in the control group. However, 10 months after the reconstitution, their frequency had increased 2.5-fold and half a year later it became 4-fold higher than the group with untreated cells. The data seems to prove a functional advantage of long-term repopulating HSC from the bone marrow of PTH treated mice. This could be a consequence of either an increase in the number of clones accomplishing hematopoiesis at the moment or an expansion of the clone size due to an increase in the proliferative potential of PTH treated HSC. The increase in frequency of CRU from the PTH treated bone marrow became more apparent with time, meaning the PTH influence had affected the very early HSC. This time-prolonged increase is especially important concerning potential therapeutic applications, as it can possibly bring down the frequency of the reversion of hematopoiesis and the probability of late relapses after allogeneic bone marrow transplantation. The decrease in the proportion of Y+ CFU-S in the bone marrow of mice reconstituted with untreated HSCs (Table 3) could be explained by the reversion of hematopoiesis and the beginning of function of the recipient stem cells. It was previously shown that gradual reversion of hematopoiesis to the recipient type occurs when lethally irradiated females are reconstituted with male HSC. Analogous phenomenon was observed in the case of a single sorted male stem cell injection to irradiated females. Frequency of successful engraftments in this design was 2.4-fold lower than in the case of all other combinations (female into male, female into female and male into male transplantations) (Uchida et al., 2003). Such impairment of the engraftment could be connected to immunological rejection due to sex antigens (Scott et al., 1997). At the same time, the proportion of Y+CFU-S in mice reconstituted with PTH treated marrow cells steadily increased; after 16 months, reconstitution significantly exceeded the control levels. This data demonstrates the increased contribution of PTH treated CFU-S to hematopoiesis. The ratio of Y+ CFU-S to mature cells varies through the lifespan of mice (Table 3). Variations in the proportion of Y+ precursors during the time of observation (as was observed in mouse #3, for example) could be explained by clonal succession. Taking into account the simultaneous function of more than 50 clones (Drize et al., 1996), such instability seems to be physiological. However, when mice were reconstituted with a mixture of intact female and PTH treated male HSC, there was a tendency of the Y+CFU-S proportion to increase. Clonal succession model and the dynamic balance of hematopoiesis can explain the differences in the proportion of Y+ CFU-S, bone marrow and peripheral blood cells analyzing simultaneously. In the competitive repopulation assay, only the progeny of marked cells were detected, so reversion of hematopoiesis to the recipient type could not distort the obtained results.

Therefore, treatment of mice with PTH leads to a significant increase in CRU frequency, 2.5- to 4-fold.

It was shown that the frequency of the other hematopoietic precursors CAFC 28-35 increased 3.5-fold in the bone marrow of mice treated with 80 μg/kg of PTH (Table 4). This data confirms the results obtained by Calvi et al (Calvi et al., 2003). When 10 μg/kg was used, CAFC 28-35 frequency increased almost 7-fold, while the intermediate dose of 30 μg/kg had no significant effect (Figure 3). The absence of linear dependency between dose and effect on the early hematopoietic precursors results from the different types of PTH influence on the various elements of the hematopoietic microenvironment. It is known that stimulation of MAPK and proliferation of osteoblasts in low concentrations of PTH is protein kinase C dependent. Activation of osteoblasts results in stimulation and recruitment of osteoclasts. Much of these processes are protein kinase A dependent. Protein kinase A activation leads to inhibition of MAPK and proliferation of osteoblasts (Swarthout et al., 2002). So a balance between the reciprocal effects of osteoblast and osteoclast activation could be disturbed by the slightest alteration in PTH dose. It is known that attempts to increase dosage of the hormone for osteoporosis treatment resulted in bone fractures (Finkelstein et al., 2003). Obviously, it is essential to thoroughly attenuate the dose of PTH in case of therapeutic application for HSC stimulation.

Among the doses studied, 80 μg/kg of PTH was shown to have the most stimulative effect on CAFC 28-35 proliferation. This data confirm the ability of PTH to stimulate the proliferation of not only osteoblasts but also, indirectly, early hematopoietic precursors. HSC expansion *in vivo* induced by PTH could be the consequence of an increase in the niche number caused by proliferation of osteoblasts as well as the induction of HSC proliferation caused by newly formed osteoblastic niches.

Concentration of less immature precursors such as CAFC 7 and CFU-S decreased after PTH treatment, while the frequency of more differentiated precursors like CFU-C did not change. Proliferation of CAFC 7 was not detected in the experiments performed. The CFU-C proliferation index was higher in the bone marrow of PTH treated mice than in controls. The ability of PTH to stimulate CFU-S proliferation was shown long ago using the thymidine suicide method (Gallien-Lartigue and Carrez, 1974). Various types of hematopoietic precursor respond to PTH in different ways.

Why did an increase in the number of early hematopoietic stem cells not lead to a corresponding elevation of the number of more differentiated progenitor cells? Each category of precursors is regulated specifically by the balance between growth factors and intracellular interactions. As hematopoietic cells do not bear high-affinity PTH-receptors, any regulation by this hormone is performed indirectly. Distantly regulated hematopoietic precursors might be not affected by PTH, or they may respond to it in complex, indirect way.

There is another possible explanation of the absence of expansion of late hematopoietic precursors and mature differentiated cells. When the homing of hematopoietic precursors to PTH treated marrow stroma was studied, it turned out that 24 hours seeding efficiency of early HSC (CAFC 28) did not change after PTH treatment (Figure 4b), while it dramatically decreased for CFU-S (Figure 4a). Interaction of HSC with the niche cells is a dynamic process (Yin and Li, 2006), so the impairment of such relations for CFU-S might affect the maturation process. Obviously, the CFU-S niche in the bone marrow was changed by the hormone, although there is still no direct evidence of PTH influence on the vascular niche. Homing to the spleen was not affected by PTH, probably because of the absence of PTH receptors on spleen stroma.

Despite large errors in the data of these experiments, it can be stated that PTH does not negatively affect the ability of early HSC to adhere to bone marrow stroma. Taking into consideration the results of Calvi et al. (Calvi et al., 2003a), increase in the adherence of immature hematopoietic precursors to the PTH treated marrow stroma was to be expected. Therefore, the increase in the number of osteoblasts forming hematopoietic niches is not connected directly with ability of HSC to home to the bone marrow stroma. Probably the molecules of homing, chemoattractants and adhesion molecules are located on different cells. The seeding efficiency for CAFC 28 to the spleen of mice treated with low doses of PTH did not differ from controls (Figure 4b). Dramatic decreases in the seeding efficiency of CAFC 28 to the spleen of mice treated with 80 μg/kg of PTH could not be explained with fatal changes in the stromal microenvironment. The distribution of hematopoietic cells throughout all bone marrow territories has been calculated (Colvin et al., 2004), and femurs and tibias were shown to contain about 14% of all HSC. Taking this fact into consideration, one is able to estimate the number of HSC homed to the hematopoietic microenvironment within the first 24 hours after irradiation and injection of bone marrow cells. In the conditions used, about 60% of injected CAFC 28 did not get to the hematopoietic territory in both untreated and PTH treated animals. The differences among the cells that did get there were mainly in the pattern of distribution among the territories tested (34% of CAFC 28 in the bone marrow and 7% in the spleen of untreated animals and 39% and 1% correspondingly in the bone marrow and spleen of mice injected with 80 μg/kg of PTH). So, PTH could affect some signaling pathways; on the other hand, it changes the stromal components essential for the homing of hematopoietic precursors that are regulated not in the osteoblastic, but in the vascular niche (Yin and Li, 2006).

The data shown signify a deep impact of PTH on the features of the stromal microenvironment. However, whether it might affect mesenchymal stem cells (MSC) was not clear. This study has shown that PTH had no effect on MSC, neither during the process of forming the hematopoietic foci de novo (when it was injected for the first 4 weeks after the implantation of the bone marrow plug under the renal capsule of syngeneic animals) nor in case of pretreated donors of the bone marrow (Table 6). Moreover, PTH treatment did not affect the self-renewing ability of MSC tested by reimplatation of the foci formed (Svinareva et al., 2004a). These results are of the highest importance, considering the prospective use of PTH for HSC expansion and facilitation of hematopoiesis in the patients undergoing radio- and chemotherapy. In addition, it seems that osteogenic differentiation of MSC progeny was not changed, even as the weight of the bone shell formed around the foci had increased following high dose PTH (data not shown). Unlike in the bone marrow, the hematopoietic precursor frequency did not change in ectopic foci after PTH treatment (Tables 4, 5). This fact might be connected with the delayed answer of osteogenic precursors differentiating during the foci formation and injections of PTH. Proliferation of all hematopoietic precursors tested at the same time was increased (Tables 4, 5). These data correlate with the results obtained from the PTH treated bone marrow analysis.

The influence of PTH *in vitro* on stromal cells capable of transferring the microenvironment was also tested. Six-weeks of PTH treatment to the adherent cell layer turned out to not affect the MCS, judged by the size of the foci formed following implantation into intact animals. However, if the recipients were injected with PTH during foci formation, the size of the foci increased 2- to 3–fold, independent of previous culture treatment (Table 7). It is additional evidence that it is the main responsibility of more differentiated than MSC progenitors for generation of a functional stromal layer in LTBMC (Chertkov et al., 1983). The data confirm the existence of hierarchy in the MSC compartment. Undifferentiated multipotent stromal precursors form the microenvironment in culture and respond to PTH during foci formation. The insensitivity of MSC to PTH seems to be of most importance concerning its potential usage for therapeutic purposes.

In long-term bone marrow cultures, total cell production did not change during the PTH treatment (Svinareva et al., 2005a). CAFC 28-35 frequency had increased since the 3rd week of cultivation (Figure 5). CAFC 7 and CFU-C frequency had not changed during the first 6 weeks (Figure 6, 7). Nevertheless, the concentration of CFU-C increased significantly by the 10th week in culture. This could be explained by the observed increase in proliferation of these cells; on the other hand, previous expansion of less differentiated precursors was detected at the 6th week of cultivation.

All types of hematopoietic precursors tested proliferate with various intensity up to 10th week in culture. The exhaustion of hematopoiesis *in vitro* could be a consequence of a small number of early HSC surviving the procedure of bone marrow explantation in the culture flask, as well as of oppression of their proliferative potential. It takes 2-3 weeks to form a functional adherent cell layer *in vitro*, and a considerable number of HSC can not survive such a period without a regulating niche.

The relative expression level of genes regulating the proliferation and self-renewing of HSC such as Notch-1, Jagged-1 and Bmi-1 did not change following 3 weeks of PTH treatment. Prolongation of PTH addition led to upregulation of the expression level of Jagged-1 and genes of stromal differentiation pathways like COMP and osteopontin. The latter is known to be both the marker protein for osteogenic differentiation and a HSC regulator; it also might smooth over the PTH influence *in vitro*. The data confirm hematopoiesis to be a multifactorial process including apoptosis and cell cycle. Oseoblast activation results in stimulation of HSC production; simultaneously in the niche, osteopontin synthesis leads to inhibition of HSC expansion and a decrease in their number (Adams and Scadden, 2006) (Nilsson et al., 2005). The hematopoietic system has a multitude of mechanisms to maintain HSC number, their ability to self-renew, and their ability to differentiate.

In summary, the data obtained on the *in vitro* model confirm the results of *in vivo* investigations. It could be stated that PTH has no pathological effect on hematopoietic and stromal cells.

In conclusion, one can note that PTH is an attractive tool for *in vivo* HSC expansion and for improvement in the recovery of hematopoiesis in patients who have undergone chemo- and radiotherapy. PTH is able to increase the number of HSC and enhance engraftment and bone marrow repopulation. This is done first by increasing the osteoblast number and as a consequence the niche number, second, by means of transplanted HSC expansion, and third, by stimulation of proliferation of more differentiated progeny. Undoubtedly, a number of investigations should be performed before the start of clinical applications of PTH. The fine adjustment of concentrations and the injection timing are of the highest importance.

REFERENCES

Adams,G.B. and Scadden,D.T. (2006). The hematopoietic stem cell in its place. *Nat. Immunol.* 7, 333-337.

Arai,F., Hirao,A., Ohmura,M., Sato,H., Matsuoka,S., Takubo,K., Ito,K., Koh,G.Y., and Suda,T. (2004). Tie-2/angiopoietin-1 signaling regulates hematopoietic stem cell quiescence in the bone marrow niche. *Cell* 118, 149-161.

Arai,F., Hirao,A., and Suda,T. (2005). Regulation of hematopoietic stem cells by the niche. *Trends Cardiovasc. Med.* 15, 75-79.

Avecilla,S.T., Hattori,K., Heissig,B., Tejada,R., Liao,F., Shido,K., Jin,D.K., Dias,S., Zhang,F., Hartman,T.E., Hackett,N.R., Crystal,R.G., Witte,L., Hicklin,D.J., Bohlen,P., Eaton,D., Lyden,D., de Sauvage,F., and Rafii,S. (2004). Chemokine-mediated interaction of hematopoietic progenitors with the bone marrow vascular niche is required for thrombopoiesis. *Nat. Med.* 10, 64-71.

Bellido,T., Ali,A.A., Plotkin,L.I., Fu,Q., Gubrij,I., Roberson,P.K., Weinstein,R.S., O'Brien,C.A., Manolagas,S.C., and Jilka,R.L. (2003). Proteasomal degradation of Runx2 shortens parathyroid hormone-induced anti-apoptotic signaling in osteoblasts. A putative explanation for why intermittent administration is needed for bone anabolism. *J. Biol. Chem.* 278, 50259-50272.

Calvi,L.M. (2006). Osteoblastic activation in the hematopoietic stem cell niche. Ann. N. Y. Acad. Sci. 1068, 477-488.

Calvi,L.M., Adams,G.B., Weibrecht,K.W., Weber,J.M., Olson,D.P., Knight,M.C., Martin,R.P., Schipani,E., Divieti,P., Bringhurst,F.R., Milner,L.A., Kronenberg,H.M., and Scadden,D.T. (2003i). Osteoblastic cells regulate the haematopoietic stem cell niche. *Nature* 425, 841-846.

Calvi,L.M., Sims,N.A., Hunzelman,J.L., Knight,M.C., Giovannetti,A., Saxton,J.M., Kronenberg,H.M., Baron,R., and Schipani,E. (2001). Activated parathyroid hormone/parathyroid hormone-related protein receptor in osteoblastic cells differentially affects cortical and trabecular bone. *J. Clin. Invest* 107, 277-286.

Cardier,J.E. and Barbera-Guillem,E. (1997). Extramedullary hematopoiesis in the adult mouse liver is associated with specific hepatic sinusoidal endothelial cells. *Hepatology* 26, 165-175.

Chertkov,J.L., Drize,N.J., Gurevitch,O.A., and Udalov,G.A. (1983). Hemopoietic stromal precursors in long-term culture of bone marrow: I. Precursor characteristics, kinetics in culture, and dependence on quality of donor hemopoietic cells in chimeras. *Exp. Hematol.* 11, 231-242.

Chertkov,J.L. and Gurevitch,O.A. (1979). Radiosensitivity of progenitor cells of the hematopoietic microenvironment. *Radiat. Res.* 79, 177-186.

Chertkov,J.L., Gurevitch,O.A., and Udalov,G.A. (1980). Role of bone marrow stroma in hemopoietic stem cell regulation. *Exp. Hematol.* 8, 770-778.

Chomczynski,P. and Sacchi,N. (1987). Single-step method of RNA isolation by acid guanidinium thiocyanate-phenol-chloroform extraction. *Anal. Biochem.* 162, 156-159.

Colvin,G.A., Lambert,J.F., Abedi,M., Hsieh,C.C., Carlson,J.E., Stewart,F.M., and Quesenberry,P.J. (2004). Murine marrow cellularity and the concept of stem cell competition: geographic and quantitative determinants in stem cell biology. *Leukemia* 18, 575-583.

Dempster,D.W., Cosman,F., Kurland,E.S., Zhou,H., Nieves,J., Woelfert,L., Shane,E., Plavetic,K., Muller,R., Bilezikian,J., and Lindsay,R. (2001). Effects of daily treatment with parathyroid hormone on bone microarchitecture and turnover in patients with osteoporosis: a paired biopsy study. *J. Bone Miner. Res.* 16, 1846-1853.

Dexter,T.M., Allen,T.D., and Lajtha,L.G. (1977). Conditions controlling the proliferation of haemopoietic stem cells in vitro. *J. Cell Physiol.* 91, 335-344.

Drize,N.J., Keller,J.R., and Chertkov,J.L. (1996). Local clonal analysis of the hematopoietic system shows that multiple small short-living clones maintain life-long hematopoiesis in reconstituted mice. *Blood* 88, 2927-2938.

Duncan,A.W., Rattis,F.M., DiMascio,L.N., Congdon,K.L., Pazianos,G., Zhao,C., Yoon,K., Cook,J.M., Willert,K., Gaiano,N., and Reya,T. (2005). Integration of Notch and Wnt signaling in hematopoietic stem cell maintenance. *Nat. Immunol.* 6, 314-322.

Ejersted,C., Andreassen,T.T., Nilsson,M.H., and Oxlund,H. (1994). Human parathyroid hormone(1-34) increases bone formation and strength of cortical bone in aged rats. *Eur. J. Endocrinol.* 130, 201-207.

El Badri,N.S., Wang,B.Y., Cherry, and Good,R.A. (1998). Osteoblasts promote engraftment of allogeneic hematopoietic stem cells. *Exp. Hematol.* 26, 110-116.

Finkelstein,J.S., Hayes,A., Hunzelman,J.L., Wyland,J.J., Lee,H., and Neer,R.M. (2003b). The effects of parathyroid hormone, alendronate, or both in men with osteoporosis. *N. Engl. J. Med.* 349, 1216-1226.

Gallien-Lartigue,O. and Carrez,D. (1974). [In vitro induction of the S phase in multipotent stem cells of bone marrow by parathyroid hormone]. *C. R. Acad. Sci. Hebd. Seances Acad. Sci.* D. 278, 1765-1768.

Habener,J.F., Rosenblatt,M., and Potts,J.T., Jr. (1984). Parathyroid hormone: biochemical aspects of biosynthesis, secretion, action, and metabolism. *Physiol. Rev.* 64, 985-1053.

Harada,S. and Rodan,G.A. (2003). Control of osteoblast function and regulation of bone mass. *Nature* 423, 349-355.

Harrison,D.E. (1980). Lifespans of immunohemopoietic stem cell lines. *Adv. Pathobiol.* 7, 187-199.

Harrison,D.E. (1983). Long-term erythropoietic repopulating ability of old, young, and fetal stem cells. *J. Exp. Med.* 157, 1496-1504.

Haylock,D.N. and Nilsson,S.K. (2005). Stem cell regulation by the hematopoietic stem cell niche. *Cell Cycle* 4, 1353-1355.

Hurley,M.M., Okada,Y., Xiao,L., Tanaka,Y., Ito,M., Okimoto,N., Nakamura,T., Rosen,C.J., Doetschman,T., and Coffin,J.D. (2006). Impaired bone anabolic response to parathyroid hormone in Fgf2-/- and Fgf2+/- mice. *Biochem. Biophys. Res. Commun.* 341, 989-994.

Hurley,M.M., Tetradis,S., Huang,Y.F., Hock,J., Kream,B.E., Raisz,L.G., and Sabbieti,M.G. (1999). Parathyroid hormone regulates the expression of fibroblast growth factor-2 mRNA and fibroblast growth factor receptor mRNA in osteoblastic cells. *J. Bone Miner. Res.* 14, 776-783.

Iwata,M., Awaya,N., Graf,L., Kahl,C., and Torok-Storb,B. (2004). Human marrow stromal cells activate monocytes to secrete osteopontin, which down-regulates Notch1 gene expression in CD34+ cells. *Blood* 103, 4496-4502.

Kiel,M.J., Yilmaz,O.H., Iwashita,T., Yilmaz,O.H., Terhorst,C., and Morrison,S.J. (2005). SLAM family receptors distinguish hematopoietic stem and progenitor cells and reveal endothelial niches for stem cells. *Cell* 121, 1109-1121.

Kobari,L., Dubart,A., Le Pesteur,F., Vainchenker,W., and Sainteny,F. (1995). Hematopoietic-promoting activity of the murine stromal cell line MS-5 is not related to the expression of the major hematopoietic cytokines. *J. Cell Physiol.* 163, 295-304.

Kopp,H.G., Avecilla,S.T., Hooper,A.T., and Rafii,S. (2005). The bone marrow vascular niche: home of HSC differentiation and mobilization. *Physiology.* (Bethesda.) 20, 349-356.

Kulkarni,N.H., Halladay,D.L., Miles,R.R., Gilbert,L.M., Frolik,C.A., Galvin,R.J., Martin,T.J., Gillespie,M.T., and Onyia,J.E. (2005). Effects of parathyroid hormone on Wnt signaling pathway in bone. *J. Cell Biochem.* 95, 1178-1190.

Kuznetsov,S.A., Riminucci,M., Ziran,N., Tsutsui,T.W., Corsi,A., Calvi,L., Kronenberg,H.M., Schipani,E., Robey,P.G., and Bianco,P. (2004). The interplay of osteogenesis and hematopoiesis: expression of a constitutively active PTH/PTHrP receptor in osteogenic cells perturbs the establishment of hematopoiesis in bone and of skeletal stem cells in the bone marrow. *J. Cell Biol.* 167, 1113-1122.

Lane,N.E., Sanchez,S., Modin,G.W., Genant,H.K., Pierini,E., and Arnaud,C.D. (1998). Parathyroid hormone treatment can reverse corticosteroid-induced osteoporosis. Results of a randomized controlled clinical trial. *J. Clin. Invest.* 102, 1627-1633.

Li,W., Johnson,S.A., Shelley,W.C., and Yoder,M.C. (2004). Hematopoietic stem cell repopulating ability can be maintained in vitro by some primary endothelial cells. *Exp. Hematol.* 32, 1226-1237.

Lin,H. (2002). The stem-cell niche theory: lessons from flies. *Nat. Rev. Genet.* 3, 931-940.

Maniatis,T., Fritsch,E.F., and Sambrook,J. (1982). *Molecular Cloning, A Laboratory Manual.* Cold Spring Harbor Laboratory).

Marie,P.J. (2002). Role of N-cadherin in bone formation. *J. Cell Physiol.* 190, 297-305.

Mayer,P. (1983). The growth of swine bone marrow cells in the presence of heterologous colony stimulating factor: characterization of the developing cell population. *Comp. Immunol. Microbiol. Infect. Dis.* 6, 171-187.

Miller,C.L. and Eaves,C.J. (1997). Expansion in vitro of adult murine hematopoietic stem cells with transplantable lympho-myeloid reconstituting ability. *Proc. Natl. Acad. Sci. USA* 94, 13648-13653.

Moore,K.A. and Lemischka,I.R. (2004). "Tie-ing" down the hematopoietic niche. *Cell* 118, 139-140.

Moore,K.A. and Lemischka,I.R. (2006). Stem cells and their niches. *Science* 311, 1880-1885.

Neer,R.M., Arnaud,C.D., Zanchetta,J.R., Prince,R., Gaich,G.A., Reginster,J.Y., Hodsman,A.B., Eriksen,E.F., Ish-Shalom,S., Genant,H.K., Wang,O., and Mitlak,B.H. (2001a). Effect of parathyroid hormone (1-34) on fractures and bone mineral density in postmenopausal women with osteoporosis. *N. Engl. J. Med.* 344, 1434-1441.

Nilsson,S.K., Johnston,H.M., Whitty,G.A., Williams,B., Webb,R.J., Denhardt,D.T., Bertoncello,I., Bendall,L.J., Simmons,P.J., and Haylock,D.N. (2005). Osteopontin, a key component of the hematopoietic stem cell niche and regulator of primitive hematopoietic progenitor cells. *Blood* 106, 1232-1239.

Nilsson,S.K., Simmons,P.J., and Bertoncello,I. (2006). Hemopoietic stem cell engraftment. *Exp. Hematol.* 34, 123-129.

Park,I.K., Qian,D., Kiel,M., Becker,M.W., Pihalja,M., Weissman,I.L., Morrison,S.J., and Clarke,M.F. (2003). Bmi-1 is required for maintenance of adult self-renewing haematopoietic stem cells. *Nature* 423, 302-305.

Patel,N., Talwar,A., Donahue,L., John,V., and Margouleff,D. (2005). Hyperparathyroidism accompanying multiple myeloma. *Clin. Nucl. Med.* 30, 540-542.

Ploemacher,R.E., van der Sluijs,J.P., Voerman,J.S., and Brons,N.H. (1989a). An in vitro limiting-dilution assay of long-term repopulating hematopoietic stem cells in the mouse. *Blood* 74, 2755-2763.

Ponomaryov,T., Peled,A., Petit,I., Taichman,R.S., Habler,L., Sandbank,J., Arenzana-Seisdedos,F., Magerus,A., Caruz,A., Fujii,N., Nagler,A., Lahav,M., Szyper-Kravitz,M., Zipori,D., and Lapidot,T. (2000). Induction of the chemokine stromal-derived factor-1 following DNA damage improves human stem cell function. *J. Clin. Invest.* 106, 1331-1339.

Potts,J.T. (2005). Parathyroid hormone: past and present. *J. Endocrinol.* 187, 311-325.

Prestidge,R.L., Watson,J.D., Urdal,D.L., Mochizuki,D., Conlon,P., and Gillis,S. (1984). Biochemical comparison of murine colony-stimulating factors secreted by a T cell lymphoma and a myelomonocytic leukemia. *J. Immunol.* 133, 293-298.

Reya,T., Duncan,A.W., Ailles,L., Domen,J., Scherer,D.C., Willert,K., Hintz,L., Nusse,R., and Weissman,I.L. (2003). A role for Wnt signalling in self-renewal of haematopoietic stem cells. *Nature* 423, 409-414.

RIXON,R.H. and Whitfield,J.F. (1961). The radioprotective action of parathyroid extract. *Int. J. Radiat. Biol.* 3, 361-367.

RIXON,R.H., Whitfield,J.F., and YOUDALE,T. (1958). Increased survival of rats irradiated with x-rays and treated with parathyroid extract. *Nature* 182, 1374.

Rodan,G.A. and Martin,T.J. (2000). Therapeutic approaches to bone diseases. *Science* 289, 1508-1514.

Sauvageau,G., Iscove,N.N., and Humphries,R.K. (2004). In vitro and in vivo expansion of hematopoietic stem cells. *Oncogene* 23, 7223-7232.

Schiller,P.C., D'Ippolito,G., Roos,B.A., and Howard,G.A. (1999). Anabolic or catabolic responses of MC3T3-E1 osteoblastic cells to parathyroid hormone depend on time and duration of treatment. *J. Bone Miner. Res.* 14, 1504-1512.

Schofield,R. (1978). The relationship between the spleen colony-forming cell and the haemopoietic stem cell. *Blood Cells* 4, 7-25.

Scott,D.M., Ehrmann,I.E., Ellis,P.S., Chandler,P.R., and Simpson,E. (1997). Why do some females reject males? The molecular basis for male-specific graft rejection. *J. Mol. Med.* 75, 103-114.

Selye H. (1932). On the stimulation of new bone formation with parathyroid extract and irradiated ergosterol. *Endocrinology* 16, 547-558.

Staal,F.J. and Clevers,H.C. (2005). WNT signalling and haematopoiesis: a WNT-WNT situation. *Nat. Rev. Immunol.* 5, 21-30.

Stier,S., Ko,Y., Forkert,R., Lutz,C., Neuhaus,T., Grunewald,E., Cheng,T., Dombkowski,D., Calvi,L.M., Rittling,S.R., and Scadden,D.T. (2005). Osteopontin is a hematopoietic stem cell niche component that negatively regulates stem cell pool size. *J. Exp. Med.* 201, 1781-1791.

Strewler,G.J., Stern,P.H., Jacobs,J.W., Eveloff,J., Klein,R.F., Leung,S.C., Rosenblatt,M., and Nissenson,R.A. (1987). Parathyroid hormonelike protein from human renal carcinoma cells. Structural and functional homology with parathyroid hormone. *J. Clin. Invest.* 80, 1803-1807.

Svinareva,D.A., Nifontova,I.N., Chertkov,I.L., and Drize,N.I. (2005a). Ex vivo expansion of hemopoietic precursor cells on a sublayer treated with parathyroid hormone. *Bull. Exp. Biol. Med.* 140, 334-337.

Svinareva,D.A., Nifontova,I.N., Chertkov,I.L., and Drize,N.I. (2005b). Ex vivo expansion of hemopoietic precursor cells on a sublayer treated with parathyroid hormone. *Bull. Exp. Biol. Med.* 140, 334-337.

Svinareva,D.A., Nifontova,I.N., and Drize,N.I. (2004a). Effect of parathyroid hormone PTH (1-34) on hemopoietic and stromal stem cells. *Bull. Exp. Biol. Med.* 138, 571-574.

Swarthout,J.T., D'Alonzo,R.C., Selvamurugan,N., and Partridge,N.C. (2002a). Parathyroid hormone-dependent signaling pathways regulating genes in bone cells. *Gene* 282, 1-17.

Teitelbaum,S.L. (2004). Postmenopausal osteoporosis, T cells, and immune dysfunction. *Proc. Natl. Acad. Sci. USA* 101, 16711-16712.

Till,J.E.M.E.A. (1961). A direct measurement of radiation sencitivity of normal mouse bone marrow. *Radiation Research* 14, 213-221.

Tobimatsu,T., Kaji,H., Sowa,H., Naito,J., Canaff,L., Hendy,G.N., Sugimoto,T., and Chihara,K. (2006). Parathyroid hormone increases beta-catenin levels through Smad3 in mouse osteoblastic cells. *Endocrinology* 147, 2583-2590.

Uchida,N., Dykstra,B., Lyons,K.J., Leung,F.Y., and Eaves,C.J. (2003). Different in vivo repopulating activities of purified hematopoietic stem cells before and after being stimulated to divide in vitro with the same kinetics. *Exp. Hematol.* 31, 1338-1347.

Van Den Berg,D.J., Sharma,A.K., Bruno,E., and Hoffman,R. (1998). Role of members of the Wnt gene family in human hematopoiesis. *Blood* 92, 3189-3202.

van der,H.G., Farih-Sips,H., Lowik,C.W., and Karperien,M. (2005). Multiple mechanisms are involved in inhibition of osteoblast differentiation by PTHrP and PTH in KS483 Cells. *J. Bone Miner. Res.* 20, 2233-2244.

Visnjic,D., Kalajzic,Z., Rowe,D.W., Katavic,V., Lorenzo,J., and Aguila,H.L. (2004). Hematopoiesis is severely altered in mice with an induced osteoblast deficiency. *Blood* 103, 3258-3264.

Wagers,A.J. (2005). Stem cell grand SLAM. *Cell* 121, 967-970.

Weber,J.M., Forsythe,S.R., Christianson,C.A., Frisch,B.J., Gigliotti,B.J., Jordan,C.T., Milner,L.A., Guzman,M.L., and Calvi,L.M. (2006). Parathyroid hormone stimulates expression of the Notch ligand Jagged1 in osteoblastic cells. *Bone* 39, 485-493.

Whitfield,J., Bird,R.P., Morley,P., Willick,G.E., Barbier,J.R., MacLean,S., and Ross,V. (2003). The effects of parathyroid hormone fragments on bone formation and their lack of effects on the initiation of colon carcinogenesis in rats as indicated by preneoplastic aberrant crypt formation. *Cancer Lett.* 200, 107-113.

Whitfield,J.F. (2005). Parathyroid hormone and leptin--new peptides, expanding clinical prospects. Expert. Opin. Investig. Drugs 14, 251-264.

Whitfield,J.F. (2006). Parathyroid hormone: A novel tool for treating bone marrow depletion in cancer patients caused by chemotherapeutic drugs and ionizing radiation. *Cancer Lett.*

Willert,K., Brown,J.D., Danenberg,E., Duncan,A.W., Weissman,I.L., Reya,T., Yates,J.R., III, and Nusse,R. (2003). Wnt proteins are lipid-modified and can act as stem cell growth factors. *Nature* 423, 448-452.

Yin,T. and Li,L. (2006a). The stem cell niches in bone. *J. Clin. Invest.* 116, 1195-1201.

Feng,J.Q., Harris,S., Wiedemann,L.M., Mishina,Y., and Li,L. (2003). Identification of the haematopoietic stem cell niche and control of the niche size. *Nature* 425, 836-841.

Zhu,J. and Emerson,S.G. (2004). A new bone to pick: osteoblasts and the haematopoietic stem-cell niche. *Bioessays* 26, 595-599.

In: Stem Cell Research Trends
Editor: Josse R. Braggina, pp. 187-208
ISBN: 978-1-60021-622-0

Chapter VI

MICROBIOLOGICAL ISSUES IN STEM CELL RESEARCH: NEW TECHNOLOGIES FOR DIAGNOSIS

Fernando Cobo*
Stem Cell Bank of Andalucía (Spanish Central Node)
University Hospital Virgen de las Nieves
Avenida Fuerzas Armadas, 2
18014 Granada, Spain

ABSTRACT

In the field of medicine, stem cell research is now the most exciting opportunity for the treatment of some diseases that until now have been incurable. The discovery of the regenerative properties of stem cells, both embryonic and adult, has been one of the main advances in Biology in the past years. This fact has contributed to the increase in the number of projects of human stem cell line derivation.

In the process of obtaining stem cell lines for research or clinical application purposes, there are still some problems that should be resolved. Apart from the difficulties of cell line generation, the problems with the culture media and the possibility of tumoral degeneration, one of the main complications of stem cell cultures is the possibility of contamination. The main sources for this to occur are the stem cells themselves, the laboratory workers and the environment in which these cell lines are being handled.

Due to the large quantity of microorganisms that could contaminate these cultures (e.g. bacterias, fungi, viruses and prions), all the stem cell research centres should incorporate an exhaustive microbiological and environmental program in order to both detect the possible pathogens and to avoid the transmission to the recipients.

Any microbial contamination during the manufacturing process presents a serious hazard to recipients. In order to establish the sterility of stem cell cultures, there are

* E-mail: fernando.cobo.sspa@juntadeandalucia.es.

several methods to study the possible presence of pathogens both in cell cultures and in the environment of the manufacturing cell factories.

Until now, the methods used for this purpose are mainly standard protocols based on the culture in liquid or solid growth media and incubated at different temperatures, reflecting conditions for human pathogen culture and environmental microorganisms. Other techniques used are PCR and RT-PCR methods, electron microscopy and in vitro inoculation and in vivo animal inoculation. However, some of these tests are long standing techniques and can take a number of weeks and can prove difficult to implement in the scheduling for release testing of time critical products. In this sense, the technologies based on the use of hybridization chips, using microarrays of immobilized oligonucleotides or antigens/antibodies/genes for microorganisms can provide a rapid and useful methodology for the identification of contaminants.

This review will discuss the methodology that could be used in stem cell research centres to assure the quality and biosafety of cell lines and biotechnological products and avoid the transmission of pathogens like bacteria, virus and prion particles.

INTRODUCTION

The recent advances with respect to the regenerative capacity of stem cells is one of most important facts in the past years in Biology and Medicine. The discovery of the regenerative properties [1] of both, adult and embryonic stem cells, to provide a solid base in research for the future treatment of some diseases that until now have been incurable (e.g. Parkinson disease, degenerative osteoarticular diseases, myocardial infarctation, etc). Since the main purpose of these kind of cells is the application in human beings, the quality and safety of cultures for transplantation should be assured. This should be achieved by means of the standardization of processes involved and the implementation of quality control programs and methods which reflect best practices.

However, these possible new treatments still have several problems such as the possibility of chromosomal alterations, the difficulties of obtention, culture and differentiation, ethical problems in embryonic stem cells and potential for tumorigenicity [2]. In this sense, one of the main problems of these cultures is the possibility of contamination and the transmission of pathogens to the recipients of these cell products. The most common potential forms of contamination are bacterias (including mycoplasma), yeasts and fungi. These microorganisms can be diagnosed on a routine basis [3, 4]. Moreover, some stem cell lines may contain endogenous viruses or can be contaminated with exogenous viruses or prion particles. Viral and prion contamination of cell cultures and feeder cells is the most challenging and serious outcome to diagnose, due to the difficulty involved in virus and prion detection and the potential to cause serious disease in the recipients.

Thus, to ensure the provision of safe and reliable cells for these purposes, it is necessary to regulate the obtention, processing, testing, storage and distribution of all the cells that will be transplanted in the human body [5, 6]. Stem cell research and in particular stem cell banks must assure the quality and safety of the cell products in order to avoid transmissible diseases caused by all the above mentioned microorganisms. These centres should implement adequate quality assurance programs like the microbiological control program. So, accredited stem cell research centres should standardize procedures and protocols used for microbiological testing and in validation programs to assure both the quality and safety of the cells.

The microbiological control program should be dynamic and should include upgraded protocols related to the testing of all microorganisms that could be found in stem cell cultures. The three key areas in this microbiological control are:

1) Clean room environmental control and plans for processing and storage areas.
2) Microbiological control of stem cell lines, "feeder" cells, and other biotechnological products and reagents used in cell culture manufacturing.
3) Microbiological screening of donors of biological material (e.g. embryos, bone marrow, umbilical cord, etc) in order to avoid the transmission of infectious agents.

This chapter will discuss the most appropriate methodology that could be used in stem cell research centres in order to assure the safety and quality of cell and biotechnological products, and intends to give an overview of the microbiological controls that should be carried out on the cell lines in order to avoid the transmission of pathogens to the recipients.

Microbiological Contamination Sources in Stem Cell Cultures

Bacterial, Yeast and Fungi Contamination

The most important source of contamination for bacterias, yeasts and fungi are the contaminated cells that are used as the primary starting material for cell culture. The cell lines recently imported and introduced in the culture laboratory represent the greatest source of contamination, depending on their culture history and past exposure to microorganisms. For this reason, providers of stem cell lines should be able to provide details of passage history and appropriate testing [7] and a "quarantine status" should be established.

Apart from other stem cell lines, another important source of contamination are the laboratory workers who handle the cell cultures. Thus, the most prevalent contaminants will be the microorganisms present in the human skin flora (e.g. *staphylococcus*, grampositive bacilli, *Candida*). Moreover, these microorganisms are dispersed into the air [8], so another potential source of contamination is the environment in which the cultures are carried out, because these microorganisms can be deposited into the cultures.

Finally, other potential sources of microbial contamination in stem cell cultures include culture media and reagents, biological products (bone marrow, embryos, blood umbilical cord), glassware or apparatus (e.g. storage bottles and pipettes) and break-down in aseptic procedures.

Viral Contamination

The primary sources of potential viral contamination in cell cultures are infected animal products used to prepare biological reagents and culture media, contamination during manipulation by laboratory workers and biological products from donors used to obtain the stem cell lines (e.g. embryos, bone marrow, blood umbilical cord, etc.).

Until this moment, in embryonic stem cell cultures the use of "feeder" layers both of animal and human origin is necessary to maintain undifferentiated growth. This requirement provides intimate contact between the embryonic stem cells and the "feeder" cells; this contact could be used as the means to transmit pathogens or bioactive molecules in the final cell product.

With respect to the use of "feeders" of animal origin, above all of murine origin, certain mouse viruses like lymphocytic choriomeningitis virus (LCMV), reovirus-3 and Hantaan virus have been diagnosed in mouse colonies [9] and these viruses have caused serious infection and include fatalities in laboratory staff [10, 11] and may also be transmitted in cell lines and reagents [12]. Moreover, there is also evidence that other mouse viruses like Sendai virus and lactic dehydrogenase virus are capable of infecting human or primates [13]. Other murine viruses like ectromelia virus, Toolan virus, Kilham rat virus, mouse adenovirus, mouse cytomegalovirus, etc., are capable of replicating in vitro in cells of human or primate origin although is not known that cause serious human disease [13] (Table 1).

On the other hand, in the case of use of "feeder" cells of human origin [14], there are numerous microorganisms that could be transmitted to the recipient due to the fact that the human "feeder" cells can be infected by viruses and other pathogens. Some viruses can be transmitted and should be screened like HIV-I/II, HBV, HCV and CMV. Other viruses susceptible of testing are HTLV-I/II, HAV, HEV, and other potentially viral agents that could be transmitted and cause disease include human herpesviruses (HHV-6, HHV-7, HHV-8, EBV, HSV), parvovirus B19, TTV virus and human polyomaviruses (JC and BK virus).

Table 1. Murine viruses that can be transmitted to human cells (Obtained form EMEA, 1998)

Viruses with capacity for infecting humans	Viruses with no evidence for infecting humans
Hantaan virus *	Ectromelia virus *
Lactic Dehydrogenase virus *	K virus
Lymphocytic Choriomeningitis virus *	Minute virus of mice *
Reovirus type 3 *	Mouse adenovirus *
Sendai virus *	Mouse cytomegalovirus *
	Mouse encephalomyelitis virus
	Mouse hepatitis virus
	Mouse rotavirus *
	Polyoma virus
	Thymic virus
	Pneumonia virus of mice *
	Kilham rat virus *
	Rat coronavirus
	Sialoacryoadenitis virus
	Toolan virus *

* Viruses capable of replicating *in vitro* in cells of human origin.

These viruses could remain latent and later some of these viruses could be involved in oncogenic transformation [15].

Moreover, depending on the geographical origin of human "feeder" cells, some pathogens can be transmitted to the recipients and could be susceptible of testing. Several examples of these microorganisms can be the lymphocytic choriomeningitis virus [16], the coronavirus (severe acute respiratory syndrome-SARS), HTLV-3/4 [17] and rabies virus [18]. In this sense, any new pathogen should be considered as a risk factor and a testing should be performed in order to avoid the possible transmission to human biological products.

Prion Particles Contamination

The American and European regulatory agencies [19, 20] have identified the risk of prion particle transmission with the use of all products derived from rumiants (e.g. bovine foetal serum). For this reason, the introduction of techniques for testing the prionic protein (PrP^{Sc}) is necessary. Cell cultures could be an adequate medium to allow the replication of PrP^{Sc} and to maintain the infectivity [21].

On the other hand, several products that can be used to obtain stem cell lines (peripheral blood, bone marrow) could transmit the prion protein to the recipients of the final products [22].

Control of Donors of Biological Products

All human cells (including reproductive cells) could transmit any infectious diseases. So, the procedures and technical criteria to assess are the eligibility of the donors, the laboratory tests required and the criteria for acceptance of cells.

In the recent past, donors have been screened for the human immunodeficiency virus (HIV-1 and 2), the hepatitis B virus (HBV) and the hepatitis C virus (HCV) [23].

However, current regulations require that donor screening includes tests for a range of viruses causing serious human diseases and that are known to be transmitted by blood and tissues (Table 2). For this reason, additional viruses should be included in the screening like HTLV-I/II, CMV, hepatitis A virus (HAV), hepatitis E virus (HEV) which are considered "cell associated viruses". This list could be different in each country where the national guidelines might vary.

Although this list may be expanded even further in light of developing knowledge and technology, it is inevitable that a balance will be drawn between the associated risk of infection and the resources and time required to perform an ever-increasing list of virus tests.

It is known that some viruses, such as herpesviruses [herpes simplex virus (HSV), Epstein-Barr virus (EBV), human herpesvirus (HHV-6, 7, 8), human Polyomaviruses (JC and BK viruses), parvovirus B19 and transfusion transmitted virus (TTV)] remain latent and detectable in humans from early childhood and are potential contaminants of cells from normal, healthy individuals [24-28]. As these viruses are so ubiquitous, there may be no clinical impact of their presence in transplanted tissues and cells for the majority of patients, and in certain cases, such as parvovirus B19, contamination of blood products up to a maximum limit (currently 10^5 genome equivalents per dose for B19) is acceptable for use in humans. However, in some instances, these agents can prove to be of concern.

Table 2. Infections, screening tests and indications for donors of biological cell, tissues and organs (Modified from US Food and Drugs Administration,1999)

Microorganism	Indication	Test
HIV-1/2	Always	HIV-1/2 antibody; MT
Hepatitis B	Always	HBsAg [1, 2]; anti-core HBc; MT
Hepatitis C	Always	HCV antibody; MT
Treponema pallidum	Always	Treponemal-specific antibody (TPHA)
HTLV-I/II	Donor risk factors	Anti HTLV-I/II
CMV	For solid organ, allogenic bone marrow donors	CMV antibody
Toxoplasma	Heart, liver and bone marrow donors	Toxoplasma antibody
Epstein-Barr virus	Donor risk factors	MT
Prions	Donor risk factors	ELISA, W-B, MT
Neisseria gonorrhoeae	Donors of reproductive cells	Bacterial culture
Chlamydia trachomatis	Donors of reproductive cells	Ig G anti-*Chlamydia*

HIV: Human immunodeficiency virus; MT: Molecular techniques (nucleic acid tests); HTLV: Human T lymphocytotropic virus; TPHA: Treponemal Hemagglutination; CMV: Cytomegalovirus; ELISA: Enzyme linked immunoassay; WB: Western-blot.

Some microbial agents have marked variation in their geographical distribution, producing infectious epidemics in different areas like the lymphocytic choriomeningitis virus in USA [29] and the very recent outbreak of severe acute respiratory syndrome (SARS) virus in humans in South East Asia. Furthermore, two new retroviruses (HTLV-3 and HTLV-4) have been recently identified among African bush-meat hunters [30] and recent cases of transplant transmitted disease due to rabies virus in USA and Germany [18] have been reported, which clearly shows the potential emergence of new serious pathogens of the re-emergence of known pathogens.

Accordingly, all infectious agents should be considered as potential pathogens and any new entity that arises should be considered as a contamination risk factor, and a risk balance for the use of contaminated products established, which should include consideration of the recipient`s prior exposure and competence to fight infection. Moreover, specific tests may be required, and these may be developed for surveillance initiatives, and in the case of SARS, for which detection methods are being developed for the causative coronavirus agent [31].

With respect to non-viral agents, currently donor screening may include *Treponema pallidum* [32], and it is inevitable that there will be a requirement to test for a transmissible spongiform encephalopathies (TSE_s), including Creutzfeldt-Jakob Disease (CJD), as sensitive and clinically validated assays become available [23, 33]. Moreover, the Food and Drugs Administrtion is proposing to require that donors of reproductive cells and tissue be tested for *Neisseria gonorrhoeae* and *Chlamydia trachomatis* (which are known to have been transmitted through artificial insemination) and screened for other sexually transmitted and genitourinary disease that could contaminate reproductive cells and tissue during recovery and then be transmitted to the recipient of those cells or tissues (Table 2).

The screening of the majority of microorganisms has been carried out using currently available tests based on the detection of donor antibodies to viral infection, but more recent research has shown that the detection of antibodies exclusively runs the risk of samples for tests being taken during an antibody-negative window period of these infections, where an individual has been exposed to viral infection and indeed can be viraemic [34]. This finding has led to the introduction of nucleic acid amplification techniques such as polymerase chain reaction (PCR) and the retrotranscriptase polymerase chain reaction (RT-PCR) in which the presence of viruses can be observed by means of the amplification of sequences of viral genome. The addition of nucleic acid test methods to the screening of tissue donors, and to the testing of cell lines, could reduce the risks of these infections among recipients of stem cells [35].

MICROBIOLOGICAL DIAGNOSIS IN STEM CELL LINES

Cell lines may become contaminated with a large variety of bacterial, yeasts and fungal agents. Potential sources of contamination include other cell lines, laboratory conditions and workers poorly trained in core areas such as aseptic techniques and good laboratory practice. Most of these contaminants become grossly obvious if the cells are grown in medium without antimicrobial agents. However, the routine use of such agents should be avoidable with good sterile cell culture techniques and clean cultures.

The use of cells and reagents of known origin and quality alone is not sufficient to guarantee the quality of product (cell stock or culture products); it is necessary to demonstrate quality throughout the production process and also in the final product. Screening aids the early detection of contamination since all manipulations are a potential source of contamination. For any reasonable quality control program involving cell lines it is absolutely necessary to include routine culture tests for bacteria, yeast, fungi, virus and prion particles. However, the fundamental element in avoiding contamination and laboratory infection is aseptic technique and this should be a major component of the training in any tissue and cell culture laboratory. Moreover, one of the most important approaches to avoid contamination problems is to avoid the routine use of antimicrobials. These are not a substitute for aseptic techniques and may only mask contamination which may reappear at a later date.

Finally, providers of cell lines should be able to provide details of passage history and appropriate testing [36]. Once the cell lines have been obtained from a reliable source, it is important at the earliest stage to establish a master bank and apply appropriate tests to rule out microbiological contaminants and to confirm the authenticity of cell culture [7].

Bacterial, Yeast and Fungal Contamination

All these contaminants are generally visible to the naked eye and detected by an increase in turbidity and colour change of the culture medium; this change of colour is due to a change in pH. Early detection of contamination is possible due to daily microscopic observation of cell cultures as part of a routine screening in the laboratory. The presence of contamination enable appropriate action to be taken as soon as the signs of contamination become apparent to avoid contamination of other cultures.

In addition to daily culture observation, specific tests for the detection of bacteria and other contaminants must be used as part of a routine quality control screening procedure. It is necessary to detect low level of contamination, so samples from the stem cell cultures and their products should be inoculated either into liquid (e.g. fluid thioglycollate medium, tryptone soya broth) or onto solid (e.g. blood agar, Sabouraud`s dextrose agar, malt extract agar) growth media. European and United States Pharmacopeia have given standard protocols for such testing [3, 4, 37]. These media should be incubated at different temperatures, reflecting conditions for human pathogen culture (e.g. 32-35º C) and environmental microorganisms with lower growth temperature optima (e.g. 25-30º C). These, should be incubated for 7-14 days in microbiological culture incubators, depending on the specific testing standard used. Moreover, the media should be tested using reference strains of potential contaminants.

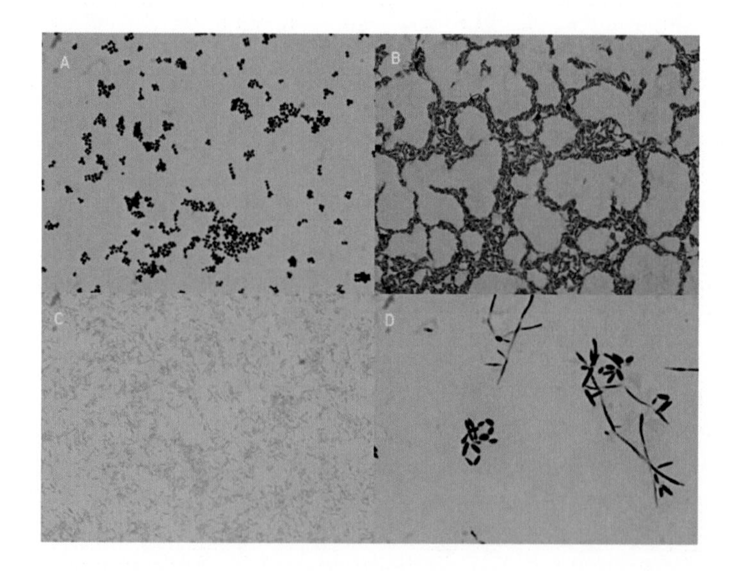

Figure 1. Gram stains in which it can observe: A.- Gram positive cocci (*Staphylococcus epidermidis*). B.- Gram positive bacilli (*Corynebacterium* spp). C.- Gram negative bacilli (*Escherichia coli).* D.- Yeast (*Candida albicans*).

Later, all these microorganisms isolated should be identified and confirmed using confirmatory tests (e.g. Gram stain) (Figure 1). Microbial identification systems are either manual (conventional procedures) or automated.

Automated systems offer the advantage of hands-off approach, allowing more time for the laboratory technician to carry out other duties; manual methods offer the advantage of using the analytical skills of the technologists for reading and interpreting the tests. These procedures should be carried out in the microbiology laboratory isolated from the cell culture laboratory.

Mycoplasma Contamination

Mycoplasma is a generic term given to microorganisms of the order Mycoplasmatales that can infect cell cultures. Those that belong to the families Mycoplasmataceae (*Mycoplasma*) and Acholeplasmataceae (*Acholeplasma*) are of particular interest. This microorganisms are the smallest free-living self-replicating prokayotes (0.3 μm in diameter) and can be observed as filamentous or coccal forms. They lack a cell wall and lack the ability to synthesize one.

The first observation of mycoplasma infection of cell cultures was by Robinson et al in 1956 [38]. Although mycoplasma contamination of primary cultures and continuous cell lines has been known since this date and for several decades, it still represents a significant problem in cell culture. This might be due to the inability of workers to detect these contaminants by microscopic observation. Mycoplasma contamination will also fail to be detected during routine sterility testing for other bacterial, fungi or yeast contaminations due to their fastidious growth requirements.

The incidence of such infection has since been found to vary from laboratory to laboratory. At present, the surveys of cell culture laboratories and cell banks substantiate that on average 15-35% of all cell cultures may be contaminated with mycoplasma [39, 40]. Mycoplasma contamination is usually caused in 98% by seven species: *Mycoplasma hyorhinis*, *Mycoplasma arginini*, *Mycoplasma orale*, *Mycoplasma salivarium*, *Mycoplasma fermentans*, *Mycoplasma hominis* and *Acholeplasma laidlawii* [41].

In cell cultures, Mycoplasmas have shown several effects, including induction of chromosome aberrations [42], induction of morphological alterations (including cytopathology) [43], interference in the rate of growth of cells [44], influence of nucleic acid and amino acid metabolism [45, 46] and induction of membrane alteration and even cell transformation [47, 48].

The contamination mainly spreads from one culture to another, transmitted by aerosols or by poor cell culture practice. Therefore, good laboratory practice and frequent monitoring of the cell lines is mandatory for every laboratory engaged in research using cell cultures [49, 50]. A range of assay techniques is available for the detection of mycoplasma contamination, and it is usually recommended to use at least two techniques for testing cell banks to ensure optimum sensitivity and specificity. These include culture, PCR, indirect DNA staining and non-isotopic detection system.

Mycoplasma Culture

This method of mycoplasma detection is the reference technique and has a theoretical level of detection of 1 colony-forming unit. The culture is carried out in specific solid agar plates (e.g. PPLO, pig serum) and/or liquid media (e.g. PPLO broth, pig serum broth) for 4 weeks in an aerobic and anaerobic atmosphere [4]. Internal negative and positive controls should be carried out by means of using of reference strains. The colonies are observed on agar plates, with some species exhibiting a characteristic "fried egg" appearance (Figure 2). However, there are several strains of mycoplasmas that are non-cultivable (e.g. *M. hyorhinis*).

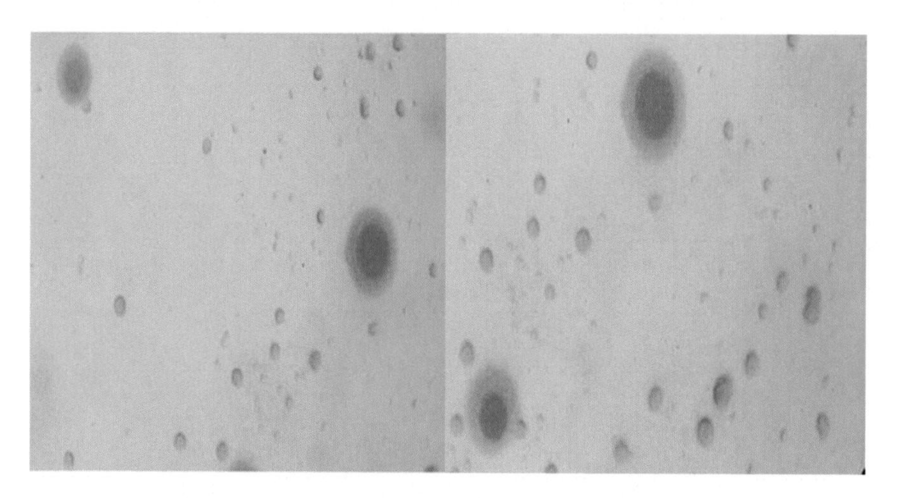

Figure 2. In this microphotograph, some strains of *Mycoplasma* spp with a characteristic "fried egg" appearance are shown (Magnification 10x).

PCR for Mycoplasma Detection

The PCR technology combines a number of advantages such as the sensitivity and specificity; moreover, this is a rapid procedure (above all, real time PCR) which detects defined DNA sequences by amplification of the target DNA sequences and by visualization of the fragment on an ethidium bromide stained gel or by simultaneous amplification-detection in real-time PCR.

PCR technology can be used routinely in both the cell culture and molecular biology laboratory without any training in classical mycoplasma detection [51].

However, this technique can be affected by some inhibitors which may be introduced from cell culture media. The bacterial contamination of the PCR reagents can themselves lead to an amplification that is obviously not backed up in cell culture contaminations, especially in nested PCR methods [52].

For this method, the regions of the genome which have been conserved can be used as targets for the primers for all species detection, as the 16S rRNA coding region [53].

Before testing for mycoplasma, the use of antibiotics in cell culture should be minimized, and the cells should be cultured without antibiotics for several passages, or at least two weeks, to allow the mycoplasmas to grow to detectable amounts.

Non-Isotopic Mycoplasma Detection System

This technique has been tested for a wide range of species of *Mycoplasma*, *Acholeplasma*, *Spiroplasma* and *Ureaplasma*. The sensitivity of detection is low (> 10^5) and it is possible to carry out the detection in 75 minutes. This method uses an all-bacterial probe that detects all species of *Mycoplasma* and *Acholeplasma* which commonly infect the cell cultures, because it uses the principle of nucleic acid hybridization and of ribosomal RNA (rRNA) detection [54]. This technique uses a single stranded DNA probe with a DNA chemiluminescent label which is complementary to the rRNA of the target microorganism. Thus, stable DNA-RNA hybrids are formed and are measured in a luminometer [55].

Indirect DNA Staining for Mycoplasma Detection

Hoechst is a stain that binds to any DNA, and by using a UV fluorescence microscope, the fluorescent nuclei and an extranuclear fluorescence (small cocci or filaments) that correspond to the mycoplasma DNA can be observed. This method provides results within 24 hours, but the sensitivity is very much reduced (10^6 cfu/ml). This sensitivity might be improved by co-culturing the test cell line (indicator cell line such as Vero), which provides a surface to adhere to and grow the mycoplasma (sensitivity of 10^4 cfu/ml).

Viral Diagnosis in Stem Cell Culture

Some stem cell lines and feeder cells might contain endogenous virus or could be contaminated from several sources with exogenous viruses, both from human or animal origin. This type of contamination is one of the most potentially serious problems in stem cell cultures, due to the difficulty involved in virus detection and the potential to cause serious disease in recipients of these cell lines. The establishments which have stem cell lines and feeder cells should assure the quality and safety of these cells and should introduce some protocols in order to detect viral products.

Electron Microscopy

This method allows the direct visualization of viral particles in cell cultures (Figure 3). Electron microscopy became essential in characterizing several new isolates detected in cell cultures [56]. This technique also allows a rapid detection of viruses and other agents if a sufficient amount of particles exists.

The main advantages of electron microscopy are the lack of requirement for viral viability and several kinds of viruses can potentially be seen. Disadvantages include the cost and complexity of maintaining an electron microscope, the need for a skilled operator and relative lack of sensitivity related to the amount of viral particles; a concentration of 10^6 particles/ml is required for visualization [57].

With direct electron microscopy the specimens might be used directly or the viruses might be concentrated before negative staining. Immunoelectron microscopy increases the sensitivity and specificity and is very useful if the number of the virus particles present is small.

Viral Culture

Viruses require living cells in order to replicate. The viral culture is a method that increases the amount of virus, facilitating diagnosis and characterization. This technique provides viable viruses that can be stored for future studies. Moreover, this method allows the detection of several viruses, including those not previously known.

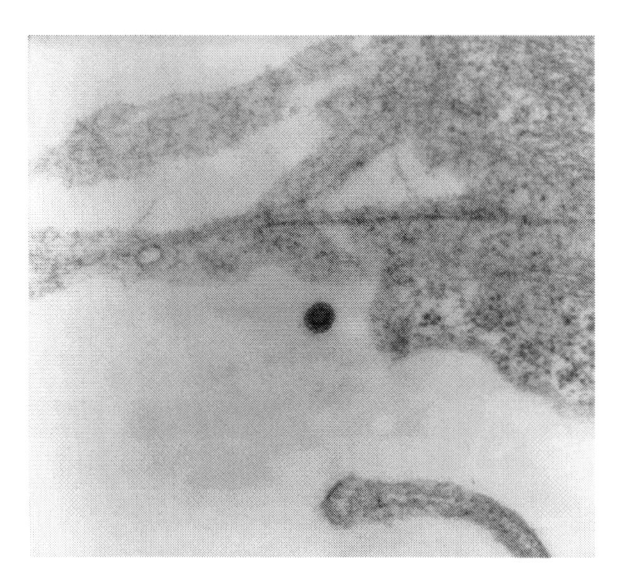

Figure 3. Electron microscopy of murine "feeder" cells, that shows a viral particle corresponding to Orthoretrovirus (possible Murine Leukemia Virus) (Magnification 50,000x).

A cell lysate can be introduced into the cell culture capable of detecting wide ranges of viruses. Usually, a minimum of three cell lines (e.g. MRC-5, Vero, cell used for production) is recommended [58]. Additional kinds of cells might be required depending on several circumstances like the passage history and the cell source. After inoculation, cultures should be incubated at 35-37° C for 14-28 days and observed periodically (e.g. weekly) in order to demonstrate the cytopathic effect. However, in many viruses like cytomegalovirus, the incubation period can be greater than 28 days.

Additional tests (e.g. hemadsorption, hemagglutination) should be used for viruses with minimal visible cytopathic effect [59].

Viral Antigen Detection

This method could provide diagnostic information within a few hours after the cell culture processing. An important advantage is the lack of requirement of viral viability.

Enzyme immunoassay (EIA) is a method that can be applied to diverse specimens; viruses for which antigen EIAs have been widely used are herpes simplex virus, hepatitis B virus and human immunodeficiency virus.

Direct fluorescent antibody staining (DFAs) is a method in which a fluorescent label (usually fluorescein isothiocyanate) is conjugated directly to the antibody that recognizes the viral antigen. In the indirect format, the antiviral antibody is unlabeled and is detected by a second antibody that recognizes immunoglobulins from the animal species of origin of the antiviral antibody. This second antibody carries the fluorescent label. The visualization is by means of a fluorescent light microscope. These methods are widely used for the detection of herpes simplex virus and human cytomegalovirus.

Finally, immunoperoxidase staining is similar to FAs except that peroxidase is used in place of a fluorescent label. In this case, the visualization can be carried out by means of light microscope.

Molecular Methods

These methods detect specific nucleic acid sequences and could be applied to the detection of any virus. Depending on the target, the assays can be specific for a single virus species or for a group of related viruses. These techniques are of special interest for viruses that are difficult or impossible to culture, viruses that grow slowly in culture and viruses for which antigen detection cannot be applied.

Polymerase chain reaction (PCR) is the main method of target amplification assays, and was invented in the eighties by Kary Mullis [60]. In this method, DNA polymerase copies a strand of DNA by elongation of complementary strands initiated from a pair of closely spaced chemically synthesized oligonucleotide primers and includes repeated cycles of amplifying selected nucleic acid sequences. Finally, the PCR product is detected by gel electrophoresis or one of several probe-hybridization techniques. At the moment, a PCR assay (e.g. Real time quantitative PCR) has been developed in which the synthesis of the PCR product is detected in real-time [61]. This method includes both amplification and analysis in the same time and with no need for gels, radioactivity or sample manipulation. Reaction products are detected with a fluorescence detection system. The fluorescence of DNA dyes or probes is monitored each cycle. Thus, the possibility of contamination by amplicons is decreased because the systems are closed, with no handling of the reaction contents after completion of PCR.

Several modifications of the standard PCR assay have been developed. RT-PCR was developed to amplify RNA targets that are first converted to complementary DNA (cDNA) by reverse-transcriptase (RT), and then amplified by PCR. This new technique has played an important role in the diagnosis of RNA viruses [62].

Other molecular techniques have been introduced for the viral diagnosis like multiplex PCR [63], nested PCR [64], nucleic acid probes [65] and Branched DNA signal amplification [66].

The advantages of molecular methods are their high sensitivity, they are easy to set up and have a fast turnaround time. However, the main inconvenience is that for each virus one PCR is necessary, so it is not feasible for the laboratory.

Diagnosis of Prion Particles

The use of bovine foetal serum is the main source of transmission of prion particles to stem cell cultures. Other sources are the use of murine feeder cells and the use of biological products from donors (e.g. bone marrow, embryos, umbilical cord blood) to establish stem cell lines. In stem cell research centres some tests for the diagnosis of the agents of the Transmission Spongiform Encephalopathy (TSE) such as the Creutzfeldt-Jakob disease in cell cultures should be applied. There are several kinds of assays that could be used for this diagnosis.

Prion Particles Antibody Detection

The detection of the protein PrP^{Sc} is essential for the diagnosis of prion diseases. This can be carried out by means of the study of antibodies of the monoclonal IgG1 subtype, anti-PrP 6H4 [67]. This antibody recognizes a specific sequence in the prion protein (human PrP: amino acids 144-152); this sequence is conserved in the majority of mammalian PrP. The anti-PrP 6H4 can be diagnosed using ELISA and/or Western-blot (e.g. Prionics ®). However, other monoclonal antibodies can be used, such as 34C9 that recognize a specific sequence in the bovine prion protein and a polyclonal antibody R029 that also recognizes bovine PrP.

Recently, a new monoclonal anti-prion antibody has been developed by Sanquin ® (clone 1E4). This antibody has the advantage that it is reactive against several species. The antibody 1E4 has been tested by means of some methods such as ELISA, RIA, Western-blot, FACS, EliBlot and immunohistochemistry.

Transgenic Mice In Vivo Assays

This method consists of the injection in transgenic mice of tissues and/or fluids that have unknown infectivity. After this, it can measure the antibody reactions against PrP^{Sc}. This method improves the sensitivity of the above mentioned methods.

Cyclic Amplification of Protein Misfolding

This technique is similar to polymerase chain reaction, involving cyclic amplification of protein misfolding PrP^{Sc} [68]. In this method, new PrP^{Sc} is formed and after cyclic amplification more than 97% of the protease resistant PrP present in the sample correspond to newly converted protein. This method can be applied to diagnose the presence of prion protein in tissues and biological fluids such as blood [69]. This method has a 89% sensitivity and 100% specificity. The implementation of a similar procedure for culture prion detection will contribute to minimizing the risk of contamination with prions.

New Methods on the Future: Microarrays, Protein Arrays and Biosensors

DNA arrays consist of nucleic acid targets immobilised on a substrate of glass, nitrocellulose or nylon membrane. The arrays of high density can have thousands of probes per cm^2 and are known as microarrays. The microarrays synthesised on silicon surface are known as DNA chips.

Microbial diagnostic microarrays consist of nucleic acid probes, and each probe is specific for a strain, species or genus. These platforms can be used as a complement of culture methods for identification of microorganisms [70].

There are two kinds of microarrays: the PCR product based DNA microarrays and the oligonucleotide-based DNA microarrays. In the microarrays based on the PCR product, the first step is the design of primers to amplify specific regions of interest. All probes on a microarray should be highly specific for their target, should bind efficiently to target sequences to allow the detection of low targets, and should display a similar hybridization behaviour. After this, a whole genome PCR amplification should be carried out. The purified PCR products are spotted onto membranes or coated glass slides.

The DNA microarrays based on oligonucleotides use oligonucleotides synthesized on a glass surface. In this case, no reverse transcription or amplification steps are involved. The main advantages of these last microarrays are that there is less likelihood for contamination due to non specific amplification and mishandling, that there is a reduction in cross-hybridization, that it is easier to normalize the oligonucleotide concentrations and that high density oligonucleotide arrays enable high coverage of the genome.

From the microbiological point of view, the main advantage of DNA microarrays for stem cell cultures is the possibility of detection and analysis of hundreds or thousands of microorganisms in a single experiment. Aside from their costs and the difficulties associated with designing and making a suitable array, the majority of the problems with their use are related to quality control, due to the difficulties of standardisation and reproducibility associated with the large number of probes on an array. Moreover, there may be contamination if PCR amplicons are used, as large amounts of these products will be made. There may also be problems with the hybridization reaction.

Other similar new technologies include protein arrays and biosensors. Proteomics include different methods to identify all the proteins present in a cell or tissue at a given time. Protein arrays are prepared with antigens or antibodies bound to a solid phase and used to capture specific antibodies or antigens, respectively [71]. The slides are incubated with serum samples and later with fluorescently labelled secondary antibodies.

A microarray platform of oligosaccharides on nitrocellulose has been developed for capturing carbohydrate proteins, and it may be possible to develop this for microbial diagnosis [72].

Biosensors are defined as small devices which use biological reactions to detect targets [73]. The target bind to a ligand immobilized on a solid phase, and the hybridization of the probe and target is detected by electronic means. Biocatalytic arrays involve an immobilized enzyme being used to recognize the substrate of the enzyme which is the target of the array. The reaction might be recognized either by colorimetric means or by electronic transducer.

Finally, the amount of data generated by microarray experiments is very large. These data require specialised software to assess the patterns. There is no single universally accepted method of statistically analysing microarray data and each method has advantages and disadvantages.

Table 3. Summary of the main contaminants and diagnosis methods in stem cell cultures

Microorganisms	Bacterias	Mycoplasma	Fungi	Yeasts	Viruses	Prion particles
Diagnosis methods	Culture in solid and liquid media Microarrays *	Culture in specific solid and liquid media (PPLO) PCR Indirect DNA staining (Hoechst stain) Non-isotopic detection system Microarrays *	Culture in specific solid and liquid media (Sabouraud agar) Microarrays *	Culture in specific solid and liquid media (Sabouraud agar) Microarrays *	Electron microscopy Cell culture Antigen viral detection Molecular methods In vivo animal inoculation Microarrays *	Detection of antibodies (ELISA, WB) Cyclic amplification of protein misfolding Microarrays *

* Microarrays: need validation for sensitivity and specificity.
PPLO: pleuropneumonia-like organism.
PCR: polymerase chain reaction.
ELISA: Enzyme linked immunoassay.
WB: Western-blot.

Conclusions

Stem cell research centres are the establishments that must guarantee the existence of an appropriate source of cell lines in a standardized way for their use in research and/or human therapies through clinical trials. Moreover, these establishments must assure the quality and biosafety of biological products for use in cell therapy.

One of the main risks associated with the use of cell lines and biological products in stem cell cultures and cell therapy is related to cell contamination. Potential sources of microbial contamination include, reagents, laboratory environment and other cell lines. Routine screening of cell lines helps in the early detection of contamination, since any kind of manipulation is a potential source of contamination. The assurance of the quality of the cell lines requires authentication, characterization and accurate description and to test the possible presence of microorganisms like bacteria (include mycoplasma), virus, prions, fungi and yeast. Selecting and testing of stem cell lines and biotechnological products is one part of a strategy for establishing a microbiological safety program (Table 3).

There are still several problems to be solved before the final and routine application in humans are carried out. While the technology to avoid the animal products in the culture processing is being developed, the safety of the animal and/or human products used in the cell cultures with respect to microorganism contamination should be obtained for application of an exhaustive program of microbiological screening by means of the combination of the above mentioned diagnostic methods.

Acknowledgements

To Ms Angela Barnie for the English correction of this chapter.

References

[1] Thomson JA, Itskovitz-Eldor J, Shapiro SS, Waknitz MA, Swiergiel JJ, Marshall VS, Jones JM. Embryonic stem cell lines derived from human blastocysts. *Science* 1998; 282: 1145-1147.

[2] Takahashi K, Ichisaka T, Yamanaka S. Identification of genes involved in tumor-like properties of embryonic stem cells. *Methods Mol. Biol.* 2006; 329: 449-458.

[3] *European Pharmacopoeia*, 2004a. European Pharmacopoeia Section 2.6.1. (Sterility). Maisonneuve SA, Sainte Ruffine.

[4] *European Pharmacopoeia, 2004b.* European Pharmacopoeia Section 2.6.7. (Mycoplasma). Maisonneuve SA, Sainte Ruffine.

[5] Directive 2004 /23/CE of the European Parliament and the council of March 31st, relating to the establishment of quality and safety norms to donate, to obtain, to assess, to process, to preserve, to store and to distribute cells and human tissues.

[6] United States Department of Health and Human Services. Guidance for human somatic cell therapy and gene therapy. *Food and Drug Administration.* Center for Biologics Evaluation and Research. March 1998.

[7] Stacey GN, Masters JRW, Hay RJ, et al (2000). Cell contamination leads to inaccurate data: we must take action now. *Nature* 403:356.

[8] Whyte W, Bailey PV (1985). Reduction of microbial dispersion by clothing. *J. Parenter Sci. Technol.* 39:51-61.

[9] Kraft V, Meyer B (1990). Seromonitoring in small laboratory animal colonies. A five year survey: 1984-1988. *Z. Ver. Tierkd.* 33: 29-35.

[10] Lloyd G, Jones N (1986). Infection of laboratory workers with hantavirus acquired from immunocytomas propagated in laboratory rats. *J. Infect.* 12: 117-125.

[11] Mahy BW, Dykewicz C, Fisher-Hoch S, Ostroff S, Tipple M, Sanchez A (1991). Virus zoonoses and their potential for contamination of cell cultures. *Dev. Biol. Stand.* 75: 183-189.

[12] Nicklas W, Kraft V, Meyer B (1993). Contamination of transplantable tumors cell lines, and monoclonal antibodies with rodent viruses. *Lab. Anim. Sci.* 43: 296-300.

[13] European Medicines Evaluation Agency (1997). Note for guidance on quality of biotechnological products: viral safety evaluation of biotechnological products derived from cell lines of human or animal origin (CPMP/ICH/295/95). *European Medicines Evaluation Agency,* 7. Westferry Circus, Canary Wharf, London, E14 4H UK.

[14] Genbacev O, Krtolica A, Zdravkovic T, et al. (2005). Serum-free derivation of human embryonic stem cell lines of human placental fibroblast feeders. *Fertil. Steril.* 83: 1517-1529.

[15] Garbuglia AR, Iiezzi T, Capobianchi MR, et al. (2003). Detection of TT virus in lymph node biopsies of B-cell lymphoma and Hodgkin`s disease, and its association with EBV infection. *Int. J. Immunopathol. Pharmacol.* 16: 109-118.

[16] WNV update 2006. 2006 West Nile Virus activity in the United States. Reported to CDC as of October 10, 2006. http://www.cdc.gov/ncidod/dvbid/westnile/index.htm

[17] Wolfe ND, Heneine W, Carr JK, et al. Emergence of unique primate T-lymphotropic viruses among central African bushmeat hunters. *Proc. Natl. Acad. Sci. USA*. 2005; 102: 7994-7999.

[18] Srinivasan A, Burton EC, Kuehnert MJ, Rupprecht C, Sutker WL, Ksiazek TG, Paddock CD, Guarner J, Shieh WJ, Goldsmith C, Hanlon CA, Zoretic J, Fischbach B, Niezgoda M, El-Feky WH, Orciari L, Sanchez EO, Likos A, Klintmalm GB, Cardo D, LeDuc J, Chamberland ME, Jernigan DB, Zaki SR. Transmission of rabies virus from an organ donor to four transplant recipients. *N. Engl. J. Med.* 2005; 352: 1103-1111.

[19] U.S. Food and Drug Administration. CFR (Code of Federal Regulations) section 1271. *Subpart C-Suitability Determination for Donors of Human Cellular and Tissue Based Products,* proposed rule 64 FR 189, 30 September 1999, U.S. Food and Drug Administration, Rockville, MD.

[20] European Medicines Evaluation Agency. EMEA Workshop on application of pharmaceutical assays for markers of TSE: report to CPMP from the BIOT-WP (CPMP/BWP/257/99). *EMEA*, 1999, London.

[21] Solassol J, Crozet C, Lehmann S. Prion propagation in cultured cells. *Br. Med. Bull.* 2003; 66: 87-97.

[22] Peden AH, Head MW, Ritchie DL, Bell JE, Ironside JW. Preclinical vCJD after blood transfusion in a PRNP codon 129 heterozygous patient. *Lancet* 2004; 264: 527-529.

[23] US Food and Drugs Administration. CFR (Code of Federal Regulations) section 1271. *Subpart C-suitability determination for donors of human cellular and tissue based products*, proposed rule 64 Fr 189. US Food and Drugs Administration, 1999, Rockville, MD.

[24] Takeuchi H, Kobayashi R, Hasegawa M, Hirai K. Detection of latent infection by Epstein-Barr virus in peripheral blood cells of healthy individuals and in non-neoplastic tonsillar tissue from patients by reverse transcription-polymerase chain reaction. *J. Virol. Methods* 1996; 58: 81-89.

[25] Cassinotti P, Burtonboy G, Fopp M, Siegl C. Evidence for persistence of human parvovirus B19 DNA in bone marrow. *J. Med. Virol.* 1997; 53: 229-232.

[26] Garbuglia AR, Iezzi T, Capobianchi MR, Pignoloni P, Pulsoni A, Sourdis J, Pescarmona E, Vitolo D, Mandelli F. Detection of TT virus in lymph node biopsies of B-cell lymphoma and Hodgkin`s disease, and its association with EBV infection. *Int. J. Immunopathol. Pharmacol.* 2003; 16: 109-118.

[27] Fanci R, De Santis R, Zakrzewska K, Paci C, Azzi A. Presence of TT virus DNA in bone marrow cells from hematologic patients. *New Microbiol.* 2004; 27: 113-117.

[28] Arnold DM, Neame PB, Meyer RM, Soamboonsrup P, Luinstra KE, O`hoski P, Garner J, Foley R. Autologous peripheral blood progenitor cells are a potential source of parvovirus B19 infection. *Transfusion* 2005; 45: 394-398.

[29] Centers for Disease Control and Prevention. West Nile Virus screening of blood donations and transfusion associated transmission- United States. *MMWR Morb Mortal Wkly Rep.* 2003; 53: 281-284.

[30] Wolfe ND, Heneine W, Carr JK, García AD, Shanmugan V, Tamoufe U, Torimiro JN, Prosser AT, Lebreton M, Mpoudi-Ngole E, McCutchan FE, Birx DL, Folks TM, Burke DS, Switzer WM. Emergence of unique primate T-lymphotropic viruses among central African bushmeat hunters. *Proc. Natl. Acad. Sci. USA* 2005; 102: 7994-7999.

[31] Juang JL, Chen TC, Jiang SS, Hsiong CA, Chen WC, Cheng GW, Lin SM, Lin JH, Chiu SC, Lai YK. Coupling multiplex RT-PCR to a gene chip assay for sensitive and semiquantitative detection of severe acute respiratory syndrome coronavirus. *Lab. Invest.* 2004; 84: 1085-1091.

[32] UK MSBT. *Guidelines on the microbiological safety of human organs, tissues and cells used in transplantation.* Advisory Committee on the Microbiological Safety of Blood and Tissues for Transplantation, MSBT, 2000.

[33] European Medicines Evaluation Agency. EMEA workshop on application of pharmaceutical assays for markers of TSE: report to CPMP from the BIOT-WP (CPMP/BWP/257/99). *EMEA*, 1999, London.

[34] Hitzler WE, Runkel S. Screening of blood donations by hepatitis C virus polymerase chain reaction (HCV-PCR) improves safety of blood products by window period reduction. *Clin. Lab.* 2001; 47: 219-222.

[35] Zou S, Dodd RY, Stramer SL, Strong M. Probability of viremia with HBV, HCV, HIV and HTLV among tissue donors in the United States. *N. Engl. J. Med.* 2004; 351: 751-759.

[36] Stacey GN, Phillips P. Quality assurance for cell substrates. *Dev. Biol. Stand.* 2000; 98: 141-151.

[37] United States Pharmacopeia. Revision 23, Eighth Supplement. Section 1116. United States Pharmacopeial Convention, May 15, 1998.

[38] Robinson LB, Wichelhausen RB, Roizman B. Contamination of human cell cultures by pleuro-pneumonia-like organisms. *Science* 1956; 124: 1147-1148.

[39] Drexler HG, Uphoff CC. Mycoplasma contamination of cell cultures. Incidence, sources, effects, detection, elimination, prevention. *Cytotechnology* 2002; 39: 75-90.

[40] Uphoff CC, Drexler HG. Detection of mycoplasma contaminations. *Methods Mol. Biol.* 2005; 290: 13-23.

[41] McGarrity GJ. Detection of mycoplasmal infection of cell cultures. In: Maramorosch K (ed). *Advances in cell cultures*, 1982, vol. 2, Academic, New York, pp 99-131.

[42] Aula P, Nichols WW. The cytogenetic effects of mycoplasma in human leucocyte cultures. *J. Cell Physiol.* 1967; 70: 281-290.

[43] Butler M, Leach RH. A mycoplasma which induces acidity and cytopathic effect in tissue culture. *J. Gen. Microbiol.* 1964; 34: 285-294.

[44] McGarrity GJ, Phillips D, Vaidya A. Mycoplasmal infection of lymphocyte cultures: infection with *M. salivarium*. In Vitro 1980; 16: 346-356.

[45] Levine EM, Thomas L, McGregor D, Hayflick L, Eagle M. Altered nucleic acid metabolism in human cell cultures infected with mycoplasma. *Proc. Nat. Ac. Sci. USA* 1968; 60: 583-589.

[46] Stanbridge EJ, Hayflick L, Perkins FT. Modification of amino acid concentrations induced by mycoplamas in cell culture medium. *Nature (London) New Biol.* 1971; 232: 242-244.

[47] Wise KS, Cassell GH, Action RT. Selective association of murine T lymphoblastoid cell surface alloantigens with mycoplasma hyorhinis. *Proc. Nat. Ac. Sci. USA* 1978; 75: 4479-4483.

[48] MacPherson I, Russel W. Transformations in hamster cells mediated by mycoplasmas. *Nature* 1966; 210: 1343-1345.

[49] Hartung T, Balls M, Bardouille C, Blank O, Coecke S, Gstramthaler G, Lewis D. Good cell culture practice. ECVAM Good Cell Culture Practice Task Force Report 1. *ATLA* 2002; 30: 407-411.

[50] Microbiological Control Methods. In the *European Pharmacopeia: present and future*. European Directorate for the Quality of Medicines Meeting, Copenhagen, 2003, pp 5-7, and 29-31.

[51] Toji LH, Lenchitz TC, Kwiatkowski VA, Sarama JA, Mulivor RA. Validation of routine mycoplasma testing by PCR. In Vitro Cell. *Dev. Biol. Anim.* 1998; 34: 356-358.

[52] Razin S. DNA probes and PCR in diagnosis of mycoplasma infections. *Moll. Cell Probes*. 1994; 8: 497-511.

[53] Stacey GN. Detection of mycoplasma by DNA amplification. In: Doyle A, Griffitths JB (eds). *Cell and tissue culture for medical research*, chapter 2. Wiley, New York, pp 58-61.

[54] Weisburg WG, Tully JG, Rose DL, Petzel P, Oyaizu H, Yang D, Mandelco L, Sechrest J, Lawrence TG, Van Etten J, Maniloff J, Woese CR. A phylogenetic analysis of the mycoplasmas: basis of their classification. *J. Bacteriol.* 1989; 171: 6455-6467.

[55] Nelson NC, Reynolds MA, Arnold LJ Jr. Detection of acridinium esters by chemiluminiscence. In: Kricka LJ (ed). *Non isotopic probing, plotting and sequencing.* Academic, San Diego, CA, pp 391-428.

[56] Biel SS, Gelderblom HR. Diagnostic electron microscopy is still a timely and rewarding method. *J. Clin. Virol.* 1999; 13: 105-119.

[57] Miller SE. Diagnosis of viral infections by electron microscopy. In: Lennette EH, Lennette DA, Lennett ET (eds). *Diagnostic procedures for viral, rickettsial, and chlamydial infections*. American Public Health Association, Washington, DC, 1995, pp 37-78.

[58] Schiff LJ. Review: production, characterization, and testing of banked mammalian cell substrates used to produce biological products. *In Vitro Cell Dev. Biol. Anim.* 2005; 41: 65-70.

[59] Ayala MA, Laborde J, Milocco S, Carbone C, Cid de la Paz V, Galosi CM. Development of an antigen for the diagnosis of Kilham rat parvovirus by hemagglutination inhibition test. *Rev. Argent Microbiol.* 2004; 36: 16-19.

[60] Mullis K, Faloona FA. Specific synthesis of DNA in vitro via a polymerase-catalyzed reaction. *Methods Enzymol.* 1987; 155: 335-350.

[61] Heid CA, Stevens J, Livak KJ, Williams PM. Real time quantitative PCR. *Genome Research* 1996; 6: 986-994.

[62] Young KK, Resnick RM, Meyers TW. Detection of hepatitis C virus by a combined reverse transcription-polymerase chain reaction assay. *J. Clin. Microbiol.* 1993; 31: 882-886.

[63] Dineva MA, Candotti D, Fletcher-Brown F, Allain JP, Lee H. Simultaneous visual detection of multiple viral amplicons by dipstick assay. *J. Clin. Microbiol.* 2005; 43: 4015-4021.

[64] Erlich HA, Gelfand D, Sninsky JJ. Recent advances in the polymerase chain reaction. *Science* 1991; 252: 1643-1651.

[65] Denniston KJ, Hoyer BH, Smedile A, Wells FV, Nelson J, Gerin JL. Cloned fragment of the hepatitis delta virus RNA genome: sequence and diagnostic application. *Science* 1986; 232: 873-875.

[66] Urdea MS, Horn T, Fultz TJ, Anderson M, Running JA, Hamren S, Ahle D, Chang CA. Branched DNA amplification multimers for the sensitive, direct detection of human hepatitis viruses. *Nucleic. Acids. Symp. Ser.* 1991; 24: 197-200.

[67] Enari M, Flechsig E, Weissmann C. Scrapie prion protein accumulation by scrapie-infected neuroblastoma cells abrogated by exposure to a prion protein antibody. *Proc. Natl. Acad. Sci. USA* 2001; 98: 9295-9299.

[68] Saborio GP, Permanne B, Soto C. Sensitive detection of pathological prion protein by cyclic amplification for protein misfolding. *Nature* 2001; 411: 810-813.

[69] Castilla J, Saá P, Soto C. Detection of prions in blood. *Nat. Med.* 2005; 11: 982-985.

[70] Vianna ME, Horz HP, Gomes BP, Conrads G. Microarrays complement culture methods for identification of bacteria in endodontic infections. *Oral Microbiol. Immunol.* 2005; 20: 253-258.

[71] Holt LJ, Büssow K, Walter G, Tomlinson IM. By-passing selection: direct screening of antibody-antigen interactions using protein arrays. *Nucleic. Acids Res.* 2000; 28: e72.

[72] Fukui S, Feizi T, Galustian C, Lawson AM, Chai W. Oligosaccharide microarrays for high-throughput detection and specificity assignments of carbohydrate-protein interactions. *Nat. Biotechnol.* 2002; 20: 1011-1017.

[73] Wang J. From DNA biosensors to gene chips. *Nucleic. Acids Res.* 2000; 28: 3011-3016.

In: Stem Cell Research Trends
Editor: Josse R. Braggina, pp. 209-226
ISBN: 978-1-60021-622-0

Chapter VII

EXPRESSION OF NEURONAL STEM CELLS AND NEUROGENESIS AFTER TRAUMATIC BRAIN INJURY

Tatsuki Itoh[1*], Takao Satou[1, 2, 3], Shigeo Hashimoto[4] and Hiroyuki Ito[1]

[1]Department of Pathology, Kinki University School of Medicine, Osaka, Japan
[2]Division of Hospital Pathology, Hospital of Kinki University School of Medicine, Osaka, Japan
[3]Division of Sports Medicine, Institute of Life Science, Kinki University, Osaka, Japan
[4]Department of Pathology, PL Hospital, Osaka, Japan

ABSTRACT

Traumatic brain injury (TBI) usually occurs as a result of a direct mechanical insult to the brain, and induces degeneration and death in the central nervous system (CNS). CNS disorder can be caused by widespread neuronal and axonal degeneration induced by TBI. It was originally thought that recovery from such injuries was severely limited because neuronal loss and degeneration in the adult brain were irreversible in the mammalian nervous system. However, recent studies have indicated that the mammalian nervous system has the potential to replenish populations of damaged and/or destroyed neurons via proliferation of neural stem cells (NSCs). NSCs have been identified in adult mammals and have the potential to differentiate into either glial or neural phenotypes. However the proliferation, differentiation and neurogenesis of NSCs at damaged regions after TBI remain unclear. We investigated the chronology of the occurrence and differentiation of NSCs around cerebral cortical damaged areas after traumatic brain injury in the rat.

* Corresponding author: Tatsuki Itoh, Address: 377-2, Ohno-higashi, Osakasayama-city, Osaka 589-8511, Japan. Email: tatsuki@med.kindai.ac.jp. Tel: +81-72-366-0221. Fax: +81-72-360-2028.

NSCs, isolated and cultured from damaged cerebral cortical brain tissue after rat TBI, were confirmed to have the potential to differentiate into neurons and glial cells, although it was unclear if NSCs migrated to the damaged area from the subventricular zone (SVZ) or the subependymal zone (SEZ) at the dentate gyrus (DG)-hilus interface, and if astrocytes showed blastogenesis in the reactivity after injury. Moreover, *in vivo*, there were immature neurons among the gliosis and glial scars around the damaged area at 7days, in decreasing stages of NSCs, after rat TBI. Additionally, we confirmed that some immature neurons matured and survived among the glial scars at 30days after injury. Our data suggested that promotion of maturation and differentiation of newly formed immature neurons around a damaged area may improve brain dysfunction induced by glial scars after brain injury.

Keywords: traumatic brain injury; neural stem cell; glial scar.

Introduction

Traumatic brain injury (TBI) occurs as a result of a mechanical insult to the brain and this induces degeneration and death in the central nervous system [1, 2]. Following the initial mechanical insult, secondary pathways are activated that contribute to ischemic damage induced by circulatory disturbance, blood-brain-barrier disruption and excitotoxic damage [3, 4]. These results suggest that central nervous disorder can be caused by neuronal and axonal degeneration induced by TBI [1, 2]. It was thought that recovery from these injuries was severely limited because the neuronal loss and degeneration in the adult brain was irreversible in the mammalian nervous system. However, recent studies have indicated that the mammalian nervous system has the potential to replenish the population of damaged and /or destroyed neurons, by means of proliferation of neural stem cells [5]. Neural stem cells(NSCs) have been identified in adult mammals and have the potential to differentiate into either glial or neural phenotypes [6, 7].

NSCs were confirmed at two portions in the adult rodent brain. One is the subventricular zone (SVZ) of the lateral ventricles [8]. The other location is the subgranular zone (SGZ) at the dentate gyrus (DG) -hilus interface. Thus, a constant slow rate of neurogenesis is occurring all the time in these areas of adult brain [7, 9].

Following insults to the brain such as brain trauma, seizures and ischemia, there is an increase in the number of mitotically active cells in the brain. In these models, an upregulation of dividing cells is observed in the SVZ or SGZ [1, 2, 9]. Nestin is a NSC marker [10, 11], and nestin positive cells were found to migrate into the granule cell layer where they differentiated into granule neurons, and these cells contributed to neurogenesis after injury [1, 2, 9].

Recent studies using the controlled cortical impact (CCI) and the lateral fluid percussion injury model of TBI have demonstrated an increase of activated cells and NSCs after injury and showed an increase in neurogenesis in the DG / SGZ [12]. However, it is unclear if those nestin-positive cells have the potential to differentiate into neurons or glia and contribute to neurogenesis after injury.

Doublecortin (DCX) is a microtubule-associated protein that is specifically expressed in virtually all migrating neuronal precursors of the developing CNS[13, 14]. In the adult brain, DCX expression is retained within areas of continuous neurogenesis, i.e. the SVZ and SGZ [14, 15]. Following an insult to the brain, such as brain trauma, seizures and ischemia, there is an increase in the number of mitotically active cells in the brain. In these models, upregulation of the number of dividing cells is observed in the SVZ or SGZ[1, 2, 9]. The increased number of divided cells in the SVZ or SGZ are newly formed immature neurons and express DCX [1,12,13,16]. Due to its association with the neurogenic processes, there has been recent interest in DCX as a candidate marker for neurogenesis [13].

On the other hand, TBI to the adult CNS results in a rapid response from resident astrocytes, a process often referred to as reactive astrocytes or glial scarring [17,18]. Glial scars have been reported to inhibit neurite elongation of damaged neurons and axonal regeneration, and thus prevent functional recovery [17,18]. Furthermore, neurite outgrowth of cultured rat hippocampal neurons was found to be inhibited by glial scars *in vitro* [19]. However, gliosis and glial scars protect against secondary insults that contribute to the ischemic damage induced by circulatory disturbance, blood-brain barrier disruption, excitotoxic damage and free radicals [20]. Moreover, reactive astrocytes secrete neurotrophic factor (NTF), nerve growth factor (NGF) and extracellular matrix, which induce axonal outgrowth and regeneration of the neural network [21, 22]. In addition, the formation of glial scars prevents the secreted factors from leaking to the outside and separates non-damaged areas from damaged areas, thereby maintaining the normal homeostasis of the CNS [23,24]. Therefore, it remains controversial whether gliosis and glial scars are beneficial or noxious for neurogenesis after brain injury. However the proliferation, differentiation and neurogenesis of NSCs at damaged regions after TBI remain unclear.

In this study, we investigated the chronology of the occurrence and differentiation of NSCs around cerebral cortical damaged areas after traumatic brain injury in the rat.

Material and Methods

Surgical Procedure

Male Wistar rats (10 weeks old, 200-250 g in weight) were anesthetized by intraperitoneal pentobarbital (50 mg/kg) injection. The scalp was incised on the midline and the skull was exposed. A 2-2.5-mm hole was drilled in the right parietal calvaria. Brain injury above the dura matter was then inflicted with a pneumatic control injury device [25, 26] at an impact velocity of 4 m/sec, with an impact tip diameter of 1 mm and at a fixed impact deformation of 2 mm depth from the cerebral surface. As a control, age matched no injury sham-operated rat were used.

Immunohistochemistry

At 1, 3, 7 and 30 days after the TBI, 7 rats were sequentially perfused intracardially with 300 ml of 0.1 M phosphate-buffered saline (PBS; pH 7.4-7.5), followed by 300 ml of 4% paraformaldehyde (PFA) in PBS (pH 7.4-7.5). The brains were removed and stored in PFA for 3 days, and then the maximum size of the lesion was sliced into five serial coronal sections (50 μm thick) using a Micro slicer (Dousaka EM, Kyoto, Japan). The first section was used for nestin immunostaining, followed by counterstaining with hematoxylin to count the number of positive cells. The second section was used for DCX immunostaining, followed by counterstaining with hematoxylin. The third section was used for double-immunofluorescence staining for DCX and glial fibrillary acidic protein (GFAP). The fourth section was used for double- immunofluorescence staining for DCX and NeuN. The fifth section was used for double-immunofluorescence staining for nestin and GFAP.

Immunostaining for Nestin

Each first section was treated with 3% H_2O_2 in Tris-buffered saline (TBS; 0.1M Tris-HCl, pH 7.5, 0.15 M NaCl) containing 0.1% Triton X-100 (TBS-T) for 30 min. Next, the sections were washed three times with TBS-T, blocked with 3% bovine serum albumin (BSA; Sigma, St. Louis, MO) in TBS-T for 30 min, and incubated with a monoclonal antibody against rat nestin (1:1000 dilution in blocking solution; BD Biosciences Pharmingen, San Diego, CA), an NSC marker, overnight at room temperature. Following extensive washing, the sections were further incubated with a HISTIFINE Rat-PO (multi)-kit (Nichirei, Osaka, Japan), consisting of a mixed solution of peroxidase-conjugated anti-mouse and rabbit IgG as the secondary antibody, for 60 min at room temperature. Labeling was visualized using diaminobenzidine (DAB; Vector Peroxidase Substrate Kit; Vector Laboratories, Burlingame, CA) for 5 min, and the sections were counterstained with hematoxylin to count the number of positive cells.

Immunostaining for DCX

Each second serial section was subjected to DCX immunostaining using the above-described method. The primary antibody was a polyclonal goat anti- DCX antibody (1:1000 dilution; Santa Cruz Biotechnology, Santa Cruz, CA) and the secondary antibody was a HISTIFINE Rat-PO (goat)-kit (Nichirei), consisting of a peroxidase-conjugated anti-goat IgG solution.

Counting the Numbers of Nestin- and DCX-Positive Cells

To determine the total numbers of nestin- and DCX-positive cells per section, the DAB-labeled cells were counted around the damaged cortical regions of the sections under a microscope with an objective magnification of 20x.

Double- Immunofluorescence Staining for DCX and GFAP

Each third serial section was washed with TBS-T, blocked with 3% BSA in TBS-T for 30 min, and incubated with a polyclonal goat anti-DCX antibody (1:300 dilution; DAKO) overnight at room temperature. Following extensive washing, the sections were further incubated with a polyclonal fluorescein isothiocyanate (FITC) -conjugated anti-goat IgG antibody (1:300 dilution; DAKO) for 80 min at room temperature. Next, the DCX-stained sections were washed extensively and incubated with a polyclonal rabbit anti-GFAP antibody (1:300 dilution; DAKO, Kyoto, Japan) overnight at room temperature. Following extensive washing, the sections were further incubated with a polyclonal rhodamine-conjugated anti-rabbit IgG antibody (1:300 dilution; DAKO) for 80 min at room temperature. Subsequently, the sections were observed by fluorescence microscopy (Nikon E-800; Nikon, Tokyo, Japan).

Double- Immunofluorescence Staining for DCX and NeuN

Each fourth serial section was subjected to DCX immunofluoresecnce staining as described above. Next, the DCX-stained sections were washed extensively and sequentially incubated with a monoclonal mouse anti-NeuN monoclonal antibody (1:300 dilution; Chemicon, Temecula, CA) followed by an a polyclonal rhodamine -conjugated anti-mouse IgG rabbit polyclonal antibody (1:300 dilution; DAKO). Subsequently, the sections were observed by fluorescence microscopy (Nikon E-800; Nikon).

Double-Immunofluorescence Staining for Nestin and GFAP

Each fifth serial section was subjected to GFAP immunofluoresecnce staining as described above. Following washing in TBS, the sections were sequentially incubated with a monoclonal mouse anti-rat nestin antibody (1:300 dilution; BD Biosciences Pharmingen), followed by a FITC-conjugated anti-mouse IgG (1:300 dilution; DAKO) for 80 min at room temperature. All antibodies were diluted in blocking solution. Subsequently, the sections were observed by fluorescence microscopy (Nikon E-800; Nikon).

Isolation and Culturing of NSCs

NSC culture methods were as described by Weiss et al. [6] and Yanagisawa et al. [27] . At 1 day, 3days and 7 days after TBI, only cerebral cortex the size of 2mm diameter from the center of lesion (Figure.1) was separated by gross dissection under a dissecting microscope. Care was taken to remove and discard the meninges and blood vessels. Two cerebral tissue samples from two individual rats were used as a single experimental group for one time. The tissue was then cut into small pieces and dissociated by incubation in Hanks' Balanced Salt Solution (HBSS) containing 0.1 % (w/v) trypsin and 0.01 % (w/v) DNase 1 at 37°C for 30 min. Subsequently, an equal volume of Fetal Calf Serum was added to the tissue suspension and centrifuged at 1300 rpm for 5 min. The supernatant was removed, HBSS was added, the cells were dissociated by trituration, and centrifuged at 1300 rpm for 5 min. They were plated as a single-cell suspension on ornithine and fibronectin-coated 60mm culture dishes in a plating medium (N2/DF) consisting of Dulbecco's modified Eagle's medium (DMEM) / F-12 medium supplemented with bFGF (20 ng/ml), EGF (20 ng/ml), insulin (25 μg/ml), transferrin

(100 μg/ml), and progesterone (20 nM), and maintained at 37°C in a humidified 5% CO_2 atmosphere for 3 days.

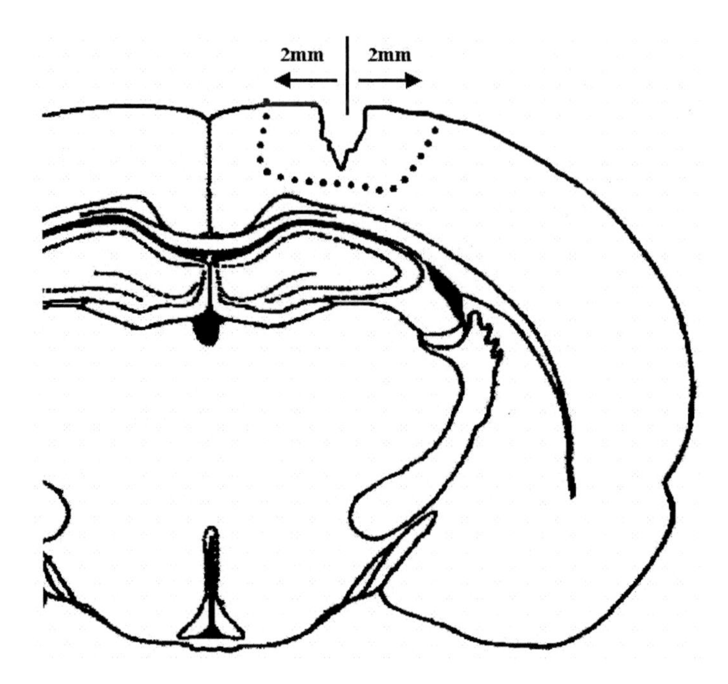

Figure 1. Schematic representation indicating the area used for isolation. The area used for isolation is surrounded by the dotted line.

After 4 days, the cells attached to the bottom of the culture dishes were desquamated by trituration then collected and dissociated by trituration in N2/DF medium. The single-cell suspension was replated on non-treated culture dishes, and after 10 to 14 days of culture, the first spheres formed. First spheres were dissociated by trituration in PBS then washed twice in PBS. The single-cell suspension was replated on non-treated culture dishes, and after 7 to 10 days of culture, secondary spheres formed. The secondary spheres were cultured on a ornithine-coated round shaped cover glass ($1cm^2$) in each wells of 24 well culture plates at 37°C in a humidified 5% CO_2 atmosphere for 4 days without bFGF and EGF.

Immunostaining of Cultured Cells

Secondary spheres and 4-day cultures were fixed using 4% PFA in PBS for 20 minutes and immunostaining was performed. The secondary spheres were stained with anti rat nestin monoclonal antibody and anti monoclonal Tuj1 antibody (Chemicon, dilution 1:300), anti polyclonal GFAP antibody (DAKO, dilution 1:300) or anti monoclonal O4 antibody (Chemicon, dilution 1:300), which is marker of oligodendrocyte, followed by a secondary antibody for FITC conjugated anti-mouse IgG, M or rhodamine conjugated anti-rabbit IgG (DAKO, dilution 1:500). All antibodies were diluted in PBS containing 3% bovine serum albumin. The 4 day cultured cells were used for fluorescence double-labeling with Tuj1 and GFAP. Culture experiments were performed three times.

Statistical Analysis

The numbers of nestin- and DCX-positive cells were expressed as means ± SD. Statistical analysis was performed using one-way analysis of variance (ANOVA) with Fisher's post-hoc test. $P < 0.05$ was considered statistically significant.

Results

Expression of Nestin and the Number of Nestin-Positive Cells after TBI

The immunostaining results for nestin expression around the damaged area after TBI are shown in Figure 2A-C. In the sham operated cerebral cortex, there were no nestin-positive cells (Figure 2A). A few small nestin-positive cells were present around the damaged area at 1 day after TBI. At 3 days after TBI, there were many nestin-positive cells around the damaged area and these cells possessed a nestin-immunopositive cytoplasm and projections similar to the shape of small astrocytes (Figure 2B).

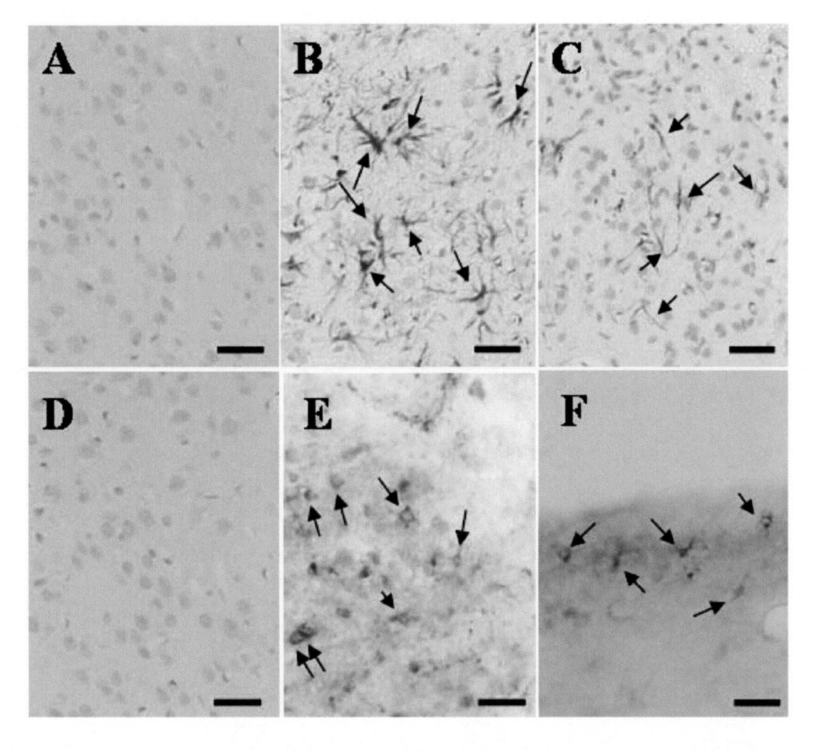

Figure 2. Immunostaining for nestin and DCX with DAB colorization around the damaged cerebral cortex after traumatic brain injury (TBI) in the rat. In the sham operated control, there are no cells immunopositive for nestin (A) or DCX (D). At 3 days after TBI, abundant nestin immunoreactivity is mainly present in the cytoplasm and projections (B, arrows). At 7 days after TBI, nestin-immunopositive elongating fibers are seen (C, arrows). DCX-immunoreactivity is mainly present in the cytoplasm and projections at 7 (E, arrows) and 30 (F, arrows) days after TBI. Scale bars = 50 μm.

There were a few nestin-positive fibers among the cells around the damaged area at 7 days after TBI (Figure 2C). However, there were no nestin-positive cells around the damaged area at 30 days after TBI.

Figure 3A shows the numbers of nestin-positive cells. The sham operated cerebral cortex did not contain any nestin-positive cells. However, there were a few positive cells (3.6 ± 2.4) at 1 day after TBI. Moreover, the number of nestin-positive cells was significantly higher at 3 days after TBI (69.2 ± 25.8) than at 1 day after TBI and reached the maximum number ($P < 0.001$). The number of nestin-positive cells was significantly decreased at 7 days after TBI (12.2 ± 3.3) compared with 3 days after TBI ($P < 0.001$). The cerebral cortex did not contain any nestin-positive cells at 30 days after TBI.

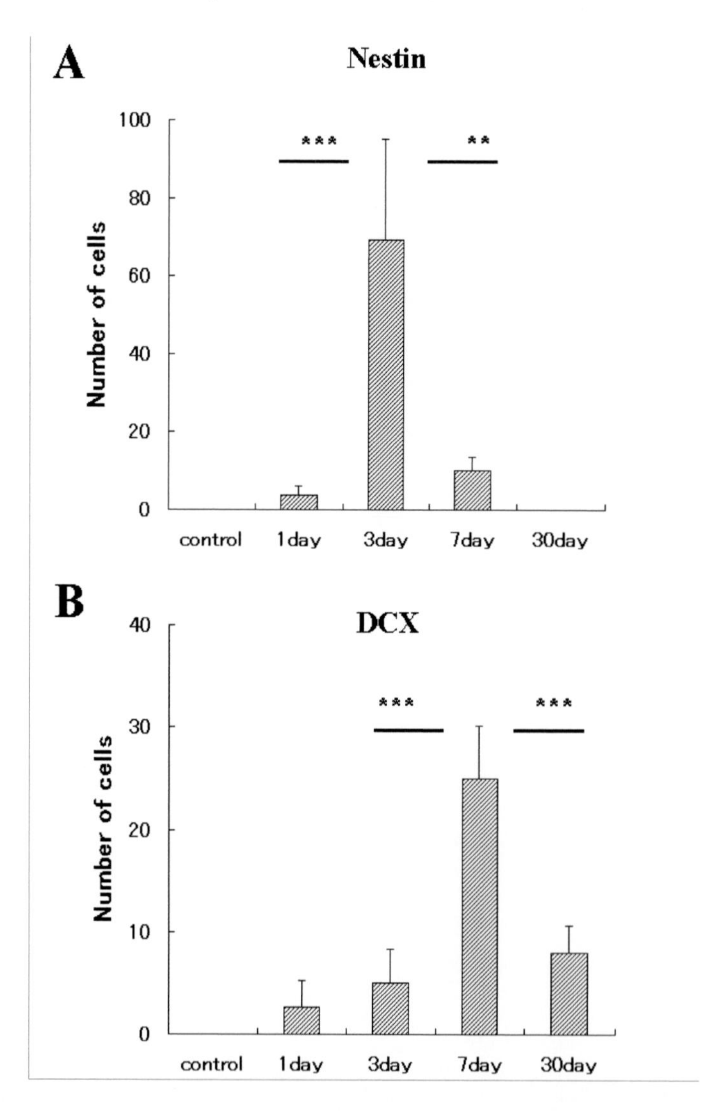

Figure 3. Graphs showing the numbers of cells immunopositive for nestin (A) and DCX (B) around the damaged cerebral cortex after traumatic rat brain injury. The results are shown as means ± SD. **: $P < 0.01$; ***: $P < 0.001$; n = 5.

NSC Isolation and Immunostaining

At 1 day and 7days after TBI, the isolation of spheres was impossible from around the cerebral cortical damaged area (Figure 4A and C). However, at 3 days after injury, spheres were isolated and cultured (Figure 4B). A phase-microscopic image of the secondary sphere is shown in Figure 4D. Almost all aggregated cells in the secondary sphere showed nestin immunopositivity (Figure 4E). In addition, the secondary spheres were not immunopositive for Tuj1 and vimentin.

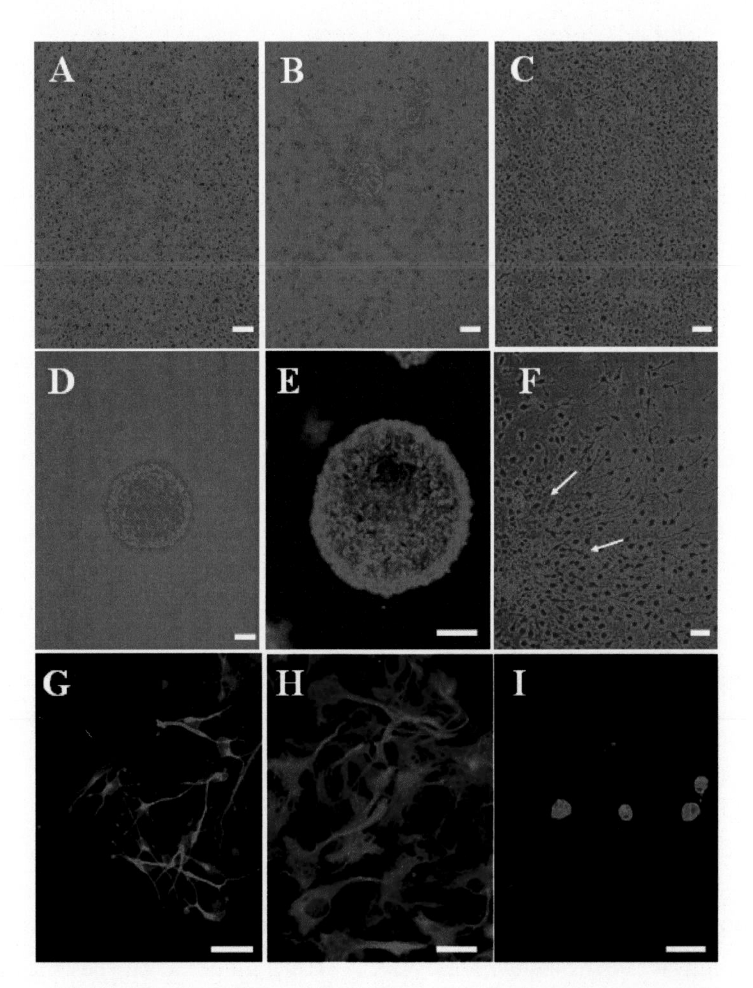

Figure 4. Photomicrographs of isolated and cultured cells originated from surrounding tissue of traumatic brain. A, B, C, D and F were a phase-contrast microscopic image. E, G, H and I were a FITC(E, G, I) or rhodamine(H) fluorescent microscopic image. There were no spheres in the culture derived from the 1 day tissue (A). There were a few spheres in the culture derived from the 3 days tissue (B). There were no spheres in the culture derived from the 7 days tissue (C). Isolated the secondary spheres from around damaged cerebral cortex after 7 days of culture (D). Isolated secondary sphere showing nestin immunoreactivity (green, E). Neurosphere showing cells differentiated after 4 days of culture in medium without bFGF and EGF. The arrow shows a neurosphere (F). A double-labeling fluorescence immunostaining showing many Tuj1 immunoreactive cells (green, G), GFAP immunoreactive cells (red, H) and O4 immunoreactive cells (green, I). G and H were taken from the same field of view which shows no co-localization. Scale bar = 50 μm.

Differentiation of the Cultures

Phase-microscopic images of cultured and differentiated cells after 4 days in culture are shown in Figure 4F. The secondary spheres immediately attached to the bottom of the cell culture dishes. Fibers from the secondary sphere elongated chronologically. After 4 days in culture, there were many cells and elongated fibers from the secondary sphere (Figure 4F). Tuj1 immunopositive labeling was found in the cytoplasm and elongated fibers in the cultures (Figure 4G). In addition, there were also cells with GFAP immunopositive labeling in the cytoplasm (Figure 4H). However, Tuji1 immunopositive labeling did not co-localize with GFAP immunopositive labeling (Figure 4G and H). Furthermore, there were cells with O4 immunopositive labeling in the cytoplasm (Figure 4I).

Expression of DCX and the Number of DCX-Positive Cells after TBI

The immunostaining results for DCX expression around the damaged area after TBI are shown in Figure 2D-F. In the sham operated cerebral cortex, there were no DCX-positive cells (Figure 2D). There were a few DCX-positive cells with DCX staining in their cytoplasm around the damaged area at 1 and 3 days after TBI. There were many DCX-positive cells with positive staining of their cytoplasm and projections around the damaged area at 7 days after TBI (Figure 2E). Furthermore, there were still DCX-positive cells with DCX staining in their cytoplasm and projections in the vicinity of lesion at 30 days after TBI (Figure 2F).

Figure 3B shows the numbers of DCX-positive cells. The sham operated cerebral cortex did not contain any DCX-positive cells. There were small numbers of positive cells at 1 and 3 days after TBI (2.6 ± 2.7 and 5 ± 3.3, respectively). Moreover, the number of DCX-positive cells was significantly higher at 7 days after TBI (25 ± 5.1) than at 3 days after TBI and reached the maximum number ($P < 0.001$). The number of DCX-positive cells was significantly decreased at 30 days after TBI (8.0 ± 2.6) to one-third the number at 7 days after TBI ($P < 0.001$). The number of DCX-positive cells was significantly higher in the TBI group than in the sham operated group at 30 days after TBI ($P < 0.001$).

Double-Immunofluorescence Staining for GFAP and Nestin

There were many GFAP-positive reactive astroglia around the damaged area at 3 and 7 days after TBI (Figure 5A and D). There were many small GFAP-positive cells with a nestin-positive cytoplasm and fibers around the damaged area at 3 days after TBI (Figure 5B and C). A small number of nestin-positive cells also showed GFAP-positivity at 7 days after TBI (Figure 5E and F).

Double- Immunofluorescence Staining for DCX and GFAP

There were many larger GFAP-positive cells with GFAP staining in their cytoplasm and long elongated projections around the damaged area at 7 days after TBI (Figure 6B). DCX-positive cells (Figure 6A) coexisted among the GFAP-positive cells (Figure 6B) around the damaged area, but the DCX staining did not co-localize with the GFAP staining (Figure 6C).

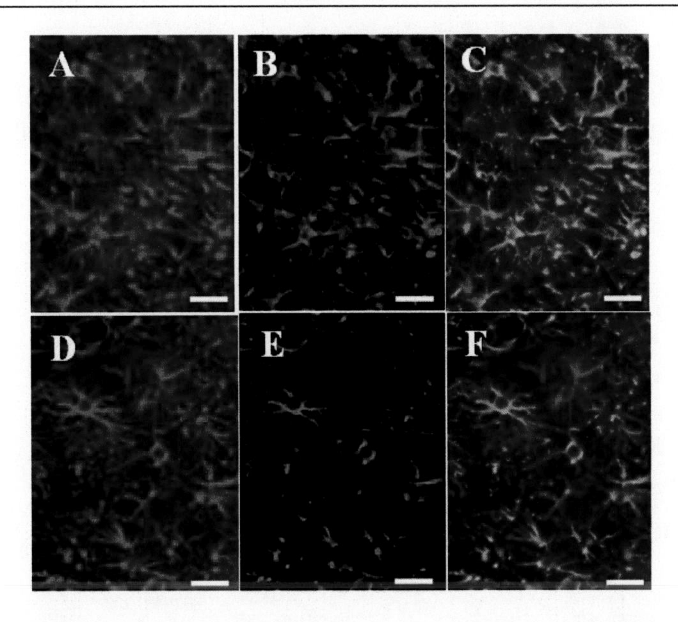

Figure 5. Double-immunofluorescence staining showing GFAP localization with rhodamine colorization and nestin with FITC colorization around the damaged area after traumatic brain injury (TBI). There are many GFAP-immunopositive cells at 3 (red, A) and 7 (red, D) days after TBI. There are a few nestin-immunopositive cells at 3 (green, B) and 7 (green, E) days after TBI. GFAP and nestin co-localize. Merged images of A and B, and D and E are shown C and F (yellow). Scale bars = 50 μm.

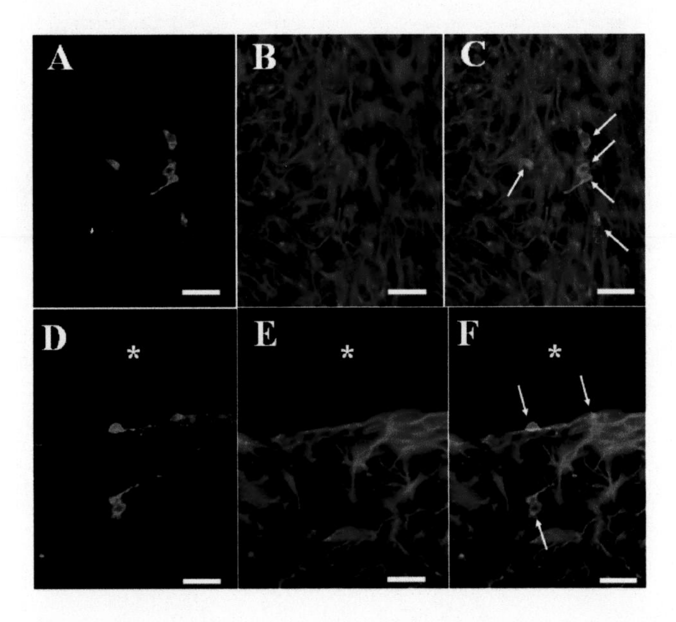

Figure 6. Double- immunofluorescence staining for DCX with FITC colorization and GFAP with rhodamine colorization around the damaged area after traumatic brain injury (TBI). At 7 days after TBI, a few DCX-immunopositive cells (green, A) coexist among the gliosis induced by reactive astrocytes (red, B). DCX and GFAP do not co-localize. A merged image of A and B is shown C. At 30 days after TBI, a few DCX-immunopositive cells (green, D) coexist among the glial scars (red, E). DCX and GFAP do not co-localize. A merged image of D and E is shown F. * shows the damaged region. Scale bars = 50 μm.

There were many GFAP-immunopositive fibers that formed glial scars at 30 days after TBI, and these fibers were enriched at the damaged region (Figure 6E). DCX-positive cells (Figure 6D) existed among the glial scars (Figure E), but the DCX staining did not co-localize with the GFAP staining (Figure 6F).

Double- Immunofluorescence Staining for DCX and NeuN

DCX immunostaining (Figure 7A) did not co-localize with NeuN immunostaining (Figure 7B) at 7 days after TBI (Figure 7C). A small number of nuclei in DCX-immunopositive cells (Figure 7D) co-localized with NeuN immunostaining (Figure 7E) at 30 days after TBI (Figure 7F).

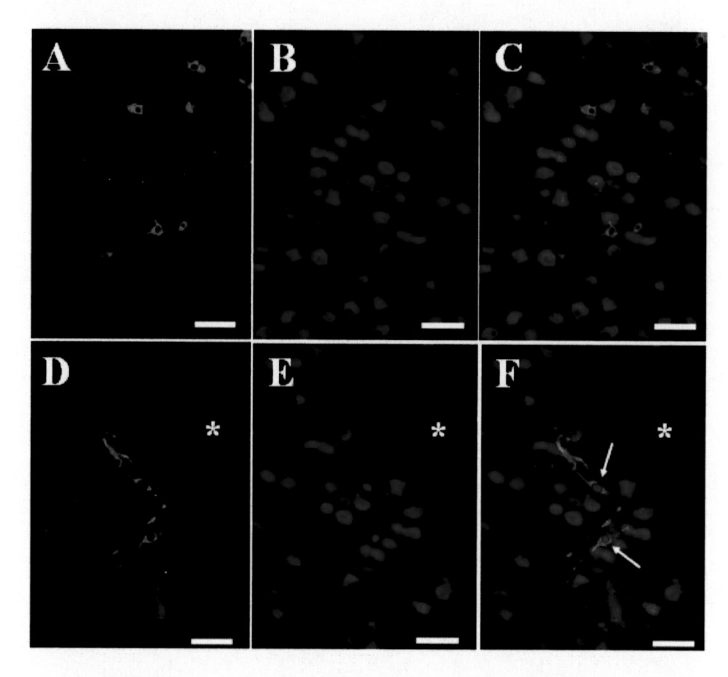

Figure 7. Double- immunofluorescence staining for DCX with FITC colorization and NeuN with rhodamine colorization around the damaged area after traumatic brain injury (TBI). At 7 days after TBI, there are a few DCX-immunopositive cells (green, A) and many NeuN-immunopositive cells (red, B). A and B represent the same field of view and show no co-localization. A merged image of A and B is shown C. At 30 days after TBI, two DCX-immunopositive cells (green, D) are co-localized with NeuN-immunopositivity (red, E). D and E represent the same field of view and show co-localization. A merged image of D and E is shown F. * shows the damaged region. Scale bars = 50 μm.

Discussion

In this study, there were nestin-positive cells in an early stage at 1 day to 7days after TBI. The nestin-positive immunoreactive cells showed a maximum number at 72 hrs after TBI. Nestin-positive cells were increased around cerebral damaged cortex at 1 day to 4 days after cryoinjury and then showed a decrease [28]. In addition, the number of nestin-positive cells around damaged cerebral cortex was at a maximum 3 days after ablation injury [29]. Nestin-positive cells were shown around cerebral damaged cortex at 1 day to 7 days after CCI, and the number of nestin-positive cells showed a maximum at 4 days after CCI [30]. These results indicated that the number of nestin-positive cells showed a maximum around the damaged area at 3 or 4 days after injury. At 24 hrs after TBI, the nestin-positive number did not show a significant increase compared with control, because the number of nestin immunopositive cells was small. Moon also reported that there were not many nestin-positive cells at 24 hrs after injury and the nestin-positive area did not change compared with the control group [28].

It is reported that nestin-positive cells around the damaged area after TBI showed co-localization with GFAP or vimentin [11, 28, 31, 32]. Furthermore, after injury, expression of both nestin and vimentin induced migration of activated cells and contributed to neurogenesis [1, 28, 32]. The astrocytes showed blastogenesis in the reactivity after injury. In this study, a few of nestin-positive cells showed GFAP. This meant that astrocytes could change to nestin-positive neural stems and then give rise to new neuronal cells [33-35]. However, it is unclear if NSC migrated to damage area from the SVZ and SEZ, or that astrocytes showed blastogenesis in the reactivity following injury and differentiated into nestin-positive cells.

At 1 day, 3 days and 7days after TBI, only the cerebral cortex was separated by gross dissection under a dissecting microscope and an attempt was made to culture NSCs. Spheres could not be isolated at 1 day and 7days after TBI, but spheres could be isolated and cultured at 3 days after TBI. Isolated and cultured secondary spheres primarily composed of nestin immunopositive cells, but these spheres did not show Tuj1 and vimentin immunopositivity. These results suggest that secondary spheres were neurospheres, because isolated and cultured secondary spheres showed nestin immunopositivity. It was reported that there were many nestin-positive cells in the SVZ and SGZ, and isolation and culture of NSCs from these area was possible [36]. However, isolation and culture of NSC from cerebral cortex should be impossible, because there are no NSCs in the adult cerebral cortex. In this study, the nestin-positive cells around the damaged area showed a maximum at between 1 day and 7days after TBI. Nestin protein showed a maximum in volume at 3 days after TBI with western blotting method[29]. These results suggest that increased numbers of nestin-positive cells is important for the isolation of NSCs from around the damaged brain.

Neurosphere comprising of NSCs could differentiate into neurons or glia by removing basic fibroblast growth factor (bFGF) or epidermal growth factor (EGF), which are mitogen factors in the culture medium [37, 38]. Neurospheres, which originated from rat brain, differentiated to Tuj1-positive neurons and GFAP-positive astrocytes after 2-4 days of culture without bFGF in the culture medium [39]. In this study, the secondary spheres differentiated to Tuj1-, GFAP- and O4-positive cells after 4 days of culture without bFGF in the culture medium. This result indicates that nestin positive neural stem cells from around the damaged brain had the possibility to differentiate into neurons and glia. Recent studies have indicated that immature neurons migrated from SVZ or SGZ, where there were neural stem cells in the adult rodent brain, after injury and neurogenesis was induced around damaged area [39]. However, our data confirmed there were neural stem cells, which can differentiate to neuron or glia, in addition to immature neurons as pointed out in the literature [39] around damaged area after TBI.

The double-immunofluorescence staining for nestin and GFAP indicated that the generation of nestin-positive cells around the damaged area coincided with some of the GFAP-positive cells. Although GFAP has been recognized as an astrocyte marker, recent studies have demonstrated that it is also expressed in pluripotent NSCs and neural precursor cells [40,41]. In the SVZ or SGZ after brain injury, astrocytes show blastogenesis and differentiate into neural precursor cells that are immunopositive for both nestin and GFAP. Subsequently, these cells differentiate into new immature neurons [33, 36]. Some reactive astrocytes also showed nestin-immunopositivity around the damaged area after TBI [11, 28, 31, 32]. Although the origin of these nestin-positive cells around the damaged area remains controversial, we showed that nestin-positive cells around the damaged area after TBI have the characteristics of NSCs, which can proliferate and differentiate to neurons or glia. Therefore, it is considered that the generation of endogenous immature cells, which have the possibility of differentiating into neurons, around the damaged area in the early stages after TBI is a very important fact for neurogenesis after brain injury.

The DCX-immunopositivity did not co-localize with the GFAP-immunopositivity around the damaged area after TBI. The DCX-immunopositive cells that did not co-localize with GFAP-positive cells represent newly formed immature neurons, rather than pluripotential NSCs or astrocytes[14, 33 ,42, 43].

The number of DCX-immunopositive cells reached its maximum at 7 days after TBI. Nestin-positive cells were increased around the damaged area of the cerebral cortex at 1 to 3 days after injury and then decreased. Expression of nestin and DCX is not present in both NSCs and neurons simultaneously, because their phases of cellular differentiation differ[14, 42, 43]. Pluripotential NSCs localized with nestin-immunopositivity mature into neuroblasts, and these neuroblasts are localized with DCX-immunopositivity, but not nestin-immunopositivity [14, 42, 43]. These results indicate that nestin-positive NSCs mature into immature neurons that are localized with DCX-immunopositivity chronologically and thus the number of DCX-positive cells increased at 7 days after TBI. Furthermore, the number of DCX-positive cells at 30 days after TBI was significantly decreased to one-third the number at 7 days after TBI, although the number of DCX-positive cells in the TBI group was significantly higher than the number in the sham operated group at 30 days after TBI. These results confirm that newly formed immature neurons survived among the glial scars, even at 30 days after glial scar formation following TBI. However, the numbers of nestin-positive cells were 69.2 ± 25.8 and 12.2 ± 3.3 at 3 and 7 days after TBI, respectively, while the numbers of DCX-positive cells were 5 ± 3.3 and 25 ± 5.1 at 3 and 7 days after TBI, respectively. These results indicate that not all of the nestin-positive cells changed into DCX-positive cells. Thus, it is considered that some of the nestin-positive cells may have disappeared or changed into glial cells.

Although the expression of DCX is downregulated in DCX-positive newly formed immature neurons for maturation to neurons, the newly formed immature neurons co-localized with NeuN, a marker of mature neurons[14,42]. Furthermore, the newly formed immature neurons showed DCX-immunopositivity, while during migration, early differentiation and maturation and the final phase of maturation into mature neurons, DCX-positive newly formed neurons co-localize with the β-III-tubulin protein (Tuj1) [44] and/or NeuN [45-47]. At 7 days after TBI, the DCX-immunopositivity was not co-localized with the NeuN-immunopositivity. However, at 30 days after TBI, a small number of DCX-positive cells were co-localized with NeuN-immunopositivity. These data indicate that all the DCX-positive cells were immature neurons at 7 days after TBI, and that at 30 days after TBI, some of these newly formed neurons had become mature neurons. Furthermore, these mature neurons survived near and among the glial scars.

In conclusion, the number of nestin-positive cells around cerebral cortical damaged area showed a maximum at between 1 day and 7 days after TBI and neurospheres isolated at the same time comprising NSCs were confirmed to have the potential to differentiate into neurons and glia, although it is unclear if NSCs migrated to the damaged area from the SVZ and SEZ, or if astrocytes showed blastogenesis in the reactivity after injury and gave rise to nestin-positive cells. Furthermore, there were DCX-positive immature neurons among the gliosis and glial scars around the damaged area in the early stages after rat TBI. In addition, we confirmed that some of the newly formed immature neurons became mature neurons that survived among the glial scars. Our data suggest that promotion of the maturation and differentiation of newly formed immature neurons around the damaged area may improve the brain dysfunction induced by glial scars after brain injury.

Acknowledgements

The authors thank Mari Machino (Department of Pathology, Kinki University School of Medicine), Yoshitaka Horiuchi and Katumi Okumoto (Life Science Research Institute , Kinki University) for technical assistance. The author also thank the staff of Department of Pathology, Kinki University School of Medicine.

References

[1] Chirumamilla S, Sun D, Bullock MR,Colello RJ (2002) Traumatic brain injury induced cell proliferation in the adult mammalian central nervous system. *J. Neurotrauma* 19: 693-703.

[2] Rice AC, Khaldi A, Harvey HB, Salman NJ, White F, Fillmore H,Bullock MR (2003) Proliferation and neuronal differentiation of mitotically active cells following traumatic brain injury. *Exp. Neurol.* 183: 406-417.

[3] Azbill RD, Mu X, Bruce-Keller AJ, Mattson MP,Springer JE (1997) Impaired mitochondrial function, oxidative stress and altered antioxidant enzyme activities following traumatic spinal cord injury. *Brain Res.* 765: 283-290.

[4] Xiong Y, Gu Q, Peterson PL, Muizelaar JP,Lee CP (1997) Mitochondrial dysfunction and calcium perturbation induced by traumatic brain injury. *J. Neurotrauma* 14: 23-34.

[5] Gage FH (2000) Mammalian neural stem cells. *Science* 287: 1433-1438.

[6] Reynolds BA,Weiss S (1992) Generation of neurons and astrocytes from isolated cells of the adult mammalian central nervous system. *Science* 255: 1707-1710.

[7] Kuhn HG, Dickinson-Anson H,Gage FH (1996) Neurogenesis in the dentate gyrus of the adult rat: age-related decrease of neuronal progenitor proliferation. *J. Neurosci.* 16: 2027-2033.

[8] Lois C,Alvarez-Buylla A (1994) Long-distance neuronal migration in the adult mammalian brain. *Science* 264: 1145-1148.

[9] Parent JM, Yu TW, Leibowitz RT, Geschwind DH, Sloviter RS,Lowenstein DH (1997) Dentate granule cell neurogenesis is increased by seizures and contributes to aberrant network reorganization in the adult rat hippocampus. *J. Neurosci.* 17: 3727-3738.
[10] Johansson CB, Lothian C, Molin M, Okano H,Lendahl U (2002) Nestin enhancer requirements for expression in normal and injured adult CNS. *J. Neurosci. Res.* 69: 784-794.
[11] Nakamura T, Miyamoto O, Auer RN, Nagao S, Itano T (2004) Delayed precursor cell markers expression in hippocampus following cold-induced cortical injury in mice. *J. Neurotrauma* 21: 1747-1755.
[12] Kernie SG, Erwin TM,Parada LF (2001) Brain remodeling due to neuronal and astrocytic proliferation after controlled cortical injury in mice. *J. Neurosci. Res.* 66: 317-326.
[13] Couillard-Despres S, Winner B, Schaubeck S, Aigner R, Vroemen M, Weidner N, Bogdahn U, Winkler J, Kuhn HG,Aigner L (2005) Doublecortin expression levels in adult brain reflect neurogenesis. *Eur. J. Neurosci.* 21: 1-14.
[14] Brown JP, Couillard-Despres S, Cooper-Kuhn CM, Winkler J, Aigner L,Kuhn HG (2003) Transient expression of doublecortin during adult neurogenesis. *J. Comp. Neurol.* 467: 1-10.
[15] Nacher J, Crespo C,McEwen BS (2001) Doublecortin expression in the adult rat telencephalon. *Eur. J. Neurosci.* 14: 629-644.
[16] Dash PK, Mach SA,Moore AN (2001) Enhanced neurogenesis in the rodent hippocampus following traumatic brain injury. *J. Neurosci. Res.* 63: 313-319.
[17] Davies SJ, Goucher DR, Doller C,Silver J (1999) Robust regeneration of adult sensory axons in degenerating white matter of the adult rat spinal cord. *J. Neurosci.* 19: 5810-5822.
[18] Jurynec MJ, Riley CP, Gupta DK, Nguyen TD, McKeon RJ,Buck CR (2003) TIGR is upregulated in the chronic glial scar in response to central nervous system injury and inhibits neurite outgrowth. *Mol. Cell Neurosci.* 23: 69-80.
[19] Rudge JS,Silver J (1990) Inhibition of neurite outgrowth on astroglial scars in vitro. *J. Neurosci.* 10: 3594-3603.
[20] Pekny M,Nilsson M (2005) Astrocyte activation and reactive gliosis. *Glia* 50: 427-434.
[21] Bechmann I,Nitsch R (2000) Involvement of non-neuronal cells in entorhinal-hippocampal reorganization following lesions. *Ann. NY Acad. Sci.* 911: 192-206.
[22] Deller T, Haas CA,Frotscher M (2000) Reorganization of the rat fascia dentata after a unilateral entorhinal cortex lesion. Role of the extracellular matrix. *Ann. NY Acad. Sci.* 911: 207-220.
[23] Gallo V,Chittajallu R (2001) Neuroscience. Unwrapping glial cells from the synapse: what lies inside? *Science* 292: 872-873.
[24] Silver IA, Deas J, Erecinska M (1997) Ion homeostasis in brain cells: differences in intracellular ion responses to energy limitation between cultured neurons and glial cells. *Neuroscience* 78: 589-601.
[25] Cherian L, Robertson CS, Contant CF, Jr.,Bryan RM, Jr. (1994) Lateral cortical impact injury in rats: cerebrovascular effects of varying depth of cortical deformation and impact velocity. *J. Neurotrauma* 11: 573-585.

[26] Goodman JC, Cherian L, Bryan RM, Jr.,Robertson CS (1994) Lateral cortical impact injury in rats: pathologic effects of varying cortical compression and impact velocity. *J. Neurotrauma* 11: 587-597.

[27] Yanagisawa M, Nakashima K, Arakawa H, Ikenaka K, Yoshida K, Kishimoto T, Hisatsune T,Taga T (2000) Astrocyte differentiation of fetal neuroepithelial cells by interleukin-11 via activation of a common cytokine signal transducer, gp130, and a transcription factor, STAT3. *J. Neurochem.* 74: 1498-1504.

[28] Moon C, Ahn M, Kim S, Jin JK, Sim KB, Kim HM, Lee MY,Shin T (2004) Temporal patterns of the embryonic intermediate filaments nestin and vimentin expression in the cerebral cortex of adult rats after cryoinjury. *Brain Res.* 1028: 238-242.

[29] Douen AG, Dong L, Vanance S, Munger R, Hogan MJ, Thompson CS,Hakim AM (2004) Regulation of nestin expression after cortical ablation in adult rat brain. *Brain Res.* 1008: 139-146.

[30] Chen S, Pickard JD,Harris NG (2003) Time course of cellular pathology after controlled cortical impact injury. *Exp. Neurol.* 182: 87-102.

[31] Shibuya S, Miyamoto O, Auer RN, Itano T, Mori S,Norimatsu H (2002) Embryonic intermediate filament, nestin, expression following traumatic spinal cord injury in adult rats. *Neuroscience* 114: 905-916.

[32] Sahin KS, Mahmood A, Li Y, Yavuz E,Chopp M (1999) Expression of nestin after traumatic brain injury in rat brain. *Brain Res.* 840: 153-157.

[33] Seri B, Garcia-Verdugo JM, McEwen BS,Alvarez-Buylla A (2001) Astrocytes give rise to new neurons in the adult mammalian hippocampus. *J. Neurosci.* 21: 7153-7160.

[34] Laywell ED, Rakic P, Kukekov VG, Holland EC,Steindler DA (2000) Identification of a multipotent astrocytic stem cell in the immature and adult mouse brain. *Proc. Natl. Acad. Sci. USA* 97: 13883-13888.

[35] Itoh T, Satou T, Nishida S, Hashimoto S,Ito H (2006) Cultured Rat Astrocytes Give Rise to Neural Stem Cells. *Neurochem Res.* 31: 1381-1387.

[36] Picard-Riera N, Nait-Oumesmar B,Baron-Van EA (2004) Endogenous adult neural stem cells: limits and potential to repair the injured central nervous system. *J. Neurosci. Res.* 76: 223-231.

[37] Yamamoto S, Yamamoto N, Kitamura T, Nakamura K,Nakafuku M (2001) Proliferation of parenchymal neural progenitors in response to injury in the adult rat spinal cord. *Exp. Neurol.* 172: 115-127.

[38] Zhu G, Mehler MF, Mabie PC,Kessler JA (1999) Developmental changes in progenitor cell responsiveness to cytokines. *J. Neurosci. Res.* 56: 131-145.

[39] Xu Y, Kimura K, Matsumoto N,Ide C (2003) Isolation of neural stem cells from the forebrain of deceased early postnatal and adult rats with protracted post-mortem intervals. *J. Neurosci. Res.* 74: 533-540.

[40] Doetsch F, Caille I, Lim DA, Garcia-Verdugo JM,Alvarez-Buylla A (1999) Subventricular zone astrocytes are neural stem cells in the adult mammalian brain. *Cell* 97: 703-716.

[41] Sanai N, Tramontin AD, Quinones-Hinojosa A, Barbaro NM, Gupta N, Kunwar S, Lawton MT, McDermott MW, Parsa AT, Manuel-Garcia VJ, Berger MS,Alvarez-

Buylla A (2004) Unique astrocyte ribbon in adult human brain contains neural stem cells but lacks chain migration. *Nature* 427: 740-744.

[42] Cooper-Kuhn CM,Kuhn HG (2002) Is it all DNA repair? Methodological considerations for detecting neurogenesis in the adult brain. *Brain Res. Dev. Brain Res.* 134: 13-21.

[43] Lendahl U, Zimmerman LB,McKay RD (1990) CNS stem cells express a new class of intermediate filament protein. *Cell* 60: 585-595.

[44] Menezes JR,Luskin MB (1994) Expression of neuron-specific tubulin defines a novel population in the proliferative layers of the developing telencephalon. *J. Neurosci.* 14: 5399-5416.

[45] Palmer TD (2002) Adult neurogenesis and the vascular Nietzsche. Neuron 34: 856-858.

[46] Mullen RJ, Buck CR,Smith AM (1992) NeuN, a neuronal specific nuclear protein in vertebrates. *Development* 116: 201-211.

[47] Rao MS,Shetty AK (2004) Efficacy of doublecortin as a marker to analyse the absolute number and dendritic growth of newly generated neurons in the adult dentate gyrus. *Eur. J. Neurosci.* 19: 234-246.

In: Stem Cell Research Trends
Editor: Josse R. Braggina, pp. 227-256
ISBN: 978-1-60021-622-0

Chapter VIII

TRANSDIFFERENTIATION OF MESENCHYMAL STEM CELLS – AN ALTERNATIVE APPROACH IN REGENERATIVE MEDICINE OF DEMYELINATING DISEASES

***Gerburg Keilhoff*[1] *and Hisham Fansa*[2]**
[1]Institute of Medical Neurobiology, Otto-von-Guericke-University of Magdeburg, Leipziger Strasse 44, D-39120 Magdeburg, Germany
[2]Dept. of Plastic, Reconstructive and Aesthetic Surgery – Hand Surgery, Staedtische Kliniken Bielefeld; Teutoburger Strasse 50 D-33604 Bielefeld, Germany

ABSTRACT

Bone marrow stromal cells (MSCs) are multipotent stem cells that differentiate into cells of the mesodermal lineage like bone, cartilage, fat, and muscle, and though adult, they are able to transdifferentiate. We aimed to evaluate their transdifferentiation potential into myelinating "Schwann cell-like" cells to offer new therapeutic strategies for a wide range of diseases and injuries of the nervous system.

Schwann cells are of special interest not only as central players in peripheral nerve regeneration, where they act as pathfinders for the outgrowing fibers, but also as therapeutic option in multiple sclerosis. They are not attacked by the immune cells and they are able to break down devastated myelin and to clear debris by phagocytosis, an important prerequisite for remyelination. Human Schwann cells can be obtained from nerve biopsies for autologous transplantation without the need for immunosuppression. However, the method has inevitable disadvantages, to mention only e few: limitations of nerve material, decreased yield of cells due to their restricted mitotic activity, and sacrificing one or more functioning nerves. Hence, alternative cell systems are desirable and stem cells may be such an alternative cell source.

[1] Phone: +49 391 6714368; Fax: +49 391 67 14365; gerburg.keilhoff@medizin.uni-magdeburg.de
[2] Phone: +49 521 5813950; Fax: +49 521 5813997; hisham.fansa@sk-bielefeld.de

We transformed cultivated rat MSCs into Schwann cell-like cells by using different cytokine cocktails. Treated MSCs changed morphologically into cells resembling typical spindle shaped Schwann cells with enhanced expression of $p75^{NTR}$, Krox-20, CD104 and S100ß proteins and decreased expression of BMP receptor-1A. As final proof of successful transdifferentiation the functionality, i.e. the myelinating capacity was checked. Therefore, Schwann cells and transdifferentiated or untreated MSCs cultured from male rats were grafted into autologous muscle conduits bridging a 2cm-gap in a female sciatic nerve. PCR of the SRY gene and S100 immunoreactivity of pre-labeled cells confirmed their presence in the grafts. After three and six weeks, regeneration was monitored clinically, histologically and morphometrically. Autologous nerves and cell-free muscle grafts were used as controls.

Revascularization studies suggested that transdifferentiated MSCs facilitated neo-angiogenesis and did not negatively influence macrophage recruitment. Autologous nerve grafts demonstrated the best results in all regenerative parameters. An appropriate regeneration was noted in the Schwann cell-groups and, albeit with restrictions, in the transdifferentiated MSC-groups, while regeneration in the MSC-group and in the cell-free group was impaired.

Although the results must be interpreted with caution, we want to speculate that the transdifferentiation technique provides a tool to manipulate adult stem cells for cell-based approaches in regenerative medicine of demyelinating diseases.

Introduction

Myelinating cells, oligodendrocytes in the central nervous system (CNS), and Schwann cells in the peripheral nervous system (PNS), play a crucial role in neurodegenerative and regenerative processes. Demyelinating diseases and traumatic disorders lead to heavy functional impairment for the patients. Multiple sclerosis (MS) and spinal cord injury both present variable amounts of axonal demyelination and transection. MS, a chronic autoimmune disease of the CNS, has a relatively high incidence and causes severe disabilities. The destruction of myelin sheaths coincides with a loss of oligodendrocytes, and subsequently with an axonal loss (Kornek and Lassmann, 1999; Lucchinetti et al., 1999) leading to a wide range of motor, sensory, visual and cognitive symptoms. Remyelination and subsequent restoration of neuronal function can be achieved by either promoting endogenous repair mechanisms or providing an exogenous source of myelinating cells via transplantation. Oligodendrocytes and Schwann cells, neural stem cells or stem cell derived oligodendrocytes, transplanted into the injured areas, can encourage axonal regeneration. Substantial regeneration, however, occurs only in the PNS.

Peripheral nerve lesions are often connected with myelin defects. To bridge larger nerve defects, autologous nerve grafting offers the best outcome at present. Limited availability of donor tissue, however, represents a major problem. Additionally, donor nerve harvesting causes a neurological deficit and, the quality and quantity of donor nerves are limited. Nerve allografts, on the other hand, are antigenic, undergo rejection and thus require the continuous use of immunosuppressive drugs (Evans et al., 1994). Tissue engineering of peripheral nerves is the alternative. The goal of peripheral nerve tissue engineering is to create a scaffold that guides the sprouting fibers to their endorgan and offers nutritional support. To date, basically two concepts have been established: (1) a scaffold combined with certain growth factors to enhance regeneration (Schmidt and Leach, 2003; Gravvanis et al., 2004; Myckatyn and Mackinnon, 2004), and (2) a scaffold with cultured Schwann cells, as, at least partially, regenerative capacity of the PNS has been linked to several abilities of Schwann cells, including the production of extracellular matrix and trophic factors for the sprouting nerve and the lack of growth inhibitors that are expressed in oligodendrocytes (Ide, 1996). As transplantation of viable Schwann cells offers better results than the release of just one or two growth factors, we, amongst others (Wiberg and Terenghi, 2003; Lundborg, 2004), have focused on the second concept (Fansa and Keilhoff, 2004; Stang et al., 2005).

Key point of this concept is to find the ideal scaffold, that minimizes fibrosis and allows an organized regeneration, and to harvest and expand Schwann cells. The generation of sufficient quantities of Schwann cells for transplantation from the patient's own peripheral nerve biopsy, however, requires at least 3 – 6 weeks according to established protocols (Morrissey et al., 1991; Calderon-Martinez et al., 2002). Moreover, one or more functioning nerves must be sacrificed with the consequence of loss of sensation, scarring and, possibly, neuroma formation. Although special techniques to cultivate adult Schwann cells have been established, in animal models the cells were harvested from predegenerated nerves (Keilhoff et al., 1999); in humans activated Schwann cells were available from the stump of the injured peripheral nerve, the so called neuroma (Keilhoff et al., 2000; Bachelin et al., 2005), alternative cell systems are desirable. That all the more, as Schwann cells are of special interest not only as central player in peripheral nerve regeneration; also in MS therapy they seem to be a therapeutic option. Although Schwann cells contain myelin basic protein, one of the main targets of the immune cells in MS (Deber and Reynolds, 1991), they are not affected by this disease. Furthermore, Schwann cells are able to break down devastated myelin and to clear debris by phagocytosis (Liu et al., 1995), an important prerequisite for successful remyelination (Stoll and Muller, 1999), and they can remyelinate axons in the CNS (Blakemore, 1977; Honmou et al., 1996).

Stem cells may be such an alternative source for Schwann cells. The use of embryonic stem cells, however, causes ethical problems and, in addition, their carcinogenic potential is a serious risk factor (Bjorklund et al., 2002). Thus, their clinical application is improbable in the near future. Several populations of stem cells exist in adult tissues that offer the possibility of circumventing such problems. Most promising are the mesenchymal stem cells (MSCs) (Bianco et al., 2001). They reside in the bone marrow and differentiate mainly in cell lineages of mesodermal origin to form, for example, muscle, bone, cartilage, fat and tendon (Prockop, 1997; Pittenger et al., 1999). MSCs are easily accessible through aspiration of bone marrow. They readily adhere in plastic culture dishes (Phinney et al., 1999). With appropriate stimuli and environmental conditions, MSCs have been shown to exhibit transdifferentiation and plasticity (Tao and Ma, 2003; Abderrahim-Ferkoune et al., 2004). MSCs may also differentiate into non-mesenchymal lineages, including astrocytes (Kopen et al., 1999), myocardium (Pittenger and Martin, 2004), endothelial cells (Oswald et al., 2004), neurons (Woodbury et al., 2000; Deng et al., 2001) and myelinating cells of the PNS (Dezawa et al., 2001; Tohill et al., 2004). Recently, however, transdifferentiation has been debated and several other biological explanations, e.g. cell fusion, have been put forward (Rutenberg et al., 2004; Kashofer and Bonnet, 2005).

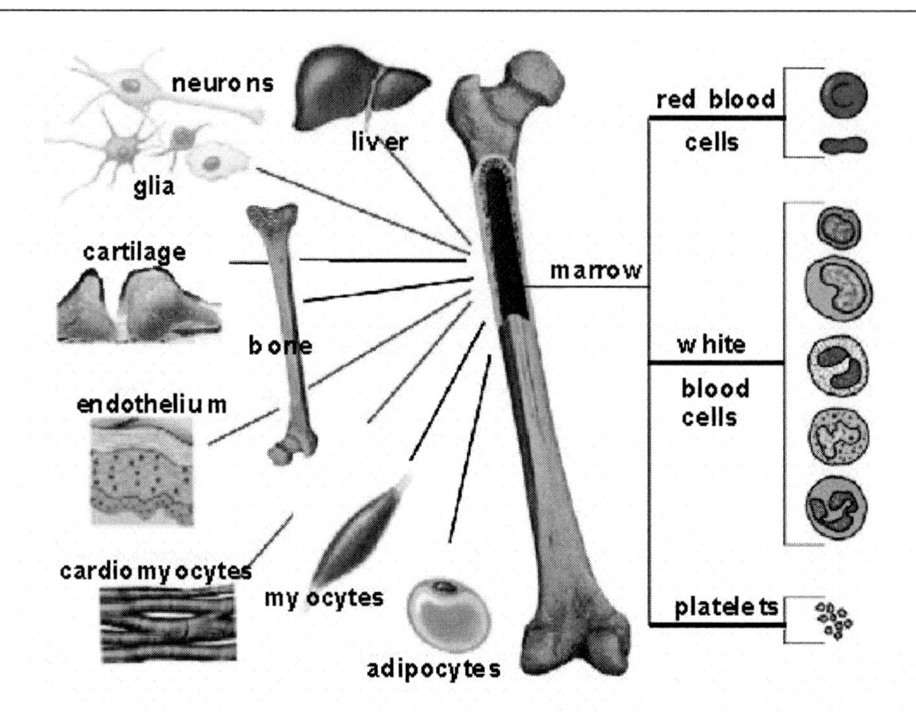

Schemata 1. Bone marrow-derived cells, modified after www.healthsystem.virginia.edu.

We aimed to evaluate the transdifferentiation potential of MSCs into myelinating "Schwann cell-like" cells. This would enormously facilitate the harvesting of the required cells from the patient, thus offering new therapeutic strategies for myelin defects. A few reports describe the possible impact of bone marrow cells on PNS, all using different approaches to lesion and regeneration characteristics. Tohill and Terenghi for example have studied the effect of grafting glial-differentiated bone-marrow stromal cells into a rat model of a peripheral nerve injury. They report a beneficial effect on Schwann cell growth within regenerating nerves (Wiberg and Terenghi, 2003; Tohill and Terenghi, 2004; Tohill et al., 2004). Cuevas et al. (2002) were able to induce peripheral nerve regeneration using undifferentiated bone marrow stromal cells, and Kocsis et al. (2002) examined the remyelinating potential of "peripheral-myelin-forming" cells derived from bone marrow to repair the injured spinal cord. The group of Dezawa, one of the pioneers in the field of stem cell derived therapies for peripheral nerve injuries, presented an artificial nerve graft made with bone marrow stromal cell-derived Schwann cells as an alternative for the difficult reconstruction of peripheral nerves (Dezawa et al., 2001; Dezawa, 2002; Mimura et al., 2004; Kamada et al., 2005).

We determined the conditions necessary to differentiate MSCs into Schwann cell-like cells and evaluated the stability of transdifferentiation. To characterize and distinguish non-differentiated, differentiated and re-differentiated MSCs, we selected several markers and showed their expression patterns with RT-PCR and immunofluorescence. Moreover, the myelinating capacity was studied in an in vitro model of normally not myelinated PC12 cells, and in our well established in vivo model of nerve tissue engineering in rats (Fansa and Keilhoff, 2004). E.g., undifferentiated or transdifferentiated (Schwann cell-like) MSCs, or, alternatively, genuine Schwann cells were implanted into devitalized muscle graft to bridge a 2-cm gap in the rat sciatic nerve. Muscle grafts were preferred since they are free of endogenous Schwann cells, thus excluding the possibility that regeneration is assisted by endogenous Schwann cells or that implanted MSCs fuse with the remaining host Schwann cells imitating a transdifferentiation effect. Devitalized muscle grafts provide an effective matrix for implanted cells to adhere to and the remaining muscle basal lamina tubes offer spaces for the regenerating axons to grow through (Hall, 1997; Fansa et al., 1999). Within such basal lamina tubes, cell columns (Büngner bands) are formed as a pathway through which regenerating axons grow to reach their target organ (Fansa et al., 1999; Evans, 2000). Moreover, the gap length of 2 cm minimizes the effect of spontaneous regeneration in rats, that is known to be highly effective and reaches 2 – 3 mm per day (Near et al., 1992).

Chemicals

α-MEM (Biochrom; Berlin, Germany; www.biochrom.de), Dulbecco´s modified Eagle´ medium (DMEM; Biochrom), fetal bovine serum (FBS; Biochrom), L-glutamine (Sigma; Saint Louis, Missouri; www.sigmaaldrich.com), penicillin/ streptomycin (Invitrogen; Carlsbad, CA; www.invitrogen.com), trypsin/EDTA (Invitrogen), laminin (Sigma), poly-D-lysine (Sigma), dexamethasone (Sigma), dispase (Boehringer-Mannheim; Mannheim, Germany; www.roche.com/diagnostics), collagenase (Sigma), hyaluronidase (Sigma), L-ascorbic acid (Merck; Darmstadt, Germany; www.merck.de), ß-glycerol phosphate (Sigma), indomethacin (Sigma), insulin (Sigma), 3-isobutyl-1-methyl-xanthine (IBMX; Merck), ß-mercaptoethanol (Merck), all-trans-retinoic acid (Sigma), forskolin (Merck), recombinant human basic fibroblast growth factor (bFGF; Chemicon; Temecula, CA; www.chemicon.com), human recombinant heregulin-β EGF domain (Her-ß; Upstate Biotechnology; Lake Placid, NY; www.upstate.com), PKH linker kit (Sigma), platelet derived growth factor (PDGF; Sigma), Toluidine blue (Chroma; Munster, Germany; www.chroma.de), Durcupan (Sigma), TRIzol® reagent (Invitrogen), TURBO DNA-free™-Kit (Ambion; Austin, Texas; www.ambion.com), RevertAidTM-H Minus First Strand cDNA Synthesis Kit (Fermentas; Burlington, Canada; www.fermentas.com), Taq-DNA-polymerase (Peqlab; Erlangen, Germany; www.peqlab.de)

Isolation and Culture of Mesenchymal Stem Cells

Male Wistar rats (Harlan-Winkelmann; Borchen, Germany; www.harlan-winkelmann.de) were used at the age of 25 to 30 days. All animal experiments were approved by the Government Committee on Animal Care of the State of Saxony-Anhalt. Animals were anaesthetized by an overdose of isoflurane, and bone marrow stromal cells were isolated from the femur. Tissue was removed and the ends of the bone were cut. Marrow was flushed out with 5 ml phosphate-buffered saline (PBS, pH 7.4) using a syringe. After centrifugation, supernatant was discarded and the cells were resuspended in growth medium that consists of α-MEM supplemented with 20 % FBS, 2 mM L-glutamine, 100 U/ml penicillin and 100 μg/ml streptomycin. To dissolve cell clusters the suspension was flushed several times with a 21-gauge needle. Cells were seeded at 1.5 to 2.0×10^6 cells/cm^2 and maintained at 37°C and 5 % fully humidified CO_2. After 72 h non-adherent cells were removed and the medium was replaced. This stage was termed passage 0 (P0). Cultures were refed every three to four days. At day ten cells grown to colonies were detached with 0.25 % trypsin and 1mM EDTA for 5 mins at 37°C. Cells of one flask were split into 2 flasks (passage 1). One week later cultures, again grown to confluence, were detached as described and disseminated in a density according to the experimental requirements (passage 2). The cell death rate was evaluated by propidium iodide labeling (5 μg/ml medium, 5 min). Propidium iodide interacts with DNA to yield a red fluorescence of dead cell nuclei.

Assays for Osteogenic, Chondrogenic and Adipogenic (Self)-Differentiation

In order to examine whether stem cells differentiate without induction into mesodermal cell lineages, MSCs of P0 to P2 were maintained in growth medium for two weeks and then stained with Sudan red (adipogenic differentiation) or Toluidin-Alizarin (osteogenic and chondrogenic differentiation). Cells containing lipids were counted and related to the total of cells in culture.

To induce osteogenic and adipogenic differentiation of MSC, cells of P2 were disseminated at a density of 5000 cells/cm^2 and maintained in growth medium for 3 days. Then medium was replaced by differentiation medium according to the literature (Phinney et al., 1999) with slight modifications. Differentiation medium contains α-MEM, 10 % FBS, 2 mM L-glutamine, 100 U/ml penicillin and 100 μg/ml streptomycin and additionally either 10 nM dexamethasone, 50 μg/ml L-ascorbic acid and 10 mM ß-glycerol phosphate (osteogenic differentiation) or 10 nM dexamethasone, 200 μg/ml indomethacin, 5 μg/ml insulin and 0.5 μM IBMX (adipogenic differentiation). Media change was performed every three to four days. After 21 days osteogenic deposits and adipocytes were visualized as described above. Additionally, adipocytes were counted and related to the total number of cells.

Transdifferentiation of MSC to SC-Like Cells

MSC were plated at 500 cells/cm^2 and expanded in growth medium for three days and a further day with an additional 1 mM β-mercaptoethanol. Then medium was replaced with transdifferentiation medium consisting of α-MEM, 10 % FBS, 2 mM L-glutamine, 100 U/ml penicillin and 100 μg/ml streptomycin and additionally 35 ng/ml all-trans-retinoic acid. For the final transdifferentiation step, cultures were incubated in transdifferentiation medium supplemented with 5 μM forskolin, 10 ng/ml bFGF, 200 ng/ml Her-β and 5 ng/ml PDGF-AA for eight days with a media change every other day. That cytokine cocktail was used after preliminary experiments.

Cultivation and Characterization of Schwann Cells

The sciatic nerve of 5 homologous adult male rats was bilaterally exposed via a dorsal incision and transected distally to the dorsal root ganglia for predegeneration. After 7 days the distal part of the predegenerated nerve, 20 mm in length, was resected. The epineurium was removed and the segments were incubated in DMEM in addition to 10% FCS, 1.25 U/ml dispase, 0.05% collagenase and 0.1% hyaluronidase for 12 h at 37°C. Nerve fascicles were dissociated mechanically, then centrifuged at 1200 rpm for 10 minutes, washed in DMEM, centrifuged again at 1200 rpm and resuspended in DMEM containing 10% FCS, 50 U/ml penicillin and 50 μg/ml streptomycin. Aliquots of cell suspension (2 ml) were spread over laminin-coated Petri dishes at a density of 1.6 x 10^6 cells per 8 cm^2, resulting in a final density of adherent cells of about 2 x 10^4 per cm^2 after discarding dead cells and cell debris.

Before implantation random samples of the cultures were characterized by antibodies directed against S100 protein, specific for Schwann cells (polyclonal, rabbit, 1:500; Dako, Hamburg, Germany, www.dakocytomation.com), and antibodies against fibronectin (monoclonal, mouse, 1:10; Boehringer Mannheim), specific for fibrocytes. In all cases the DAKO LSAB kit was used. The cultures were fixed for 30 minutes in 4% paraformaldehyde, incubated with the respective antibodies for 3 h, processed for the peroxidase-antiperoxidase technique and visualized with 3.3'-diaminobenzidine.

Immunocytochemistry

Cells in culture dishes were fixed with a solution of 4 % paraformaldehyde (PFA) in PBS for 30 min. After washing 3x5 min with PBS, non-specific antigens were blocked with 3 % horse serum in PBS for 45 min and incubated overnight with the following primary antibodies: polyclonal anti-BMPR-1A (rabbit; 1:100; Santa Cruz Biotechnology; Santa Cruz, California; www.scbt.com), monoclonal anti-Stro-1 (mouse; 1:100; RandD Systems; Minneapolis, Minnesota; www.RnDSystems.com), monoclonal anti-NGF receptor (mouse, 1:10, Chemicon), monoclonal anti-CD104 (mouse; 1:200; Pharmingen; San Jose, California; www.Pharmingen.com). Cells were washed 3x5 min with PBS and then incubated with the secondary antibody Alexa Fluor® anti-mouse IgG (goat; 1:500; Molecular Probes; Eugene, Oregon; www.probes.com) or Alexa Fluor® anti-rabbit IgG (goat; 1:500; Molecular Probes), at room temperature for 3 h. Cultures were examined using a fluorescence microscope (Axiophot; Zeiss; Jena, Germany; www.zeiss.de) equipped with phase contrast, fluoresceine and rhodamine optics and documented with a color camera AxioCam MRc (Zeiss, Jena).

RT-PCR

Total RNA was isolated using TRIzol reagent according to the manufacturer's instructions followed by DNAse treatment. For RT-step 5 μg RNA were used for first strand cDNA synthesis. PCR was performed under the respective conditions with 0.5 μg cDNA, a Taq-DNA-polymerase and the respective primers (see Table1). The respective mRNA signals were quantified by densitometric analysis using a Biometra BioDocAnalyzer and the ratio of their expression to the housekeeping gene expression (GAPDH) was calculated.

Table 1. BMPR-1A: bone morphogenetic receptor-1A; FABP4: fatty acid binding protein 4; ErbB2: v-erb-b2 erythroblastic leukemia viral oncogene homolog 2 (= Her2: human epidermal growth factor receptor-2); LNGF-R: low affinity nerve growth factor receptor; IGF-1R: insulin-like growth factor-1 receptor; S100b: S100 protein, beta polypeptide; Krox-20 = Egr2: early growth response protein 2, CD104: β4-integrin; L1: neural cell adhesion molecule Primer sequences for RT-PCR

Gene	Sequence	Product size (bp)	Cycle -No.	Reference Gene bank No
BMPR-1A	5'-CAGCCCTACATCATGGCTGAC-3' 5'-GCTTCAAAACGGCTCGAAGAC-3'	229	40	NM_030849
FABP4	5'- AAAGAAGTGGGAGTTGGCTTC -3' 5'- ACCATCCAGGGTTATGATGCT -3'	204	33	NM_053365
Osteopontin	5'- TCCGATGAATCTGATGAGTCC -3' 5'- GCAACTGGGATGACCTTGATA -3'	236	22	NM_012881
IGF-IR	5'-TCCCAAGCTGTGTGTCTCTGAA-3' 5'-GTGCCACGTTATGATGATGCG-3'	178	36	NM_052807
ErbB2	5'-AATGCCAGCCTCTCATTCCTG-3' 5'-GACTTCGAAGCTGCAGCTCC-3'	235	40	NM_017003
LNGF-R	5'-CGACAACCTCATTCCTGTCTATTGC-3' 5'-CAGTCTGCGTATGGGTCTGCTG-3'	227	40	NM_012610
S100b	5'-GAGAGAGGGTGACAAGCACAA-3' 5'-GGCCATAAACTCCTGGAAGTC-3'	169	28	NM_013191
Krox-20	5'-AGATACCATCCCAGGCTCAGT-3' 5'-CTCTCCGGTCATGTCAATGTT-3'	300	40	NM_053633
CD104	5'-GCTCTGCTGGAAATACTGTGC-3' 5'-CAGGCTTCATGAGGTTCTCAG-3'	317	40	NM_013180
L1	5'- GGAAGTGGAGGAAGGAGAAT-3' 5'- AAGTGGGCATTGCAGATGTAG-3'	202	40	NM_017345
GAPDH	5'-TTAGCACCCCTGGCCAAGG-3' 5'-CTTACTCCTTGGAGGCCATG-3'	531	24	NM_017008

L1; GAPDH: glyceraldehyde-3-phosphate dehydrogenase used as house-keeping gene.

Myelin Assay

The myelinating capacity of MSC and transdifferentiated MSC compared to SC was checked in vitro using an PC12 co-culture system and in vivo using our well-established rat sciatic nerve model (Fansa and Keilhoff, 2004; Fansa et al., 2001).

In vitro myelination assay: PC12 cells (rat pheochromocytoma cell line) were cultured at 37°C and 5% CO_2 in MEM containing 10% FCS, 2 mM glutamine, 100 U/ml penicillin, and 100 µg/ml streptomycin at a density of 500 cells/cm². The cells were allowed to differentiate for 4 days. During the experiments the medium was changed every 2 days and the additives were added each time. Co-cultures were established by removing the media from the PC12 cells and seeding 500 dissociated Schwann cells, 500 MSCs, or 500 transdifferentiated MSCs, respectively, into each dish. PC12/Schwann cells and PC12/MSCs were cultured in the "PC12-medium". PC12/transdifferentiated cells were cultured in "PC 12-medium" or in transdifferentiated medium. During the experiments the medium was changed every 2-3 days. After 14 days, co-cultures were fixed for 30 min in 0.2 M cacodylate buffer containing 2.5% glutaraldehyde, then osmicated (2% OsO_4), dehydrated, en-bloc stained with 7% uranyl acetate and finally, embedded in Durcupan. All steps were carried out in the culture dishes. Ultra-thin sections (50-70 nm) were cut and mounted on Formvar-coated slot grids. For examination an E 900 transmission electron microscope (Zeiss, Germany) were used.

In vivo myelination assay: 115 Wistar rats were used (45 for evaluation of revascularization and macrophage recruitment, 70 for the ten experimental groups, Harlan-Winkelmann; Borchen, Germany; www.harlan-winkelmann.de). The rats were kept in accordance with the guidelines of the German Animal Welfare Act. The experimental protocol was approved by a review committee of the state of Saxony-Anhalt, Germany. The animals were housed under temperature-controlled conditions at 21 ± 1°C, with a 12-h light/dark cycle, with free access to standard rat chow (Altromin 1324™, Altromin GmbH, Lage, Germany; www.altromin.de) and water. All surgical procedures were performed under general anaesthesia with intraperitoneal injection of pentobarbital (60 mg/kg).

An identical operating protocol was performed for each group. The right sciatic nerve of isogenic adult Wistar female rats was exposed through a dorsal incision under aseptic conditions. A 2 cm nerve segment was completely transected with fine surgical scissors at a level just distal to the sciatic notch.

In control groups the nerve segment was reimplanted orthotopically with 10/0 monofilament nylon epineural sutures with the aid of an operating microscope. In all other groups the gracilis muscle was harvested. The gracilis muscle offers a longitudinal fiber orientation, which is anyway similar to the endoneural tube structure. The muscles were acellularized by freezing and thawing. Therefore, the muscles were placed into liquid nitrogen (-96 °C) until thermal equilibrium was achieved, and then put into PBS (pH 7.2, 22°C).

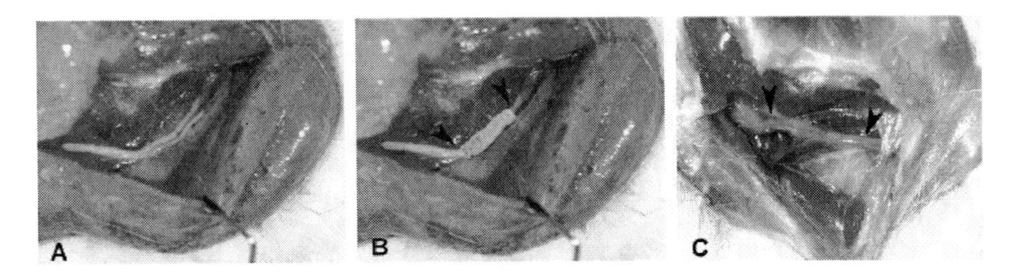

Figure 1. (A) An untreated contralateral sciatic nerve. (B) Tissue engineered SC-muscle graft (arrowheads) at time point of implantation. (C) The "nerve-like" muscle graft (arrowheads) with transdifferentiated MSCs after regeneration time of 6 weeks.

This was repeated for 3 times. A suspension of the respective cells (MSC, transdifferentiated MSC, Schwann cells, $2x10^6$ cells/ml DMEM) was transferred longitudinally into the acellular muscle with a microsyringe (29G x ½" needle) immediately before implantation.

Cell viability was assessed immediately before implantation by fluorescein staining, based on the phenomenon that living cells are able to hydrolyze fluorescein diacetate (10 µg/ml PBS) by intracellular esterases resulting in a green-yellow fluorescence.

Revascularization and Macrophage Recruitment

To evaluate revascularization and macrophages recruitment, animals were studied post-operatively at day 3, 5, and 10, respectively. At each time 3 animals per graft type were anaesthetized, the right femoral vein was exposed and 2 ml of Evans blue bound to albumin were injected (1% w/v Evans Blue with 5% w/v bovine albumin in sterile distilled water, filtered through a G-25 M Sephadex column (Pfizer/Pharmacia Corporation Austria, Wien, Austria; www.pharmacia.at)). The grafts were explanted 10 minutes after injection and the animals were killed. Specimens were fixed in 4% w/v paraformaldehyde and kept for cryoprotection for 48 h in 0.4% paraformaldehyde with 30% sucrose and frozen to -70°C. Sections of 25 µm were cut from the midpoint of the grafts in a cryostat (Leica, Bensheim, Germany) for fluorescence microscopic evaluation as described above. Evans blue bound to albumin remains in the vessels and fluoresces red (excitation wave-length 550 nm, emission 611 nm).

For visualization of invading (ED1-positive) and resident endoneurial (ED2-positive) macrophages, sections parallel to the revascularization series were incubated overnight at 4°C with monoclonal anti-rat ED1 antibodies (mouse, 1:250, Serotec GmbH, Duesseldorf, Germany, www.serotec.com) and monoclonal anti-rat ED2 antibodies (mouse, 1:250, Serotec) in PBS with 0.3% Triton X-100 and 1% FCS. After washing twice with PBS, slices were processed for the peroxidase-antiperoxidase technique and visualized with 3,3′-diaminobenzidine using the DAKO-LSAB-kit. The specificity of immunoreactions was controlled by applying a buffer instead of antibodies; all control sections remained free of any immunostaining.

Evaluation of the Regenerative Capacity

There were ten groups, each group consisted of seven animals:

(1) control, replanted nerve graft, survival time 3 weeks,
(2) control, replanted nerve graft, survival time 6 weeks,
(3) control, cell-free muscle graft, survival time 3 weeks,
(4) control, cell-free muscle graft, survival time 6 weeks,
(5) muscle graft, enriched with SCs, survival time 3 weeks,
(6) muscle graft, enriched with SCs, survival time 6 weeks,

(7) muscle graft, enriched with MSCs, survival time 3 weeks,
(8) muscle graft, enriched with MSCs, survival time 6 weeks,
(9) muscle graft, enriched with transdifferentiated MSCs, survival time 3 weeks,
(10) muscle graft, enriched with transdifferentiated MSCs, survival time 6 weeks,

Three weeks after reconstruction, specimens for histological examination were taken from the respective grafts 5 mm distal to the proximal suture. After survival time of six weeks, additionally material was taken from the distal nerve segment (5 mm distal to the distal suture). Material was fixed and prepared for light and electron microscopy as described for cultures. Semi-thin sections were stained with toluidine blue (0.1% toluidine blue dissolved in 2.5% sodium bicarbonate, 20 sek at 80°C). Ten arbitrarily selected grids from each graft or nerve segment were examined with an Axiophot microscope. For statistical analysis the groups of the nerve sections were blinded to the examiner. Five details of five cross-sections per nerve or graft segment were scanned using a CCD camera and all morphologically vital appearing axons were counted manually. Morphometric evaluation was carried out with a computer-assisted system (Image C, Imtronic, Münster, Germany). To assess the maturity of the fibers the g-ratio was calculated, expressing the relation of axonal diameter to the entire fiber diameter. Statistical analysis was performed with the non-parametric Kruskal-Wallis test. The Mann-Whitney U-test was used as a post-hoc test.

Functional Outcome

Being aware that single methods of measuring peripheral nerve regeneration give only limited data, we combined different methodologies: morphometric analysis (nerve fiber counts, g-ratio) and ultrastructural evaluation together with judging the rats ability to spread their digits actively and by measuring the sensitiveness of the digits to 55 °C hot water (rated positive if the rats retracted their legs from the water within 3 seconds). Additionally, the gastrocnemius muscle of the sacrificed animals was bilaterally removed and weighed. The ratio was calculated as indicator of the functional outcome (an indirect measurement of nerve regeneration). The clinical appearance was assessed from the beginning of the third postoperative week. From electrophysiological methods we abstained as the measurement of muscle contraction force depends on the frequency, voltage, and duration of the stimulus, which may not approximate the physiological condition. And nerve conduction velocity, measuring only the fastest conducting nerve fibers, may reveal false-positive results since a regenerated nerve may have only e few fibers that conduct very well even though a large number of remaining fibers are damaged (Kanaya et al., 1996).

Detection of Grafted Cells

To identify the donor derived male MSCs, transdifferentiated MSCs and Schwann cells within the muscle graft PCR of the sex-determining region Y gene (SRY) was used. Therefore, genomic DNA was isolated from the respective graft muscle after the respective survival time using an Invisorb Genomic DNA Kit II (Invitek, Berlin, Germany; www.invitek.de). 200 ng DNA was applied to the PCR. The following primers for SRY protein (Sry3) gene (gene bank no. X89730) were selected: forward 5'-CCCGCGGAGAGAGGCACAAGT-3' and reverse 5'-TAGGGTCTTCAGTCTCTGCGC-3'.

Alternatively, the PKH fluorescent cell linker technology was used to label viable cells before implantation. Therefore, the adherent cells (MSCs, transdifferentiated MSC) were suspended as described and a 2x suspension of the respective cells and a 2x dye solution, both in the PKH diluent supplied with the kit (PKH26-GL, Sigma), were mixed and incubated briefly at room temperature. The labeling reaction was stopped by addition of serum. Labeled cells were washed 3 times to remove unbound dye. The stable partitioning of the fluorescent dye into the membrane permits long-term monitoring while leaving the important functional surface proteins unaltered. The so labeled MSCs and transdifferentiated MSCs were implanted into muscle grafts and these were transplanted as described (3 animals per cell type). After a survival time of six weeks cryosections (20 µm) of the transplant (5 mm distal to the proximal suture) were immunostained with an antibody against myelin basic protein (MBP, monoclonal, mouse, 1:100; DakoCytomation) as described. The resulted double fluorescence labeling (PKH fluorescence – red / MBP - green) was evaluated in the Axiophot.

MSC Cultures

Bone marrow derived cultures (P0) contained a heterogeneous population of cells, where scarce MSC were first visible after several duplications at day 5 to 7 because oft their colony forming nature. At low densities, MSCs showed a spindle-like shape. When cultures reached confluence, morphology changed to a more flat and big one with seemingly torn ends. In cultures of P2, which were used for experiments, immunostaining revealed that 97.4 ± 4.5% of the total cells were immunopositive for BMPR-1A receptor and 88.3 ± 11.7% for the stem cell marker Stro-1. All Stro-1-positive cells also immunostained for BMPR-1A receptor. In all passages, the cell death rate, tested by propidium iodide, was stable by approximately 25%.

Osteogenic and Adipogenic Differentiation

In untreated cultures, adipocytes were found only in P0 and P1 in decreasing number. In P2, no more adipocytes could be observed Bone deposits were not found in any passages of untreated cultures. Incubated with adipogenic induction medium resulted in more than 30% of adipocytes in the second week. Calcification started one week after application of the osteogenic differentiation medium and showed an extensive deposition of bone material after three weeks. The differentiation could be illustrated by RT-PCR. Adipocytes-enriched fractions the FABP4 gene and the mRNA expression levels of osteopontin increased after the MSCs differentiated into osteoblasts.

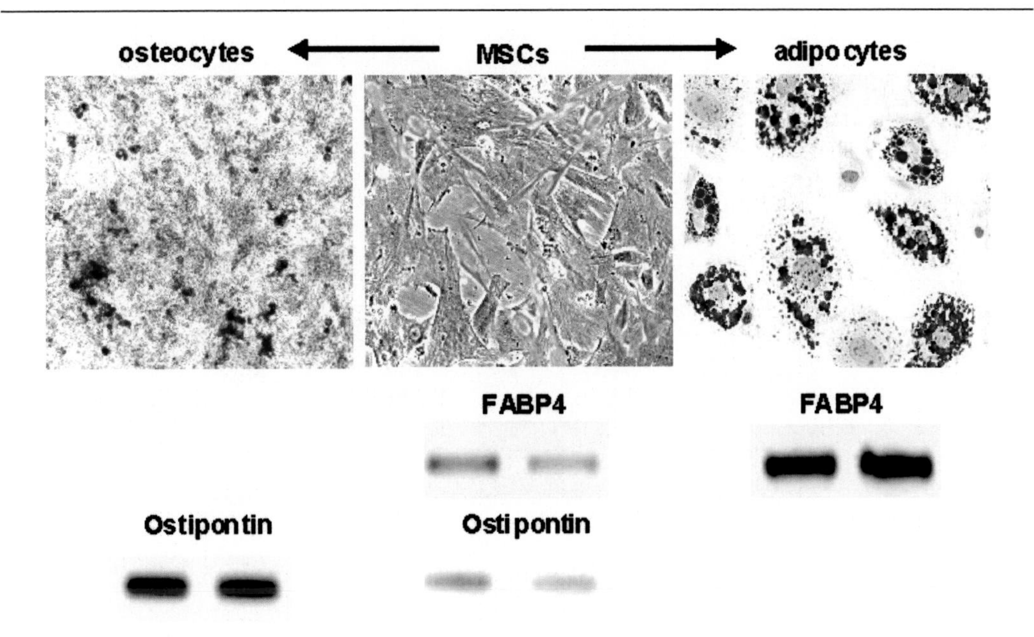

Figure 2. MSCs were able to differentiate into lipid-containing adipocytes (Sudan red-staining), expressing FABP4 and osteocytes (Toluidin-Alizarin staining), expressing ostipontin mRNA (X 400x).

Transdifferentiation of MSCs into SC-Like Cells

All transdifferentiation experiments were carried out in cultures of P2. 61.9 ± 23.7% of the total MSC adopted SC-like morphology and 56.7 ± 5.89% of them were immuno-positive for the Schwann cell-marker S100 within two days of transdifferentiation induction. Transdifferentiation efficiency was strongly dependent on the cell density. Separated cells changed their morphology faster and showed a higher S100 immunoreactivity than cells grown to a confluent layer which often did not change their morphology completely. During transdifferentiation, approximately 50% of the transforming cells, obviously dead, detached from the surface. From the remaining adherent cells about 5% showed morphological similarities with cells of other neural lineages like astrocytes, oligodendrocytes and neurons. A small subset of MSCs (less than 1%) differentiated to muscle cells when they were incubated for at least seven days.

The transdifferentiation is not stable. When the transdifferentiation medium was maintained in growth medium for three days, approximately 5% of transdifferentiated MSCs re-differentiated. The amount of these cells increased strongly with prolonged cultivation time in growth media. Three days after the medium was changed, only separate cells remained in an SC-like shape. After cells were re-differentiated, it was possible to induce transdifferentiation again by adding the respective cytokines with similar observations as described above.

Transdifferentiation was demonstrated by comparing undifferentiated and transdifferentiated MSCs with SCs using RT-PCR and immunohistochemistry. BMPR-1A, a marker for bone precursor cells, including MSC, is intensely expressed in MSCs, whereas transdifferentiation procedure reduced BMPR-1A expression.

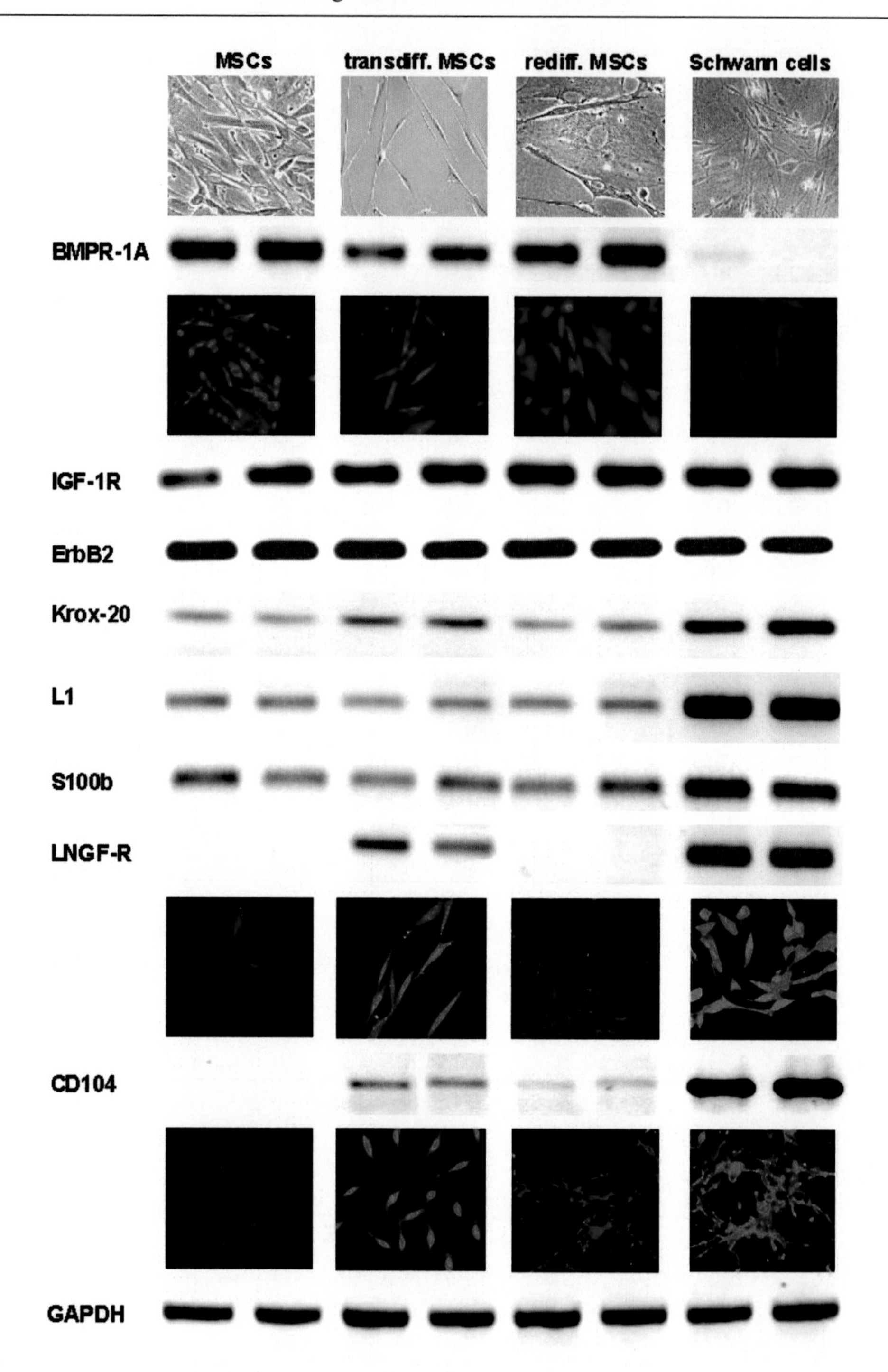

Figure 3. Phase-contrast micrographs, mRNA, and immunofluorescence pattern (magnification x 200). For mRNA sizes see Table 1.

Ratio of mRNA expression of the respective genes to the expression of GAPDH

	MSCs	transd. MSCs	rediff. MSCs	SCs
BMPR-1A	1.41 ± 0.07	0.97 ± 0.04**	1.06 ± 0.04*	0.07 ± 0.01***
IGF-1R	1.27 ± 0.14	1.38 ± 0.05	1.35 ± 0.07	1.29 ± 0.04
ErbB2	1.87 ± 0.04	1.81 ± 0.08	1.86 ± 0.12	1.86 ± 0.11
Krox20	0.19 ± 0.01	0.39 ± 0.02**	0.21 ± 0.45	0.81 ± 0.08
L1	0.44 ± 0.11	0.29 ± 0.07	0.41 ± 0.09	1.51 ± 0.07
S100b	0.71 ± 0.07	0.62 ± 0.01	0.69 ± 0.02	1.29 ± 0.04***
LNGF-R	0	1.07 ± 0.03***	0	1.45 ± 0.03***
CD104	0	0.27 ± 0.03***	0.12 ± 0.01*	1.36 ± 0.05***

Values are expressed as mean ± S.D., statistical significance was set at *$p < 0.05$, **$p < 0.005$, ***$p < 0.0005$, t-test Anova, always related to the value of MSCs.

In SCs the BMPR-1A mRNA transcript could be demonstrated only at a very low level. In correlation with that, immunostaining for BMPR-1A was highly expressed in MSCs, whereas in transdifferentiated MSCs the immunosignal was only moderate and absent in SCs. This indicates a transdifferentiation-induced silence of this precursor cell marker at both, mRNA and protein levels. mRNAs of IGF-1R and erbB2, the respective receptor for the applied growth factor Her-β, were expressed at high levels in all samples studied. Krox-20, a transcription factor for myelin genes, L1 and S100b were expressed in MSCs. Transdifferentiated MSCs transcribed more mRNA but much less than *in vitro*-cultured SCs. mRNAs of the LNGF-R and CD104 (β4-integrin), not expressed in MSCs, were induced by transdifferentiation but did not reach the levels as in SCs. Immunohistochemistry of both, LNGF-R and CD 104, demonstrates that transdifferentiation-induced mRNAs have been translated into proteins. In MSCs, immunofluorescence of this protein was absent. In transdifferentiated MSCs and SCs, LNGF-R (p75) labeling was increased indicating an enhanced number of LNGF receptors located throughout the cell surface with the highest level in SCs. Only traces of CD 104, heavily expressed in SCs, were found in MSCs. Again, the protein expression was markedly induced by transdifferentiation.

Myelinating Capacity of tMSC

In vitro myelination assay: At the beginning of cultivation, PC12 cells were round or polygonal. The great majority of the cells had none or short processes. As demonstrated in figure 4, after five days in culture, about 20% of untreated cells became spindle, triangle, or irregular shaped with noticeable processes stretching out. Addition of Schwann cells and, with curtailments, transdifferentiated MSCs induced a rapid neuron-like differentiation of the PC12 cells, as revealed by an increase in cell size and by the extension of neurites. Electron microscopy (Figure 4) confirmed myelin structures in PC12/Schwann-cultures at 14 days in vitro. This myelin was compacted and the presence of intraperiodic lines was seen. In the PC12/transdifferentiated MSCs-cultures, the grade of myelination depended on the culture medium. A remarkable myelination, like in PC12/Schwann cell cultures with a mean of 7 ± 2 layers, was noted when the cells were cultivated in transdifferentiation medium.

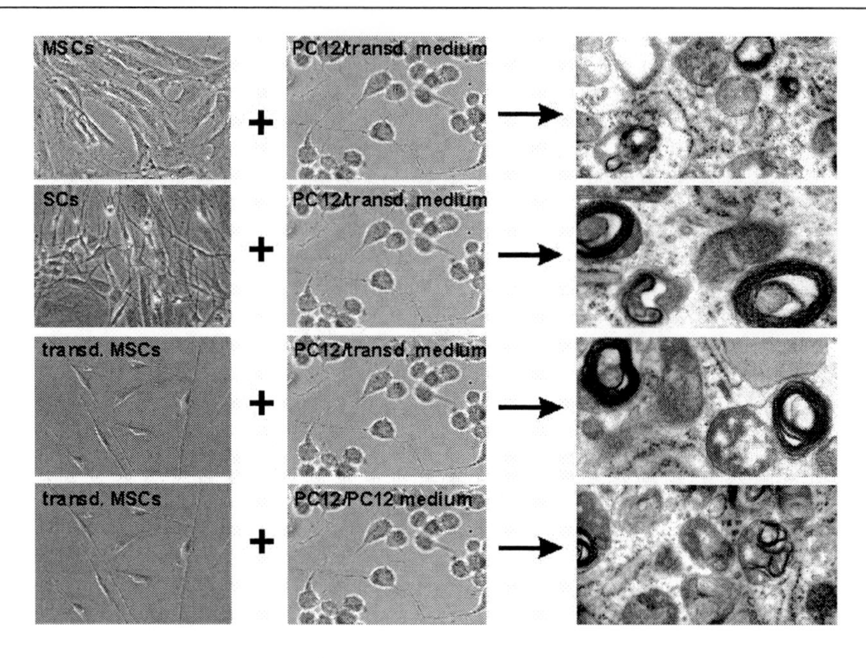

Figure 4. Phase contrast micrographs of 5 day old cultures and electron micrographs of the respective 14 day old co-cultures.

Poor myelination was found when the "PC12-medium" was used. These cultures produced myelin membranes with less layers which were not compacted. MSCs were not able to myelinize PC12 cell neurites.

In vivo myelination assay: The myelinating capacity was evaluated by the regenerative outcome three and six weeks after bridging a 2 cm gap of rat sciatic nerve with a graft tissue engineered from devitalized muscle enriched with the respective stem cells in comparison to muscle-SC grafts. Tolerance to the operations was good, all wounds healed primarily. None of the animals died. Clinical signs of pain (allodynia, hyperalgesia) or discomfort were only observed in the first two post operative weeks. Some animals, independent of treatment, showed a tendency for automutilation of their treated digits. Few of them had destruction of the entire forefoot which were healed after 4 weeks. No inflammatory changes of the reconstructed nerve were observed in any of the cases.

At both evaluation time points, best regenerative outcome, evaluated by morphometric analysis (Figure 5) and at light and electron microscopical levels (Figure 6), was demonstrable in the nerve graft group. Semi-thin sections showed a lot of freshly myelinated axons, more or less homogeneously distributed over the entire cross section of the graft (Figure 6A,B) and even in the distal nerve segment (Figure 6C). Electron microscopy demonstrated a regular axon shape, with wide cross-sections and partly thick myelin isolation in the graft (Figure 6A',B') as well as in the distal nerve segment (Figure 6C'). There was no connective tissue fibrosis. The regenerative capacity of the muscle conduits depended on the cell settlement. Poor regeneration was given in the cell-free muscle (Figure 6D-F) conduits and in ones with MSCs (Figure 6M-O). Here the prominent feature was a massive connective tissue fibrosis. A passable number of newly myelinated fibers, grouped in mini-fascicles, was seen in the muscle-SC (Figure 6G-I) conduits and in the muscle conduits enriched with transdifferentiated MSCs (Figure 6J-L).

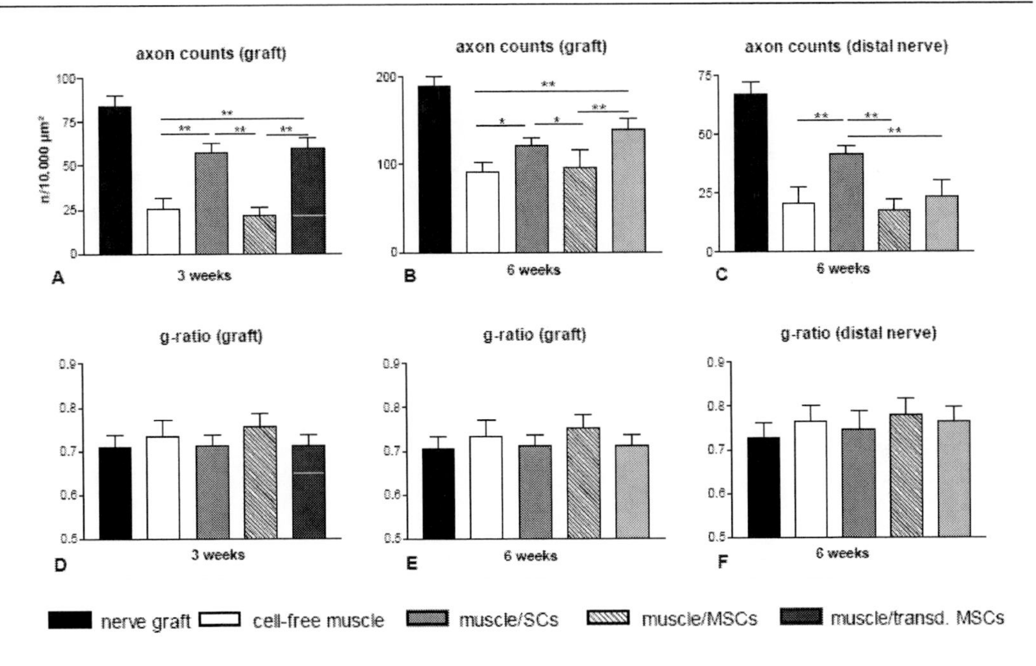

Figure 5. Three and six weeks after reconstruction, morphometric analysis of fiber counts showed a statistically significant reduction in all muscle grafts compared to the respective control nerve grafts (A,B, significances not indicated in the graph, ***p<0.0005 for cell-free and MSC-muscle grafts, *p<0.05 for the graft with Schwann cells or transdifferentiated MSCs), thereby the grafts with Schwann cells and thus with transdifferentiated MSCs developed a significant higher axon counts compared to the cell-free and MSC-muscle grafts (significances are given in the graph). (C) In the distal nerve segment, evaluated only after six weeks, the statistical significant superiority of nerve grafts was much more evident (not indicated in the graph, ***p<0.0005 for cell-free, MSC and transdifferentiated MSC grafts, **<0.005 for Schwann cell-grafts). The grafts with transdifferentiated MSCs demonstrated a reduced ability to overcome the distal suture when compared to the grafts with Schwann cells (**p<0.005, Mann-Whitney test). (D-F) The g-ratio (axon diameter/total fiber diameter, usually between 0.6 and 0.7), expressing the quality of the myelin sheaths, tended to increase in the grafts and distal nerve segments of all groups indicating thinner myelin sheaths. Data are mean values ± S.E.M.

At the electron microscope, fibers of these grafts, however, appeared to be smaller with a more irregular shape and thinner myelin sheaths. An irregular organization was visible. In contrast to SCs, transdifferentiated MSCs often myelinised two or even more axons (Figure 6K'). Myelin lamellae were regularly packed and the basal lamina seemed to be regularly developed. In the muscle-MSC group, however, the prominent feature was a massive connective tissue fibrosis. No signs of considerable regeneration could be found. Evaluation of the distal nerve segments revealed, that in SC-muscle grafts significantly more axons have crossed the distal suture (Figure 6I'). In both groups transplanted with stem cells (Figure 6L',O'), connective tissue was prominent. Isolated nerve fibers, some of which appeared to be degenerated, were seen only in the periphery.

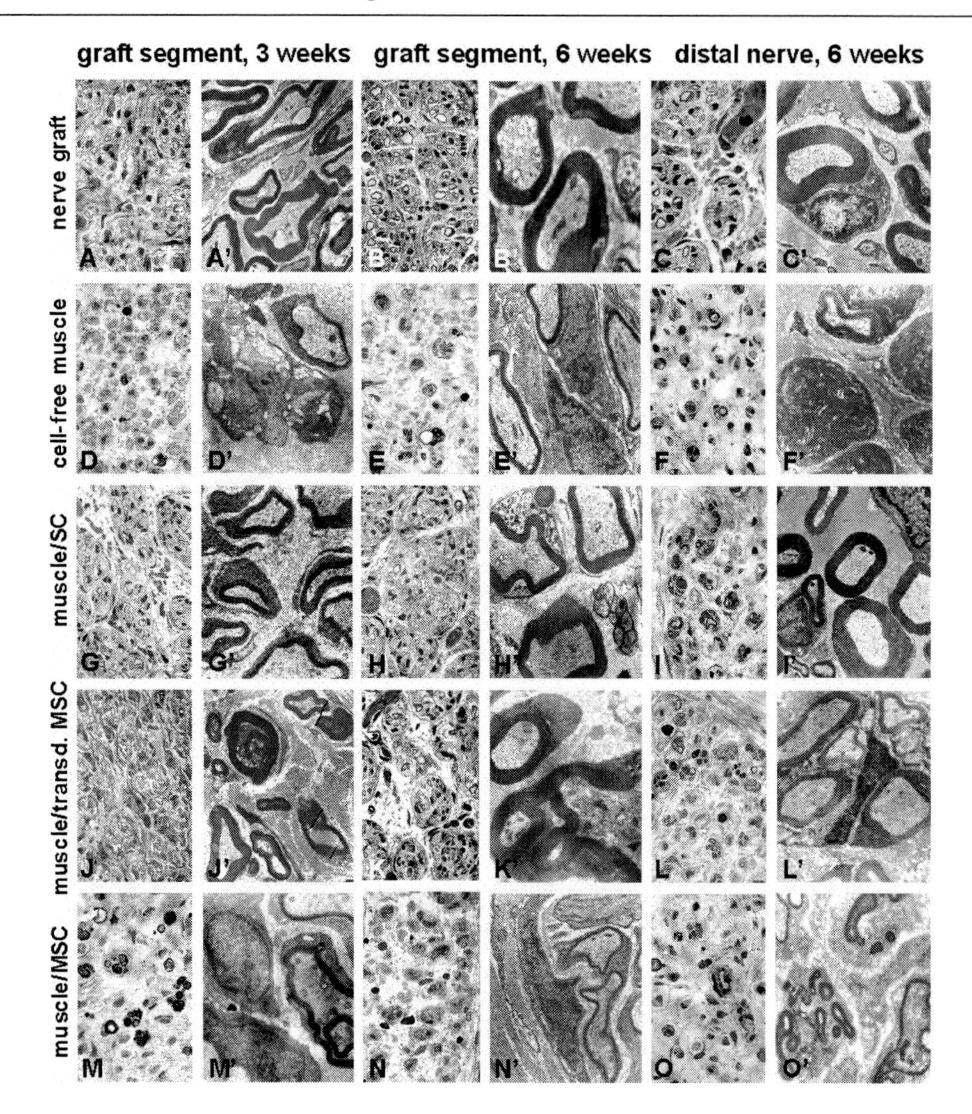

Figure 6. Light an electron micrographs of the respective graft (~10 mm distal to the proximal suture) and distal nerve segments (~5 mm distal to the distal suture) after a regeneration time of 3 and 6 weeks: (A-C) The control nerve graft showed the best regeneration indicated by a high number of axons with proper myelin sheath, a fascicular structure and no signs of cellular infiltration. Well-regenerated fibers were seen also in the distal nerve segment. (D-F) The cell-free muscle graft showed a bad regeneration. Connective tissue fibrosis is the prominent feature. Electron microscopic evaluation revealed only few deformed neurites growing into the distal nerve. (G-I) The additional implantation of Schwann cells (SC) into muscle grafts significantly improved the regeneration outcome. Axons were regularly shaped and well myelinized. The extent of connective tissue was low. Even in the distal nerve well myelinized axons, although isolated situated, were evident. (J-L) Implanted transdifferentiated MSCs were able to support regeneration throughout the muscle graft, although smaller fibers, with a more irregular shape and thinner myelin sheaths were seen compared with control and muscle-SC groups. In the distal nerve segment regeneration was bad like in the cell-free muscle graft. Connective scare tissue was prominent, indicating a problem in overcoming the distal suture. (M-O) The muscle-MSC graft showed an impaired regeneration with similar to the cell-free muscle characteristics. Semi-thin cross sections (A-O) stained with toluidine blue, x 400; electron micrographs (A'–O') x 7000.

Functional Outcome

Beginning with the 4th week, four out of seven animals in the nerve graft group, three out of seven animals in the SC-muscle group, and three out of seven animals in the muscle-transdifferentiated MSC group demonstrated thermo-sensitiveness of the treated leg. After six weeks, in all animals in these groups thermo-sensitiveness has been recovered. In the cell-free and in the MSC treated groups, only two out of seven animals revealed thermo-sensitiveness at the beginning of the 4th week and even after six weeks thermo-sensitiveness was absent in two animals in every of this groups.

Table 2. Timetable: return of thermal sensitivity/active spreading of digits

	animals	thermal sensitivity				spreading of digits			
time		3	4	5	6	3	4	5	6
nerve graft	1	-	-	+	+	-	-	-	-
	2	+	+	+	+	-	+	-	+
	3	+	+	+	+	-	-	+	+
	4	-	+	+	+	-	-	+	+
	5	+	+	+	+	-	-	+	+
	6	-	-	-	+	-	+	-	-
	7	+	+	+	+	-	-	+	+
cell-free muscle	1	-	-	-	-	-	-	-	-
	2	-	-	-	+	-	-	-	-
	3	-	-	+	+	-	-	-	-
	4	+	+	+	+	-	-	-	-
	5	+	+	+	+	-	-	-	-
	6	-	-	-	-	-	-	-	-
	7	-	+	+	+	-	-	-	-
muscle/ SCs	1	+	+	+	+	-	-	+	+
	2	-	+	+	+	-	-	+	+
	3	-	-	+	+	-	-	-	-
	4	-	+	+	+	-	-	-	-
	5	+	+	+	+	-	-	-	-
	6	+	+	+	+	-	-	-	+
	7	-	-	+	+	-	-	-	-
muscle/ MSCs	1	+	+	+	+	-	-	-	-
	2	-	-	-	+	-	-	-	-
	3	-	-	-	-	-	-	-	-
	4	-	-	-	-	-	-	-	-
	5	-	-	+	+	-	-	-	-
	6	+	+	+	+	-	-	-	-
	7	-	-	+	+	-	-	-	-
muscle/ transdiff. MSCs	1	+	+	+	+	-	-	+	+
	2	+	+	+	+	-	-	-	+
	3	+	+	+	+	-	-	-	-
	4	-	+	+	+	-	-	-	-
	5	-	-	+	+	-	-	-	-
	6	-	-	+	+	-	-	-	-
	7	-	-	-	+	-	-	-	-

(-) no reaction, (+) positive reaction, survival time in weeks

Beginning with the 5th week, none of the animals with a muscle-graft, independent on cell settlement, but two of nerve graft group were able to spread their digits actively. After six weeks, five out of seven animals in the nerve graft group, three out of seven animals in the SC-muscle group and two from the transdifferentiated MSC group demonstrated an active spreading. In the cell-free and in the MSC-muscle group this ability did not recover in the regeneration time studied (A survey is given in Table 2).

In the nerve graft group and in the SC-muscle group, the muscle weight ratio was quite good (0.52 ± 0.03, 0.43 ± 0.09). The transdifferentiated MSC group revealed a lower ratio (0.30 ± 0.04). However, this was still superior to the cell-free and the MSC-muscle groups, where the ratio reached 0.24 ± 0.07 and 0.25 ± 0.06, respectively.

Revascularization and Macrophage Recruitment

Evens-blue stained cross-sections of the contralateral nerve revealed uniform fascicular intravascular fluorescence (Figure 7A). All vessels visible by fluorescence microscopy demonstrated intraluminal red fluorescence. Extravasation of the dye was limited to epineurium. Endoneurial tissue was black. Nerve graft revascularization was established by day 3 (Figure 7B). On day 5 there was a massive distribution of endoneurial vessels throughout the entire graft (Figure 7C). Cell-free and the SC-muscle conduits showed a delayed revascula-rization. Blood vessels were evident only at day 5 (Figure 7D,F). On day 10, revascu-larization includes the whole lumen of the conduits (Figure 7E,G). Revascularization of the MSCs enriched conduits started with a further delay. Here, evens-blue staining could be seen only on day 10 (Figure 7H). Transdifferentiated MSCs, however, seemed to have the ability to induce neo-vascularization. These muscle conduits were already well revascularized by day 3 (Figure 7I).

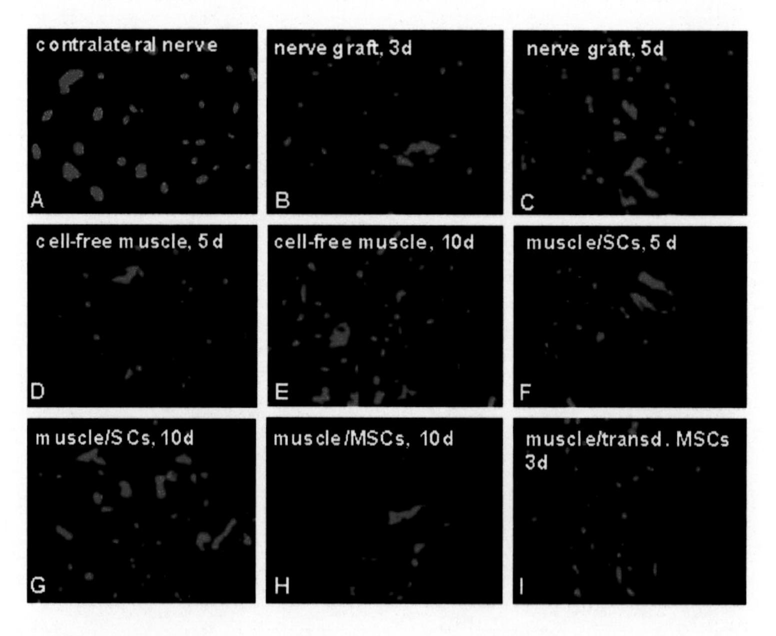

Figure 7. Revascularization pattern.

On day 3 after implantation, a moderate activation of ED2-positive resident tissue macrophages was found in the surrounding tissue of nerve and muscle conduits independent of the implanted cell type. As expected from the revascularization data, a moderate invasion of ED1-positive macrophages was demonstrated in the control nerve graft and also in the grafts enriched with transdifferentiated MSCs starting 3 days after implantation. Cell-free and SC-muscle conduits were entered by hematogenous macrophages as early as 5 days after implantation, whereas in the conduits with MSCs ED1 immunoreactivity was not detected at this time point. On day 10 ED1-positive macrophages were evident in all grafts studied.

Semi-quantitative analysis of ED1/ED2 immunostaining

graft	nerve graft		cell-free muscle		muscle/SC		muscle/MSC		muscle/transd. MSCs	
	ED1	ED2	ED1	ED2	ED1	ED2	ED1	ED2	ED1	ED2
3d border	+	+	-	+	-	+	-	+	+	+
centre	+	(+)	-	(+)	-	(+)	-	(+)	+	(+)
5d border	++	++	++	++	++	++	(+)	++	++	++
centre	++	+	+	++	+	++	-	++	++	++
10d border	++	+	++	+	++	+/++	+	+/++	++	+/++
centre	+/++	+	+/++	+	+/++	+	+	+	++	+

no immunoreactive cells,
(+) isolated immunoreactive cells,
+ isolated immunoreactive cell populations,
++ extensive immunoreactive cell populations.

Detection of Engrafted Cells

To confirm that engrafted male cells were responsible for the observed effects in the female recipient, we tested tissue for occurrence of Y-chromosomal SRY3 (Figure 8). The positive SRY3 signal indicates that the respective implanted cells were still in the graft after a regeneration period of three weeks. They had apparently not been replaced by host SCs. The negative control, a muscle of a female rat, was free of any PCR product.

Moreover, before injection stem cells were pre-labeled using a fluorescent cell linker kit, leaving the fluorogenic moiety (yellow) exposed near the outer surface of the cell. With an *in vivo* half-life, greater than 100 days, this method is ideal for cell tracking in the explanted grafts after three weeks. The combination with immunostaining for MBP (green fluorescence) showed a double staining for transdifferentiated MSCs (Figure 9A, arrows), indicating that they participate in re-myelination. A double labeling of MSCs could not be demonstrated (Figure 9B, magnification X 400).

This study clearly shows the plasticity of bone marrow derived rat MSCs by their transdifferentiation into SC-like cells with myelinating capacity. Devitalized biogenic muscle grafts enriched with transdifferentiated MSCs were able to support regeneration of the sciatic nerve. Grafts with undifferentiated MSCs lacked regenerative potency. Principally, MSCs can be induced by transdifferentiation to produce a trophic and tropic environment for regeneration processes. However, transdifferentiated MSCs did not reach the functional outcome of implanted Schwann cells.

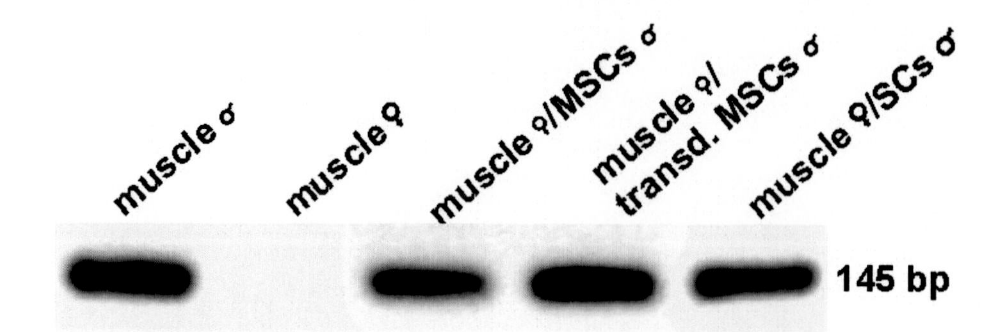

Figure 8.

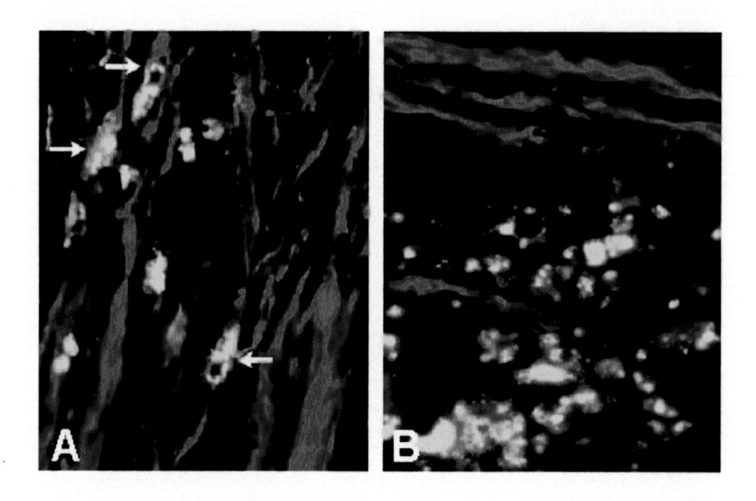

Figure 9.

In addition, in the grafts enriched with transdifferentiated Schwann cells fibers had a serious problem to cross the distal suture and to grow into the distal nerve segment.

Previous studies have shown that growth factors directly affect stem cell differentiation. However, there is no single factor which directs differentiation exclusively to one cell type (Schuldiner et al., 2000). In our hands, combinations of bFGF, PDGF-AA and Her-ß have proved to be successful. Prerequisite was the preincubation with β-mercaptoethanol and all-trans retinoic acid and co-incubation with forskolin. β-mercaptoethanol was used to promote formation of neurite-like outgrowth (Woodbury et al., 2000; Deng et al., 2001), although its effect is controversial (Lu et al., 2004). Retinoic acid is known to induce differentiation of embryonic stem cells into neural lineages cells (Fraichard et al., 1995). An increase in cAMP, and thus, an elevated expression of mitogenic genes can be achieved when cells are treated with forskolin (Fortino et al., 2002). Proliferation and differentiation effects are mediated by protein kinase C/MAP kinase pathway, which, amongst others, activates the transcription serum response factor and in the following c-fos (Eldredge et al., 1994). The anti-apoptotic pathway is mediated by phosphatidyl inositol-3-kinase (PI3-K). Downstream from PI3-K, Akt (also known as protein kinase B) is phosphorylated and multiple apoptosis preventing mechanisms are activated (Dudek et al., 1997; Kauffmann-Zeh et al., 1997). Because MSCs are likely to express many different receptor tyrosine kinases, it may also be possible for other cytokine mixtures to induce transdifferentiation into SC-like cells.

Development of an Schwann cell-like morphology cannot be taken as the only proof of commitment to an Schwann cell fate. To clearly demonstrate the success of transdifferentiation, cells should display a set of Schwann cell markers. Surprisingly, the "classic" Schwann cell gene S100b (Magnaghi et al., 2001; Jessen and Mirsky, 2002) was equally expressed in MSCs and Schwann cells, unlike expression of the LNGF receptor, CD104 and Krox20, also accepted markers for Schwann cells (Mirsky and Jessen, 1999; Keilhoff et al., 2000; Jessen and Mirsky, 2002; Chan et al., 2004) that were induced by transdifferentiation. The expression of the MSC marker BMPR-1A (Otsuka et al., 1999; Hoffmann and Gross, 2001), on the other hand, was reduced after transdifferentiation and reversed when cell were re-differentiated. The fluctuation between Schwann cell-like and MSC morphology suggests that MSCs have a dynamic, reversible response to induction conditions *in vitro*, which is also characteristic of Schwann cell differentiation *in vivo*. Fully differentiated Schwann cells retain an unusual plasticity throughout life and can readily de-differentiate to form cells similar to immature Schwann cells (Jessen, 2004).

As a "side effect" of transdifferentiation an increased cell death rate was noted. This seems to be in contrast to the presumed cytokine effect, but there is evidence indicating that either direct contact to axons or survival factors secreted by neurons are necessary for the survival and development of Schwann cells precursors as well as of mature Schwann cells (Mirsky and Jessen, 1999; Jessen, 2004).

Final proof of successful transdifferentiation is the demonstration of cell functionality, i.e. their myelinating capacity. That we have done in vitro and in vivo. In vitro, PC12 cells were used as reliable assay for detecting ensheathment of neurites. When the co-cultures were processed for electron microscopy, it could be demonstrated that Schwann cells and transdifferentiated MSCs were able to wrap P12 cell neurites normally unmyelinated. Co-cultures of PC12 and transdifferentiated MSCs cultured in "PC12-medium", were characterized by loosely compacted thin myelin sheats. Obviously, the "PC12-medium", lacking the transdifferentiation cytokine cocktail, was not able to stabilize transdifferentiation of MSCs to a needed for myelination extent.

In the present study, transdifferentiated MSCs had a good but compared with Schwann cells reduced guidance potency. These differences reflect the respective expression patterns of the calcium-binding protein S100b. As S100b that is released into the extracellular space stimulates neuronal survival, proliferation and differentiation (Donato, 2001; Shapiro and Whitaker-Azmitia, 2004), the differences in supporting extensive nerve regeneration may easily be attributed to the different levels of this neurotrophin. The expression patterns of CD 104 (beta4 integrin) and Krox20 may be also helpful to understand the differences in mylination. CD104 is a cell-surface extracellular matrix receptor, and several authors suggested a correlation between its expression and myelination, as Schwann cells from beta4 integrin-knock out mice form only rudimental myelin (Previtali et al., 2003). Krox20 (Egr2) is a transcription factor that controls Schwann cell myelination (Parkinson et al., 2004). Schwann cells in Krox20-knock out mice failed to myelinate, and unlike myelinating Schwann cells, continued to proliferate and were susceptible to death.

Moreover, the reduced ability of axons to growth through muscle grafts not enriched with Schwann cells and then to enter the distal nerve segment may be attributed to the lack of the L1 adhesion molecule, which in transdifferentiated MSCs is expressed only at a low level. L1 is a key player during peripheral nerve regeneration as axons express a preference for growing out on L1 and, particularly relevant for interpretation of our findings, to cross boundaries (Previtali et al., 2003; Rong et al., 2004).

Another explanation is the instability of MSC transdifferentiation. In this study, transdifferentiation was achieved by a cytokine cocktail leading to stable *in vitro* conditions, which cannot be assumed *in vivo*. The microenvironment of tissue plays an important role in the control of tissue-specific adult stem cells due to a combination of locally secreted factors, cell-cell interactions mediated by integral membrane proteins, and the extracellular matrix (Jan and Jan, 1998; Mehler, 2002). Normally, in a distal segment of transected peripheral nerves a microenvironment allowing regeneration is formed by reprogrammed Schwann cells aligned in Büngner bands and expressing surface molecules that guide regenerating fibers (Rong et al., 2004). It can be assumed that if transdifferentiated MSCs are transplanted into such an environment their long-term fate will be stably defined as Schwann cell-like by cell-cell interaction. In the presented experiments, however, the microenvironment for transplanted cells mainly consisted of devitalized muscle lacking most of the cues for peripheral nerve regeneration and subsequently for differentiation of MSCs into myelinating Schwann cell-like cells. The phenomenon that even terminally differentiated cells can reprogram their genome and switch their phenotype in response to extracellular cues was described by Song and Tuan (2004). Additionally, it must be taken into consideration that the pool of transdifferentiated stem cells is probably diluted by cell division. It is not necessarily conclusive that the used chemical procedure of transdifferentiation alone, without support of a respective microenvironment, results in stable gene expression patterns. Cell division is required for demethylation, which is a critical step for genome reprogramming. Without the pressure of inducing factors, fully differentiated MSCs recur into the cell cycle, modify their gene expression profile, and return to a more primitive stem cell-like stage (Song and Tuan, 2004). The reprogrammed daughter cells are then able to transdifferentiate into other cell types, here for instance into fibrocytes, in response to inductive extracellular cues or as result of communication with each other or with surrounding cells (Jan and Jan, 1998). This can explain the increased connective tissue fibrosis in the distal nerve segment of muscle grafts with transdifferentiated MSCs as well as the poor nerve regeneration in the muscle grafts enriched with undifferentiated MSCs.

Since the muscle grafts used were free of endogenous Schwann cells, it could be excluded that regeneration was assisted by endogenous Schwann cells or that implanted MSCs fused with remaining host Schwann cells pretending a transdifferentiation effect. In the grafts enriched with Schwann cells or transdifferentiated MSCs, the bundles of regenerated nerve fibers were arranged in mini-fascicles. These structures might be a result of newly formed perineurial cells isolating the regenerated clusters from the surrounding connective tissue. Such minifascicular reorganization is characteristic for nerve grafts in which the original perineurial barrier has been eliminated, as resembled by our muscle grafts (Schröder, 2001).

Interestingly, transdifferentiated MSCs were able to wrap more than one axon, a phenomenon never seen in Schwann cells but characteristic of myelination by oligodendrocytes in the CNS. Obviously, the transdifferentiation procedure used was also able to induce differentiation into other neural cell lineages, i.e. oligodendrocytes. Such a neuroglial differentiation of human MSCs has been demonstrated by other groups (Lee et al., 2004; Mimura et al., 2004; Zhao et al., 2004; Kamada et al., 2005). Unlike others (Cuevas et al., 2002; Akiyama et al., 2002; Tohill et al., 2004) we did not find any nerve regeneration in the muscle grafts enriched with undifferentiated MSCs, but rather a massive connective tissue fibrosis that suggests an arbitrary differentiation of MSCs. How can this apparently opposing data be conciliated? Experimental conditions (gap length, survival time, transplantation technique) vary a lot among different laboratories and even a sorted purified MSC population is heterogeneous *per se*.

Conclusion

We were able to transdifferentiate MSCs to Schwann cell-like cells characterized by an Schwann cell-like morphology and expression of respective biochemical markers. Functionality was confirmed by demonstrating their mylinating capacity in normally non-mylinated PC12 cell cultures and by showing a benefit for axonal regeneration after these cells were implanted into a biogenic muscle graft to bridge a sciatic nerve gap.

In the past few years research on stem cells has exploded as a tool to develop potential therapies for treatment of incurable neurodegenerative diseases. Despite promising results, significant constraints hamper the use of embryonic cells for transplantation in humans: besides ethical concerns, the viability, purity, carcinogenic potency, and final destiny of the cells have not been completely defined. Hence, adult MSCs are an attractive alternative candidate because they exhibit several important and potential advantageous features both in PNS and CNS regeneration. They can be obtained easily, and display an unorthodox plasticity. But again, significant constraints must be taken into consideration. The question of whether MSCs can transdifferentiate is only one of several aspects that must be clarified by intense laboratory experiments before grafting programs become widely applicable in clinical settings. Furthermore, it is still not known how the newly acquired differentiation program can be maintained. And it is absolutely essential to determine not only whether *in vivo* transplantation of transdifferentiated cells supports physiological processes like nerve regeneration, but it has also to be evaluated whether such transplantation induces a potential risk of tumor formation.

Nevertheless, autologous adult MSCs are a particularly attractive cellular therapy with several distinct advantages, such as: (1) They do not pose ethical problems. (2) They do not need immunosuppression. (3) They can be efficiently expanded by their selective attachment to tissue culture plastic. (4) They constitute an abundant and accessible reservoir for clinical use, overcoming the risk of obtaining cells from healthy nerves, and bone marrow harvesting is relatively non-invasive. (5) They reveal a considerable tolerance to genetic manipulations.

Although our results must be interpreted with caution, they give hope that the technique of transdifferentiation provides a tool to manipulate adult stem cells for cell-based therapy of myelin defects.

Acknowledgements

We would like to thank Karla Klingenberg, Leona Bück and Alexander Goihl for their contributions to the experiments. This work was supported by grants from the Hertie-Stiftung (Kei 1.01.1/03/011) the Klee-Stiftung and the Zinkann-Stiftung.

References

Abderrahim-Ferkoune, A., Bezy, O., Astri-Roques, S., Elabd, C., Ailhaud, G., Amri, E.Z. (2004). Transdifferentiation of preadipose cells into smooth muscle-like cells: role of aortic carboxypeptidase-like protein. *Exp. Cell Res.* 293, 219-28.

Akiyama, Y., Radtke, C., Kocsis, J.D. (2002). Remyelination of the rat spinal cord by transplantation of identified bone marrow stromal cells. *J. Neurosci.* 22, 6623-6630.

Bachelin, C., Lachapelle, F., Girard, C., Moissonnier, P., Serguera-Lagache, C., Mallet, J., Fontaine, D., Chonjnowski, A., Le Guern, E., Nait-Oumesmar, B., Baron-Van Evercooren, A. (2005). Efficient myelin repair in the macaque spinal cord autologous grafts of Schwann cells. *Brain* 128, 540-549.

Bianco, P., Riminucci, M., Gronthos, S., Robey, P.G. (2001). Bone marrow stromal cells: nature, biology, and potential applications. *Stem Cells* 19, 180-192.

Bjorklund, L.M., Sanchez-Pernaute, R., Chung, S., Andersson, T., Chen, I.Y., McNaught, K.S., Brownell, A.L., Jenkins, B.G., Wahlestedt, C., Kim, K.S., Isacson, O. (2002). Embryonic stem cells develop into functional dopaminergic neurons after transplantation in a Parkinson rat model. *Proc. Natl. Acad. Sci. USA* 99, 2344-2349.

Blakemore, W.F. (1977). Remyelination of CNS axons by Schwann cells transplanted from the sciatic nerve. *Nature* 266, 68-69.

Calderon-Martinez, D., Garavito, Z., Spinel, C., Hurtado, H. (2002). Schwann cell-enriched cultures from adult human peripheral nerve: a technique combining short enzymatic dissociation and treatment with cytosine arabinoside (Ara-C). *J. Neurosci. Methods* 114, 1-8.

Chan, J.R., Watkins, T.A., Cosgaya, J.M., Zhang, C., Chen, L., Reichardt, L.F., Shooter, E.M., Barres, B.A. (2004). NGF controls axonal receptivity to myelination by Schwann cells or oligodendrocytes. *Neuron* 43, 183-191.

Cuevas, P., Carceller, F., Dujovny, M., Garcia-Gomez, I., Cuevas, B., Gonzalez-Corrochano, R., Diaz-Gonzalez, D., Reimers, D. (2002). Peripheral nerve regeneration by bone marrow stromal cells. *Neurol. Res*. 24, 634-638.

Deber, C.M. and Reynolds, S.J. (1991). Central nervous system myelin: structure, function, and pathology. *Clin. Biochem.* 24, 113-134.

Deng, W., Obrocka, M., Fischer, I., Prockop, D.J. (2001). In vitro differentiation of human marrow stromal cells into early progenitors of neural cells by conditions that increase intracellular cyclic AMP. *Biochem. Biophys. Res. Commun.* 282, 148-152.

Dezawa, M. (2002). Central and peripheral nerve regeneration by transplantation of Schwann cells and transdifferentiated bone marrow stromal cells. *Anat. Science Int.* 77, 12-25.

Dezawa, M., Takahashi, I., Esaki, M., Takano, M., Sawada, H. (2001). Sciatic nerve regeneration in rats induced by transplantation of in vitro differentiated bone-marrow stromal cells. *Eur. J. Neurosci.* 14, 1771-1776.

Donato, R. (2001). S100: a multigenic family of calcium-modulated proteins of the EF-hand type with intracellular and extracellular functional roles. *Int. J. Biochem. Cell. Biol.* 33, 637-668.

Dudek, H., Datta, S.R., Franke, T.F., Birnbaum, M.J., Yao, R., Cooper, G.M., Segal, R.A., Kaplan, D.R., Greenberg, M.E. (1997). Regulation of neuronal survival by the serine-threonine protein kinase Akt. *Science* 275, 661-665.

Eldredge, E.R., Korf, G.M., Christensen, T.A., Connolly, D.C., Getz, M.J., Maihle, N.J. (1994). Activation of c-fos gene expression by a kinase-deficient epidermal growth factor receptor. *Mol. Cell. Biol.* 4, 7527-7534.

Evans, G.R. (2000). Challenges to nerve regeneration. *Semin. Surg. Oncol.* 19, 312-318.

Evans, P.J., Midha, R., Mackinnon, S.E. (1994). The peripheral nerve allograft: a comprehensive review of regeneration and neuroimmunology. *Prog. Neurobiol.* 43, 187-233.

Fansa, H. and Keilhoff, G. (2004). Comparison of different biogenic matrices seeded with cultured Schwann cells for bridging peripheral nerve defects. *Neurol. Res.* 26, 167-173.

Fansa, H., Keilhoff, G., Plogmeier, K., Frerichs, O., Wolf, G., Schneider, W. (1999). Successful implantation of Schwann cells in acellular muscle. *J. Reconstr. Microsurg.* 15, 61-65.

Fansa, H., Keilhoff, G., Wolf, G., Schneider, W. (2001). Tissue engineering of peripheral nerves: A comparison of venous and acellular muscle grafts with cultured Schwann cells. *Plast .Reconstr. Surg.* 107, 485-494.

Fortino, V., Torricelli, C., Gardi, C., Valacchi, G., Rossi Paccani, S., Maioli, E. (2002). ERKs are the point of divergence of PKA and PKC activation by PTHrP in human skin fibroblasts. *Cell. Mol. Life Sci.* 59, 2165-2171.

Fraichard, A., Chassande, O., Bilbaut, G., Dehay, C., Savatier, P., Samarut, J. (1995). In vitro differentiation of embryonic stem cells into glial cells and functional neurons. *J. Cell Sci.* 108, 3181-3188.

Gravvanis, A.I., Tsoutsos, D.A., Tagaris, G.A., Papalois, A.E., Patralexis, C.G., Iconomou, T.G., Panayotou, P.N., Ioannovich, J.D. (2004) Beneficial effect of nerve growth factor-7S on peripheral nerve regeneration through inside-out vein grafts: an experimental study. *Microsurgery* 24, 408-415.

Hall, S. (1997). Axonal regeneration through acellular muscle grafts. *J. Anat.* 190, 57-71.

Hoffmann, A. and Gross, G. (2001). BMP signaling pathways in cartilage and bone formation. *Crit. Rev. Eukaryot. Gene Expr.* 11, 23-45.

Honmou, O., Felts, P.A., Waxman, S.G., Koscis, J.D. (1996). Restoration of normal conduction properties in dfemyelinated spinal cord axons in the adult rat by transplantation of exogenous Schwann cells. *J. Neurosci.* 16, 3199-3208.

Ide, C. (1996). Peripheral nerve regeneration. *Neurosci. Res.* 25, 101-121.

Jan, Y.N. and Jan, L.N. (1998). Asymmetric cell division. *Nature* 392, 775-778.

Jessen, K.R. and Mirsky, R. (2002). Signals that determine Schwann cell identity. *J. Anat.* 200, 367-376.

Jessen, K.R. (2004). Glial cells. *Int. J. Biochem. Cell Biol.* 36, 1861-1867.

Kamada, T., Koda, M., Dezawa, M., Yoshinaga, K., Hishimoto, M., Koshizuka, S., Nishio, Y., Moriya, H., Yamazaki, M. (2005). Transplantation of bone marrow stromal cell-derived Schwann cells promotes axonal regeneration and functional recovery after complete transection of adult rat spinal cord. *J. Neuropathol. Exp. Neurol.* 64, 37-45.

Kanaya, F., Firrell, J.C., Breidenbach, W.C. (1996). Sciatic function index, nerve conduction tests, muscle contraction, and axon morphometry as indicators of regeneration. *Plast. Reconstr. Surg.* 98, 1264-1274.

Kashofer, K. and Bonnet, D. (2005). Gene therapy progress and prospects: stem cell plasticity. *Gene Therapy* 12, 1229-1234.

Kauffmann-Zeh, A., Rodriguez-Viciana, P., Ulrich, E., Gilbert, C., Coffer, P., Downward, J., Evan, G. (1997). Suppression of c-Myc-induced apoptosis by Ras signalling through PI(3)K and PKB. *Nature* 385, 544-548.

Keilhoff, G., Fansa, H., Schneider, W., Wolf, G. (1999). In vivo predegeneration of peripheral nerves: an effective technique to obtain activated Schwann cells for nerve conduits. *J. Neurosci. Methods* 89, 8917-8924.

Keilhoff, G., Fansa, H., Smalla, K.H., Schneider, W., Wolf, G. (2000). Neuroma: a donor-age independent source of human Schwann cells for tissue engineered nerve grafts. *Neuroreport* 11, 3805-3809.

Kocsis, J.D., Akiyama, Y., Lankford, K.L., Radtke, C. (2002). Cell transplantation of peripheral-myelin-forming cells to repair the injured spinal cord. *J. Rehabil. Res.* Develop. 39, 287-298.

Kopen, G.C., Prockop, D.J., Phinney, D.G. (1999). Marrow stromal cells migrate throughout forebrain and cerebellum, and they differentiate into astrocytes after injection into neonatal mouse brains. *Proc. Natl. Acad. Sci. USA* 96, 10711-10716.

Kornek, B. and Lassmann, H. (1999). Axonal pathology in multiple sclerosis. A historical note. *Brain Pathol.* 9, 651-656.

Lee, O.K., Ko, Y.C., Kuo, T.K., Chou, S.H., Li, H.J., Chen, W.M., Chen, T.H., Su, Y. (2004). Fluvastatin and lovastatin but not pravastatin induce neuroglial differentiation in human mesenchymal stem cells. *J. Cell. Biochem.* 93, 917-928.

Liu, H.M., Yang, L.H., Yang, Y.J. (1995). Schwann cell properties: 3. C-fos expression, bFGF production, phagocytosis and proliferation during Wallerian degeneration. J. Neuropathol. *Exp. Neurol.* 54, 487-496.

Lu, P., Blesch, A., Tuszynski, M.H. (2004). Induction of bone marrow stromal cells to neurons: differentiation, transdifferentiation, or artefact? *J. Neurosi. Res.* 77, 174-191.

Lucchinetti, C., Brück, W., Parisi, J., Scheithauer, B., Rodriguez M., Lassmann, H. (1999). A quantitative analysis of oligodendrocytes in multiple sclerosis lesions. A study of 113 cases. *Brain* 122, 2279-2295.

Lundborg, G. (2004). Alternatives to autologous nerve grafts. Handchir. Mikrochir. *Plast. Chir.* 36, 1-7.

Magnaghi, V., Cavarretta, I., Galbiati, M., Martini, L., Melcangi, R.C. (2001). Neuroactive steroids and peripheral myelin proteins. *Brain Res. Brain Res. Rev.* 37, 360-371.

Mehler, M.F. (2002). Mechanisms regulating lineage diversity during mammalian cerebral cortical neurogenesis and gliogenesis. *Results Probl. Cell. Differ.* 39, 27-52.

Mimura, T., Dezawa, M., Kanno, H., Sawada, H., Yamamoto, I. (2004). Peripheral nerve regeneration by transplantation of bone marrow stromal cell-derived Schwann cells in adult rats. *J. Neurosurg.* 101, 806-812.

Mirsky, R. and Jessen, K.R. (1999). The neurobiology of Schwann cells. *Brain Pathol.* 9, 293-311.

Morrissey, T.K., Kleitman, N., Bunge, R.P. (1991). Isolation and functional characterization of Schwann cells derived from adult peripheral nerve. *J. Neurosci.* 11, 2433-2442.

Myckatyn, T.M. and Mackinnon, S.E. (2004). A review of research endeavors to optimize peripheral nerve reconstruction. *Neurol. Res.* 26, 124-138.

Near, S.L., Whalen, L.R., Miller, J.A., Ishii, D.N. (1992). Insulin-like growth factor II stimulates motor nerve regeneration. *Proc. Natl. Acad. Sci. USA* 89, 11716-11720.

Oswald, J., Boxberger, S., Jorgensen, B., Feldmann, S., Ehninger, G., Bornhauser, M., Werner, C. (2004). Mesenchymal stem cells can be differentiated into endothelial cells in vitro. *Stem Cells* 22, 377-384.

Otsuka, E., Yamaguchi, A., Hirose, S., Hagiwara, H. (1999). Characterization of osteoblastic differentiation of stromal cell line ST2 that is induced by ascorbic acid. *Am. J. Physiol.* 277, 132-138.

Parkinson, D.B., Bhaskaran, A., Droggiti, A., Dickinson, S., D'Antonio, M., Mirsky, R., Jessen, K.R. (2004). Krox-20 inhibits Jun-NH2-terminal kinase/c-Jun to control Schwann cell proliferation and death. *J. Cell Biol.* 164, 385-395.

Phinney, D.G., Kopen, G., Isaacson, R.L., Prockop, D.J. (1999). Plastic adherent stromal cells from the bone marrow of commonly used strains of inbred mice: variations in yield, growth, and differentiation. *J. Cell. Biochem.* 72, 570-585.

Pittenger, M.F. and Martin, B.J. (2004). Mesenchymal stem cells and their potential as cardiac therapeutics. *Circ. Res.* 95, 9-20.

Pittenger, M.F., Mackay, A.M., Beck, S.C., Jaiswal, R.K., Douglas, R., Mosca, J.D., Moorman, M.A., Simonetti, D.W., Craig, S., and Marshak, D.R. (1999). Multilineage potential of adult human mesenchymal stem cells. *Science* 284, 143-147.

Previtali, S.C., Nodari, A., Taveggia, C., Pardini, C., Dina, G., Villa, A., Wrabetz, L., Quattrini, A., Feltri, M.L. (2003). Expression of Laminin Receptors in Schwann Cell Differentiation: Evidence for Distinct Roles. *J. Neurosci.* 23, 5520-5530.

Prockop, D.J. (1997). Marrow stromal cells as stem cells for nonhematopoietic tissues. *Science* 276, 71-74.

Rong, L.L., Trojaborg, W., Qu, W., Kostov, K., Yan, S.D., Gooch, C., Szabolcs, M., Hays, A.P., Schmidt, A.M. (2004). Antagonism of RAGE suppresses peripheral nerve regeneration. *FASEB J.* 18, 1812-18127.

Rutenberg, M.S., Hamazaki, T., Singh, A.M., Terada, N. (2004). Stem cell plasticity, beyond alchemy. *Int. J. Hematol.* 79, 15-21.

Schmidt, C.E. and Leach, J.B. (2003). Neural tissue engineering: strategies for repair and regeneration. *Annu. Rev. Biomed. Eng.* 5, 293-347.

Schröder, J.M. (2001). *Pathology of peripheral nerves*. Springer –Verlag, Berlin Heidelberg New York.

Schuldiner, M., Yanuka, O., Itskovitz-Elder, J., Melton, D.A., Benvenisty, N. (2000). Effects of eight growth factors on the differentiation of cells derived from human embryonic stem cells. *Proc. Natl. Acad. Sci. USA* 97, 11307-11312.

Shapiro, L.A. and Whitaker-Azmita, P.M. (2004) Expression levels of cytoskeletal proteins indicate pathological aging of S100B transgenic mice: an immunohistochemical study of MAP-2, drebrin and GAP-43. *Brain Res.* 1019, 39-46.

Song, L. and Tuan, R.S. (2004). Transdifferentiation potential of human mesenchymal stem cells derived from bone marrow. FASEB J. 18, 980-983.

Stang, F., Fansa, H., Wolf, G., Reppin, M., Keilhoff, G. (2005). Structural parameters of collagen nerve grafts influence peripheral nerve regeneration. *Biomaterials* 26, 3083-3091.

Stoll, G. and Müller, H.W. (1999). Nerve injury, axonal degeneration and neural regeneration: basic insights. *Brain Pathol.* 9, 313-325.

Tao, H. and Ma, D.D. (2003). Evidence for transdifferentiation of human bone marrow-derived stem cells: recent progress and controversies. *Pathology* 35, 6-13.

Tohill, M. and Terenghi, G. (2004). Stem-cell plasticity and therapy for injuries of the peripheral nervous system. *Bitechnol. Appl. Biochem.* 40, 17-24.

Tohill, M., Mantovani, C., Wiberg, M., Terenghi, G. (2004). Rat bone marrow mesenchymal stem cells express glial markers and stimulate nerve regeneration. *Neurosci. Lett.* 362, 200-203.

Wiberg, M. and Terenghi, G. (2003). Will it be possible to produce peripheral nerves? *Surg. Technol. Int.* 11, 303-310.

Woodbury, D., Schwarz, E.J., Prockop, D.J., Black, I.B. (2000). Adult rat and human bone marrow stromal cells differentiate into neurons. *J. Neurosci. Res.* 61, 364-370.

Zhao, L.X., Zhang, J., Cao, F., Meng, L., Wang, D.M., Li, Y.H., Nan, X., Jiao, W.C., Zheng, M., Xu, X.H., Pei, X.T. (2004). Modification of the brain-derived neurotrophic factor gene: a portal to transform mesenchymal stem cells into advantageous engineering cells for neuroregeneration and neuroprotection. *Exp. Neurol.* 190, 396-406.

In: Stem Cell Research Trends
Editor: Josse R. Braggina, pp. 257-276
ISBN: 978-1-60021-622-0

Chapter IX

WOUND HEALING ACTIVITY OF BONE MARROW STROMAL STEM CELLS

Seung-Kyu Han[1] and Chi-Ho Lee[2]
[1]Department of Plastic Surgery,
Korea University College of Medicine
[2]Department of Plastic Surgery,
Kangwon National University College of Medicine

ABSTRACT

Bone marrow stroma is the source of mesenchymal stem cells. Several research articles dealing with the use of bone marrow stromal stem cells (BSCs) for regenerating bone or cartilage have been published. However, there has been no research on the effect of BSC transplantation on wound healing. The BSCs secrete collagen and several cytokines essential for wound healing. However, it is unclear as to what extent these materials are produced and it has not yet been studied whether they are superior to fibroblasts (which have been used conventionally) in promoting the wound healing process. It has been hypothesized that BSCs and fibroblasts would differ in their ability to promote wound healing, and furthermore that BSCs should have greater activity than fibroblasts. If so, then the current cell therapy method using fibroblasts to stimulate the healing of chronic wounds could be replaced by a therapy which uses BSCs and has a superior effect. The purpose of this chapter is to compare the wound healing activity of BSCs with that of fibroblasts *in vitro* and *in vivo*. Cultured human BSCs and dermal fibroblasts taken from same patients were tested. *In vitro* study was focused on cell proliferation, synthesis of collagen, and production of three growth factors – basic fibroblast growth factor (bFGF), vascular endothelial growth factor (VEGF), and transforming growth factor beta (TGF-β) – which are considered to be important in chronic wound healing. *In vivo* study was performed to compare collagen synthesis, epithelization, and angiogenic activity of BSCs with those of fibroblasts using rat wound models. In the *in vitro* study we did not observe great differences in cell proliferation and TFG-β secretion. In contrast, the amount of collagen synthesis and the levels of bFGF and VEGF were much higher in the BSC group than in the fibroblast group. In particular, the VEGF level of the BSC group was 12 times higher than that of the fibroblast group. In the *in vivo* study, great differences were noted in collagen synthesis, epithelization, and angiogenesis among the three groups. The BSC group showed the best results, followed by the fibroblast group and then the no cell group. These results suggest that the potential of BSCs in wound healing acceleration may be superior to that of fibroblasts and that BSCs may be possibly used as a replacement for fibroblasts, which are currently being used for wound healing.

INTRODUCTION

Chronic wounds present the physician with a difficult treatment problem, and these lesions respond poorly to conventional treatment. Although there has recently been much interest in the use of topical growth factors for the treatment of chronic wounds such as diabetic foot ulcers, the effects have not generally been very dramatic [1-4]. In other attempts to deliver growth factors to wounds, fibroblast implants, which are able to adjust to a wound's environment and provide the desired growth factors and other substances that maybe be lacking in a chronic wound, have been developed as biological wound dressings [5-8].

Bone marrow stroma is the source of mesenchymal stem cells that may serve as long-lasting precursors for bone, cartilage, muscle, and connective tissues [9-11]. Mesenchymal stem cells also have a low immunity-assisted rejection rate, and they have a greater ability to divide without apoptosis than differentiated cells [12-14]. Therefore, they are drawing intense attention in the bioengineering field. Mesenchymal stem cells have been extensively studied and a lot of research articles using bone marrow stromal stem cells (BSCs) for regenerating bone or cartilage have been presented [11,15]. However, most of these studies have been performed with cells maintained under differentiating culture conditions [16-18] and there has been no research on the effect of the transplantation of BSCs on wound healing.

In this chapter, the BSC cultures were maintained in the absence of differentiation stimuli, and under these conditions they should retain the properties of mesenchymal stem cells [9, 14]. BSCs are relatively easy to isolate from other cells in bone marrow because of their tendency to adhere to tissue culture dish plastic [9]. These cells have many of the characteristics of stem cells for tissues, and they can be roughly defined as mesenchymal because they can be differentiated by culture into osteoblasts, chondroytes, adipocytes, fibroblasts and even myoblasts [9, 14]. In addition, it was demonstrated that even after 20 or 30 cell doubling in culture, they still retain stem-cell properties [14]. Therefore, BSCs present an intriguing model for examining the differentiation of stem cells, and BSCs are potentially useful for a number of cell therapies.

The cultured BSCs synthesize extracellular matrices that include type I collagen, fibronectin, type IV collagen and laminin [9]. The cells also secrete several cytokines [9]. However, it is unclear to what extent these cytokines are produced and it has not yet been studied whether they have superior characteristics in promoting wound healing process than the fibroblasts that have been used for this purpose.

It has been hypothesized that BSCs and fibroblasts would differ in their ability to promote wound healing and furthermore, BSCs should have greater activity than the fibroblasts. If this is so, then the cell therapy method using fibroblasts to stimulate healing of chronic wounds could be replaced by a therapy having a superior effect, which could be the use of BSCs in the future. The purpose of this chapter is to present potential of wound healing activity of BSCs *in vitro* and *in vivo*.

WOUND HEALING ACTIVITY OF BSCs *IN VITRO*

Previous studies of wound healing have shown that deposition of collagen, epithelization, and angiogenesis, which are fundamental processes for wound healing, are impaired in chronic wounds such as diabetic foot ulcers [19-22], and the key growth factors that are important in chronic wound healing include basic fibroblast growth factor (bFGF), vascular endothelial growth factor (VEGF), and transforming growth factor beta (TGF-β) [19, 22, 23]. Basic FGF is a angiogenic growth factor that stimulates endothelial cell proliferation and migration, and it is an efficient mediator of VEGF activity [24, 25]. It also promotes epithelization. VEGF, which is a family of platelet derived growth factors, induces the process of angiogenesis and endothelial cell proliferation, stimulates vascular permeability [25, 26], and causes vasodilatation [27]. It can also stimulate cell migration and inhibit apoptosis [27]. TGF-β is a potent stimulator of fibroblast proliferation and collagen synthesis [19, 24].

The purpose of this pilot study was to compare the essential factors for wound healing of BSCs and fibroblasts *in vitro*. This study was especially focused on cell proliferation, the synthesis of collagen which is the major component of the extracellular matrices, and the production of three growth factors – bFGF, VEGF, and TGF-β – which are considered to be important for chronic wound healing [19,29,30].

Isolation and Culture of Dermal Fibroblasts and BSCs

Leftover materials were obtained from the bone marrow and dermis of the same patients who were undergoing simultaneous bone marrow cell grafts for phalangeal bone reconstruction and skin grafts. Six normal male patients were included in this study and their age ranged from 20 to 38 years. The institutional review board approved the study, and all the patients provided an informed written consent. The patients fully understood and agreed with the purposes of the study, and they were happy and willing to take part in the study.

Freshly discarded healthy skin tissue was deepithelialized and minced into 2 X 1 mm pieces. These pieces were spread evenly over the surface of a 100-mm tissue culture plate that was precoated with 3 ml of Dulbecco's Modified Eagle Medium/Ham's F-12 (DMEM/F-12; GIBCO, Grand island, NY) containing 50% fetal bovine serum (FBS; GIBCO, Grand island, NY). The plates were incubated at 37°C for 4 hours to facilitate tissue adhesion. After this incubation, 12 ml of DMEM/F-12 containing 10% FBS and 25μg/ml gentamycin was added and the plates were returned to the incubator. All the cell cultures in this study were grown at 37°C in a 5% CO_2/95% air atmosphere. The confluent cultured fibroblasts were next trypsinized, and the dissociated cells were diluted 2.7-fold with Dulbecco's phosphate buffered saline without Mg^{2+} and Ca^{2+} (DPBS; GIBCO, Grand island, NY, U.S.A.), and then they collected by centrifugation at 450 x *g* for 17 minutes. The cells were washed twice in 40 ml DPBS, resuspended in 5 ml DPBS and filtered through a 100μm nylon mesh. The cell density was determined by counting in a hemocytometer, and viability was assessed by Trypan blue exclusion assay. Third passage cell cultures were used for the experiments in this chapter.

Adult BSCs were obtained by way of bone marrow aspiration from the posterior iliac crest of the patients. A total of 20ml of bone marrow was aspirated using a syringe coated with 5,000 IU of heparin. After the aspired cells were placed to a 50ml centrifuge tube, a Ficoll-Paque density gradient solution (Ficoll-Paque, Amersham Biotech., Stockolm, Sweden) of 1.077 was added and centrifugation was performed for 30 minutes at room temperature at 2,500 X *g*. From the separated bone marrow sample, the mononuclear cell layer of the medium's upper layer was collected to start the culture. We set up all the conditions including culture medium, culture process, and subculture to be the same with those used for the dermal fibroblast culture. Purity of the BSCs was confirmed using flow cytometer (Coulter EPICS XL, Beckman Coulter, CA). The BSCs used in the study were negative for the antibodies against hematopoietic antigens CD45 (Immunotech, Marseille, France), CD34 (Becton Dickinson Bioscience, CA), and CD14 (Immunotech, Marseille, France). Regarding the BSC characterization, there is still no specific marker for BSCs. However, the presence of the any potential hematopoietic cells can be avoided as the BSCs are maintained as primary culture for 2 or 3 weeks [14, 31, 32]. We have also confirmed that no hematopoietic cells were present in our BSC cultures.

For the monolayer cultures, the human dermal fibroblast population consisted of homogeneously spindle-shaped cells, whereas the majority of BSCs were of a larger size and more polygonal in shape (Figure 1). The BSCs and fibroblasts maintained these phenotypes throughout the experiments. As for the shape of BSCs, previous studies have documented that they were elongated, spindle shaped cells similar to the morphology of fibroblasts [12, 33, 34]. However, the shape of BSCs from the patients involved in this chapter was more polygonal, and their size was larger than the dermal fibroblasts from the same patient. To our knowledge, the data we have presented here are the first to show the differences of phenotype between BSCs and mature fibroblasts from a same human being.

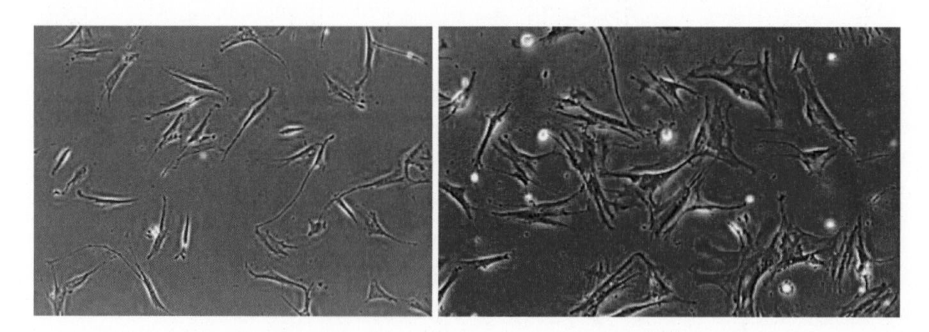

Figure 1. Human dermal fibroblasts (Left) and BSCs (Right) from a same patient during monolayer culture.

Cell Proliferation Assay

Cell proliferation was determined by 3-(4,5-dimethylthiazol-2-yl)-2,5-diphenyl tetrazolium bromide (MTT:Sigma, St. Louis) assay. Briefly, 10$\mu\ell$ MTT of 5mg/ml was added to a 100$\mu\ell$ cell monolayer in each 96-well plate, after which incubation was performed for 3 hours at 37℃. Next, 100$\mu\ell$ 0.04M HCl in propan-2-ol was added to each well and mixed thoroughly to dissolve the insoluble blue formazan crystals. The absorbance was measured at a test wavelength of 570mm and a reference wavelength of 630mm using an ELISA reader.

For statistical Analyses, all experiments were performed in triplicate on each explant from each individual and the average value was used as a datum for a subject at each time interval. Statistical comparisons were performed with the Mann-Whitney U-test, with p values < 0.05 being considered statistically significant.

There was no significant difference in cell proliferation between BSC and fibroblast groups at the first day after plating. At third and fifth day post-plating, the cell proliferation in the BSC group was 11 and 17 percent greater than the cell proliferation noted in fibroblast group, respectively. However, there was no statistical significance (Figure 2).

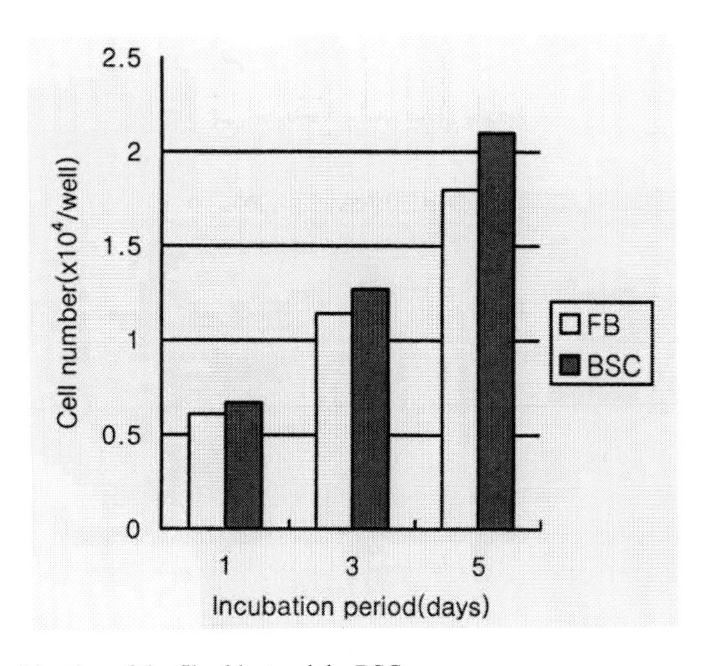

Figure 2. Cell proliferation of the fibroblast and the BSC groups.

Collagen Synthesis Assay

To measure the production of collagen, the collagen type I carboxy-terminal propeptide (CICP) enzyme immunoassay was performed using a Metra CICP kit (Quidel, CA) according to the manufacturer's instruction. Briefly, 100$\mu\ell$ diluted culture supernatants were added to each well of the monoclonal anti-CICP antibody coated plates, and then the plates were incubated at room temperature for 2 hours. 100$\mu\ell$ rabbit anti-CICP antiserums were added for 50 minutes, followed by 100$\mu\ell$ goat anti-rabbit alkaline phosphatase conjugates for 50 minutes. The reaction was stopped and the collagen synthesis was measured at 405nm.

Collagen synthesis of the BSC group significantly increased when compared to that of the fibroblast group, with increases of 28, 80 and 70 percent at 1, 3 and 5 days after plating, respectively ($p<0.05$) (Figure 3).

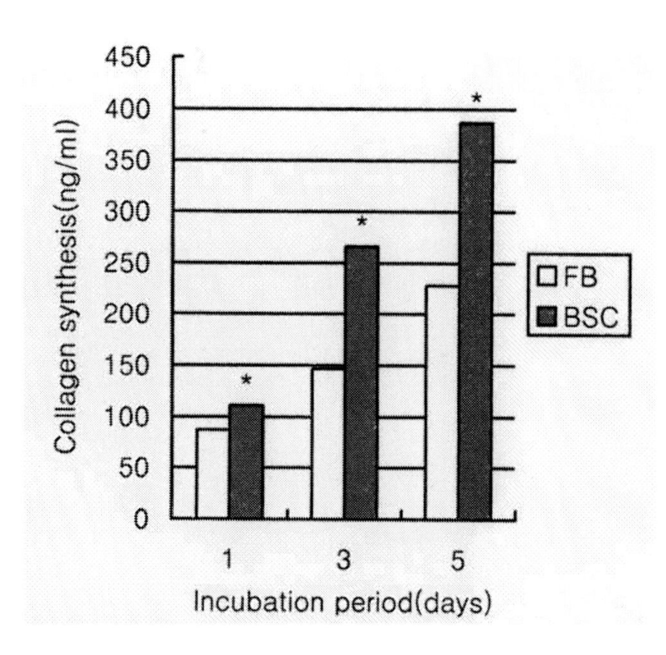

Figure 3. Collagen synthesis of the fibroblast and the BSC groups (* $p<0.05$).

Growth Factor Level Assay

The levels of bFGF, VEGF, and TGF-β1 were measured in culture media with an enzyme-linked immunosorbent assay (ELISA) kit (RandD Systems, Minneapolis, MN) according to the manufacturer's instruction. Briefly, the samples and standards were diluted appropriately in the respective assay diluent and then incubated at room temperature in wells pre-coated with the specific capturing antibodies. After washing, the respective detection antibody-conjugate was added and then incubated at room temperature before washing again. The addition of substrate solution then resulted in color development. After the addition of a stop solution, the color intensity was measured in an ELISA reader at the appropriate wavelength. The measured TGF-β1 in this chapter included both active and latent forms.

Of the three growth factors that we studied, the BSC group showed significantly higher levels for the expression of bFGF and VEGF. The bFGF levels in the BSC group were 47, 89 and 68 percent greater than those for the fibroblast group at each time interval, respectively ($p<0.05$) (Figure 4). The most profound difference was observed for the VEGF level. The VEGF levels of the BSC group were seven, twelve and twelve fold higher than the VEGF levels of the fibroblast group at each time interval, respectively ($p<0.05$) (Figure 5).

However, the TGF-β level of the fibroblast group was slightly higher than that of the BSC group by 13, 12 and 22 percent at each time interval, respectively ($p<0.05$) (Figure 6).

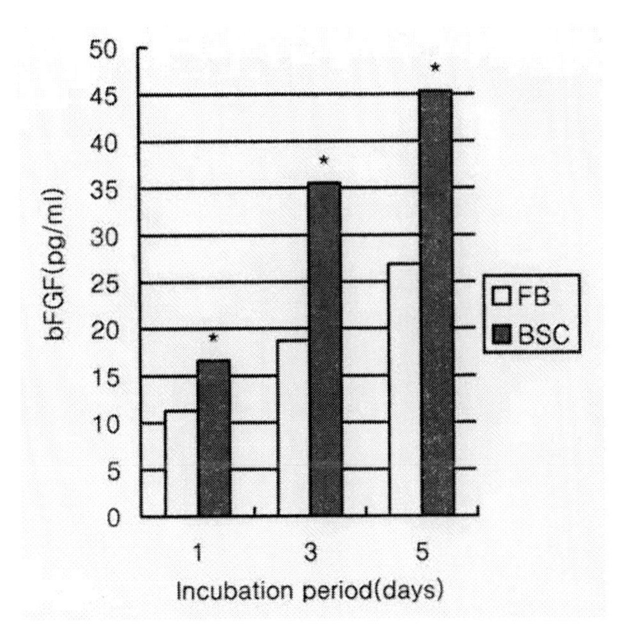

Figure 4. The bFGF levels of the fibroblast and the BSC groups (* $p<0.05$).

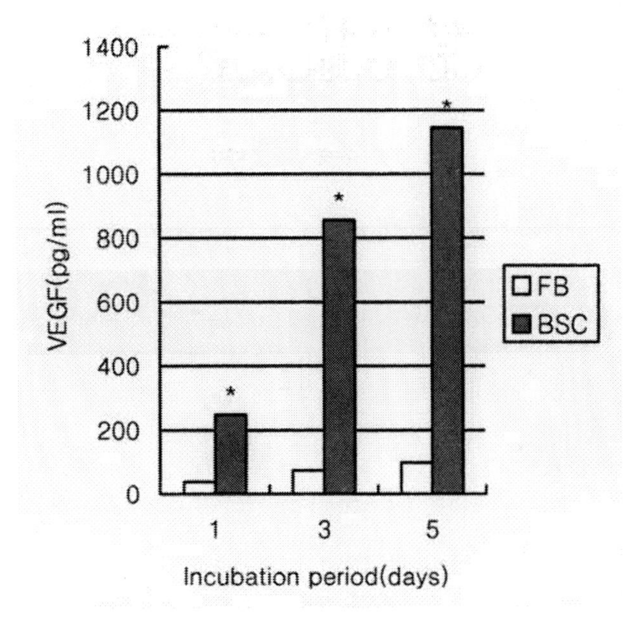

Figure 5. The VEGF levels of the fibroblast and the BSC groups (* $p<0.05$).

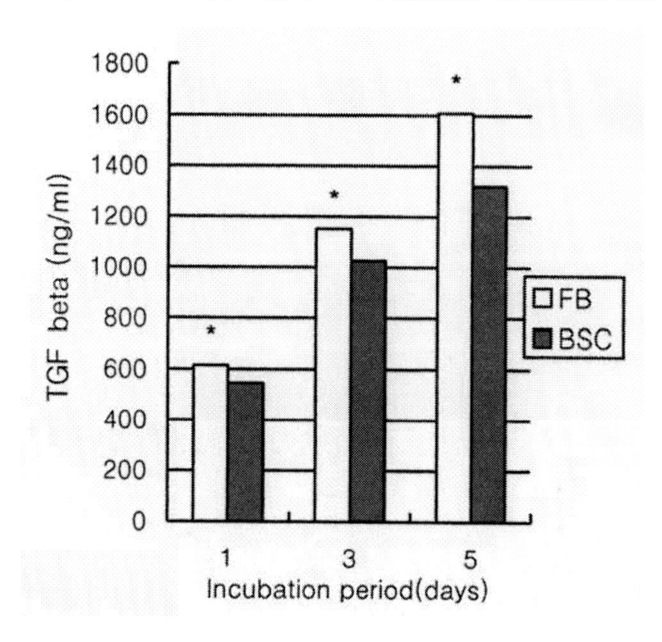

Figure 6. The TGF-β levels of the fibroblast and the BSC groups (* $p<0.05$).

As mentioned, we did not observe significant differences in cell proliferation and TFG-β secretion. In contrast, we observed a greater wound healing activity for BSCs compared with to that noted for dermal fibroblasts for collagen, bFGF, and VEGF production. An especially striking observation was seen in the VEGF level, which was twelve fold greater for BSCs.

It is interesting that the increase in type I collagen synthesis of BSCs is not related to the activity of TGF- β, since the level of TGF- β in the BSC group was lower than that in the fibroblast group. We can guess that other collagen-inducing factors probably influence the greater degree of collagen synthesis seen for the BSC. Further studies will be required to clarify the reasons for this observation.

VEGF and bFGF are potent angiogenic factors. Angiogenesis is a critical step in the wound healing process, and particularly for chronic wound healing. Therefore, we can expect that cell therapy for wound healing would benefit more from using the BSCs than by using the dermal fibroblasts. The BSCs are also capable of profoundly expanding their number while still showing a stable phenotype, as well as being immunologically tolerable, and thus, these positive characteristics open the way to using these cells during transplantation for the purpose of wound healing.

However, it is difficult to assess the definite effect of BSC transplantation on the overall process of wound healing. They could differentiate into fibroblasts or other cell types and become quiescent. In addition, cells in a wound bed are exposed to a complex milieu of growth factors and other stimuli, such as hypoxia. Future studies of how transplanted BSCs assist in healing a chronic wound was needed to examine the role of BSC transplantation on wound healing *in vivo*.

EFFECT OF BSC TRANSPLANTATION ON COLLAGEN SYNTHESIS

In the *in vitro* study, the authors showed that BSCs have a superior effect on wound healing than fibroblasts. This included improvements in cell proliferation, collagen synthesis, and in the production of three important growth factors required for chronic wound healing.

However, wound healing *in vivo* is a complex process requiring multiple factors. Thus, *in vivo* study was needed to examine the role of BSC transplantation on wound healing.

The aim of this study was to compare the effect of BSCs and fibroblasts on wound healing *in vivo*. First, the authors focused on collagen synthesis.

Preparation of Polyethylene Discs

Porous polyethylene (Medpor™; Porex, Newnan, GA) was used for the collagen synthesis experiment. Blocks of polyethylene (5mm diameter 3mm thick discs) were cut using a 5mm punch. Four discs were soaked in a 50ml centrifuge tube containing $4x10^6$ cells and 1ml of thrombin (Baxter AG, Vienna, Austria) for 10 minutes to allow the cells to infiltrate the pores of the polyethylene structure. After the porous polyethylene discs had been loaded with the cell-thrombin composite, they were moved to a 96-well culture plate. Two hundred microliters of fibrinogen (Baxter AG, Vienna, Austria) was added to each well to form a fibrin cuff that kept the infiltrated cells in the polyethylene pores. After 10 minutes, the polyethylene discs were taken from the plate and surplus fibrin was removed from disc exteriors (Figure 7). Thirty-six polyethylene discs were divided equally into three groups according to the cells with which they had been mixed. In groups I, II, and III, the discs were loaded with no cells (fibrin only, control group), fibroblasts, or BSCs, respectively.

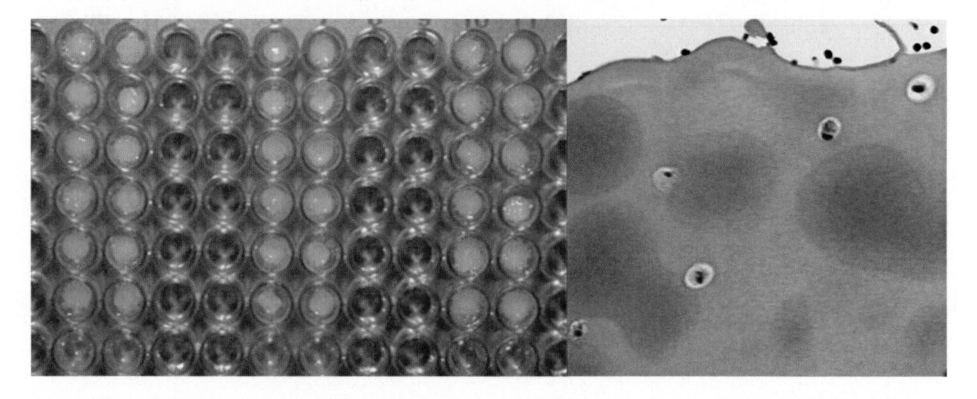

Figure 7. (Left) Porous polyethylene discs loaded with cell-thrombin composite were coated with fibrinogen in a 96-well cultured plate. (Right) Cells in the pores of a polyethylene disc (x 100).

Implanting and Harvesting Discs

Six Sprague-Dawley white rats, weighing 150 to 200g, were used for the collagen study. Three separate midline incisions approximately 5mm in length were made to create six pockets (two pockets per incision) in the dorsum of each rat, into which six discs (two discs per group) were implanted. Each pocket was closed using subcutaneous sutures to prevent interactions between pockets (Figure 8).

At 1st, 2nd, and 3rd week after implantation, discs were harvested and connective tissues around discs were carefully removed (Figure 9). Discs were then divided horizontally at disc centers, and half was used for histologic study and the other subjected to collagen assay.

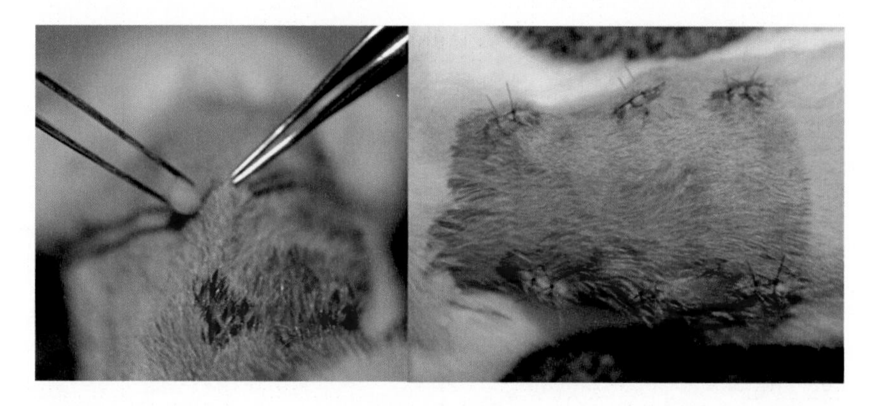

Figure 8. (Left) Polyethylene discs were implanted in the backs of white rats and (Right) the wounds were repaired.

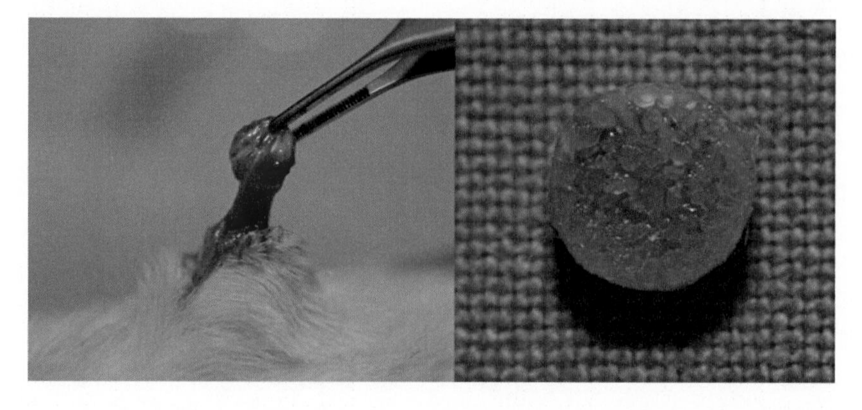

Figure 9. (Left) At one, two, and three weeks post-implantation, implanted discs were harvested. (Right) A disc after removing surrounding soft tissue.

Histological Evaluation of Collagen Growth

Disc halves were fixed in formalin and paraffin blocked. Horizontal sections (3μm) were then taken from the center of each disc. H and E staining was performed after dissolving the paraffin and polyethylene with xylene (Duksan Chem., Korea) for one minute. Extents of soft tissue growth into polyethylene disc pores were observed by one histologist under a light microscope at 40x, 100x, and 200x.

After one week, group I (the no cell group) showed little soft tissue growth in disc pores. On the other hand, group II (the fibroblast group) demonstrated soft tissue growth in disc pores at specimen peripheries, but no soft tissue was evident in the center of specimens at 100X magnification field. Only a meshwork of fibrinoid materials and inflammatory cells were observed. However, in group III (the BSC group) soft tissue growth was observed in the centers at 100X magnification field as well as peripheries of specimens. After two weeks, group I showed soft tissue growth at peripheries, but not at centers, whereas group II showed soft tissues at centers at 100X magnification field. In group III, collagen fibers were identified in specimen centers at 200X magnification field as well as 100X. After three weeks, group I still did not show soft tissues in specimen centers at 100X magnification field, whereas in groups II and III collagen fibers were observed at 100 and 200X field (Figure 10).

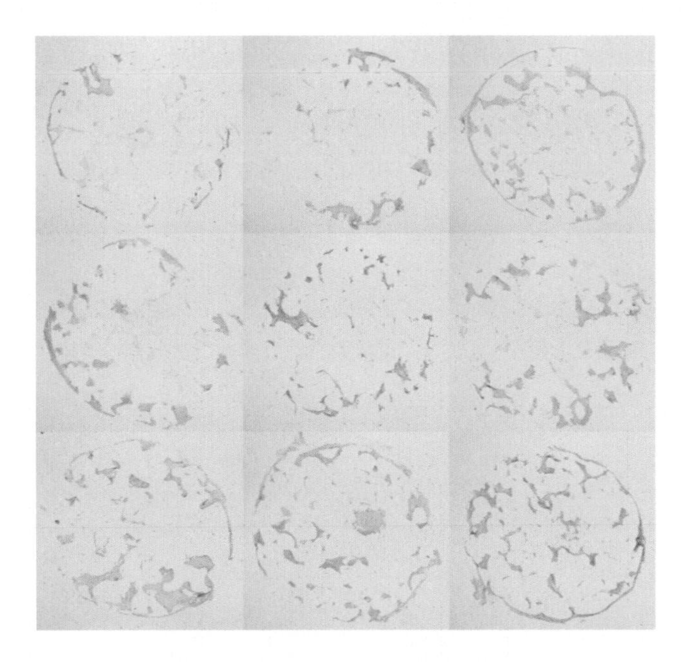

Figure 10. Histology (x 40) for collagen synthesis at 1-week (Left), 2-weeks (Center), and three weeks (Right). (Above) the non-cell treated control. (Center) Fibroblast-treated group. (Below) Bone marrow stromal stem cell-treated group.

Collagen Level Assays

ELISA was used to quantify levels of collagen in polyethylene disc pores. Rabbit anti-rat collagen type I polyclonal antibody (Chemicon International, Temecula, CA), HRP anti-rabbit IgG (Kirkegaard and Perry Laboratories Inc., Gaithersburg, MD), and Cultrex Rat Collagen I (RandD Systems, Minneapolis, MN) were used as primary and secondary antibodies, and as standard protein, respectively.

Marked differences were observed between the three groups at one and two week post-implantation. Group III showed the highest collagen level, followed groups II and I ($p<0.05$).

The average collagen levels in groups III, II and I after two weeks were 111, 82, and 60 ng/ml, respectively. Third week specimens also showed greater amounts of collagen in groups III and II than in group I ($p<0.05$). However, little difference was observed at this stage between groups III and II ($p=0.343$ and Figure 11).

Collagen is the most prevalent form in dermis, and collagen deposition is the predominant mechanism of wound healing. Collagen production is an elaborate process that involves multiple intracellular and extracellular events. Of the many factors that affect collagen synthesis, TGF-β is believed to be the most influential. The *in vitro* study in this chapter showed that fibroblast produces more TGF-β than BSC by 12% to 22%. However, in the *in vivo* study the opposite effect was observed, which indicates that the higher collagen level found in the BSC group in the present study was not directly related to TGF-β activity. Thus, it is likely that other collagen-producing factors, such as IGF, induced the greater collagen deposition observed in the BSC group. Further studies are required to explain this observation.

Histological findings show that soft tissue growth in disc pores was higher in the BSC group. In addition, collagen synthesis in the BSC group, as detected by ELISA, was much higher than those of the other two groups after one and two weeks ($p<0.05$). However, at three weeks post-implantation, no significant difference was observed between the fibroblast and BSC groups. These results suggest that collagen synthesis was near complete at two weeks post-implantation in the described model. Based on these results, the authors believe that the diameter of the Medpor^{TM} (Porex, Newnan, GA) discs used in this chapter were not large enough to evaluate the effect of BSCs beyond two weeks after implantation.

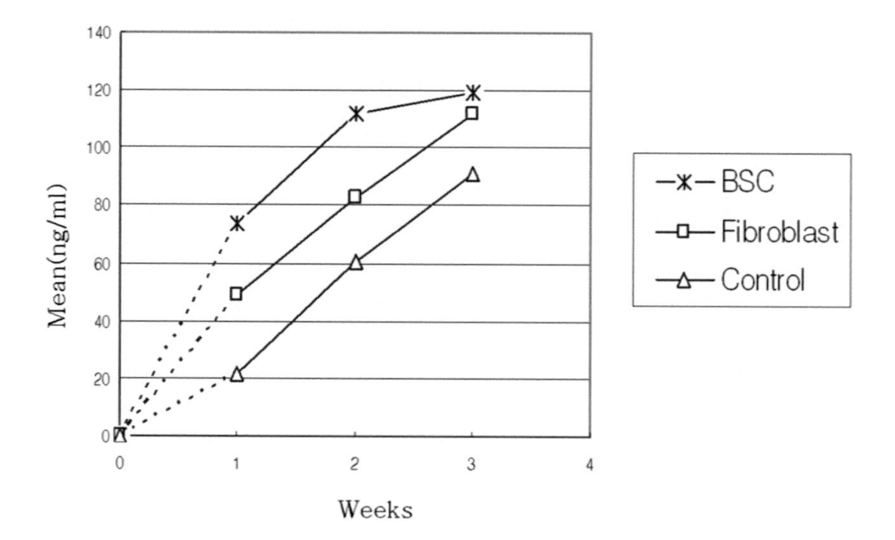

Figure 11. Amounts of collagen synthesis.

EFFECT OF BSC TRANSPLANTATION ON EPITHELIZATION

Epithelization is a vital component of successful wound healing. The effect of BSCs and fibroblasts on epithelization was compared *in vivo.*

Ten Sprague-Dawley white rats were used for the experiment. Six full-thickness 5-mm punch wounds were created through skin and subcutaneous tissue on the dorsal surfaces of each rat (two wounds per rat in each group, i.e., 20 wounds per group). The authors prepared a cell-fibrinogen (Baxter AG, Vienna, Austria) mixture ($4x10^6$ cells/ml) and applied 50μl of this to each wound. Thrombin (50μl; Baxter AG, Vienna, Austria) was then applied to all wounds to fix implanted cells. In groups I, II, and III, wounds were treated using mixtures containing no cells, fibroblasts, and BSCs, respectively. Wounds were then covered with a semiocclusive Medifoam™ (Ildong Pharm, Seoul, Korea) and OpSite™ (Smith and Nephew Ltd. UK) dressing.

Six days after application, animals were euthanized and healing wound specimens were obtained en bloc with a 6mm punch to include a margin of unwounded skin. These were then fixed in 10% neutral buffered formalin (Figure 12), and embedded in paraffin. Sections (3μm) were taken in the sagittal plane through the widest wound margin, and after removing paraffin by immersion in xylene (Duksan Chem.) for one minute, sections were HandE stained. A histopathologist, unaware of treatment history, measured the epithelial gap of each sagittal wound section at 40x. The rate of complete epithelization was calculated for each group and mean epithelial gap distances were compared (Figure 13).

Statistical comparisons were performed using the Mann-Whitney U-test and results were expressed as means±SDs. *P* values of < 0.05 were considered significant.

Complete epithelization rates averaged 20%, 45%, and 70% for groups I, II, and III, respectively. In addition, epithelial gap distances averaged 1.4mm, 0.5mm, and 0.3mm, respectively. Groups II and III showed much better results than group I ($p<0.05$), but no significant difference was found between groups II and III (Figure 14).

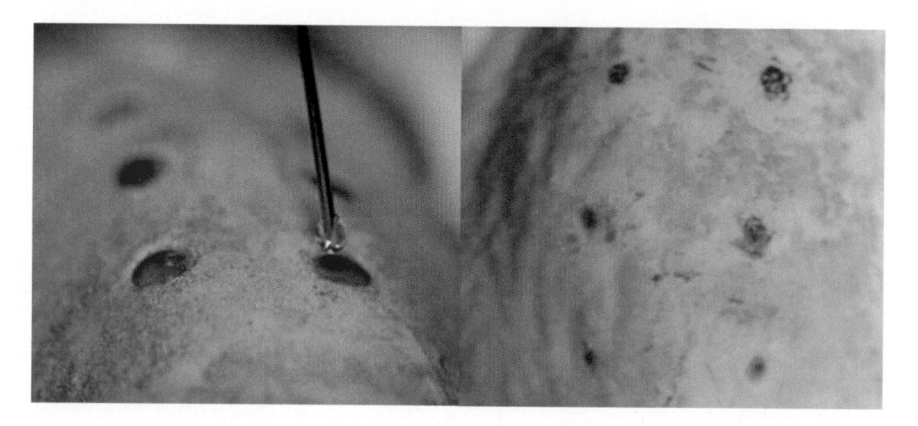

Figure 12. (Left) Transplantation of cells to wounds for the epithelization study. (Right) Six days after cell transplantation.

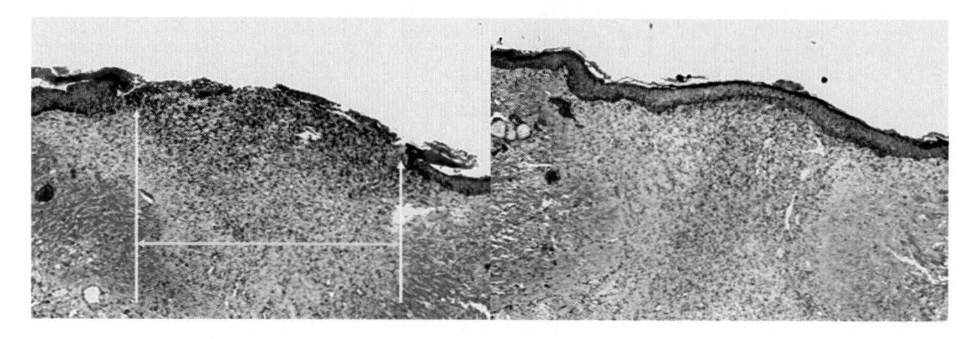

Figure 13. Microscopic view (x 40) used for measuring epithelial gaps. (Left) Epithelial gap of 1.25mm. (Right) Complete epithelization.

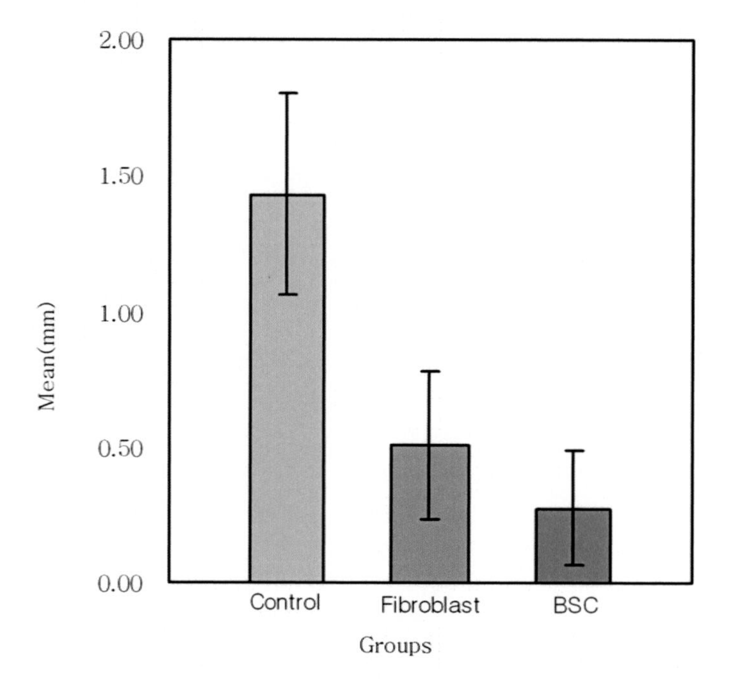

Figure 14. Epithelial gap lengths.

Wound epithelization is influenced by multiple factors, and in particular, collagen supports and promotes epithelial cell spreading and attachment. Various growth factors also affect epithelial cells, and bFGF stimulates epithelial cell migration, growth, and proliferation, whereas conversely, TGF-β inhibits proliferation [35]. In the *in vitro* study, the authors demonstrated that BSCs produce more collagen and bFGF and less TFG-β than fibroblasts. Moreover, the results of this *in vivo* study on epithelization concur with those of the former *in vitro* study.

EFFECT OF BSC TRANSPLANTATION ON ANGIOGENESIS

Angiogenesis, which is one biological mechanism for the formation of new capillaries, is fundamental to wound healing and flap or graft survival. The cell transduction signals for the stimulation of angiogenesis, including a variety of growth factors, have recently aroused considerable interest. However, there are several problems with supplying an excess of a single growth factor. First, angiogenesis requires many components, with each growth factor playing one part in this orchestrated pathway. Second, high concentrations of some growth factors may be harmful. Finally, the growth factor response may be stage-specific.

On the other hand, cell transplantation can improve the overall wound environment. This is accomplished by adjusting the secretion to obtain the optimal biological conditions for the varied growth factors essential for angiogenesis. Fibroblast cell therapy has recently been developed for wound healing and has been applied in human patients. It is well known that the transplantation of fibroblasts accelerates angiogenesis around wounds by stimulating the secretion of various growth factors. Dermal fibroblasts not only play important roles in the deposition and destruction of the matrices in a wound but also control angiogenesis in the tissue via cytokines such as VEGF, bFGF, and others.

The results of *in vitro* study suggest that BSCs transplanted into a wound can actively secrete growth factors such as VEGF and bFGF that stimulate angiogenesis, and they do so to a much greater extent than fibroblasts. However, to date, there has been no investigation into the effect on the angiogenesis of transplantation of BSCs *in vivo*.

Therefore, the aim of this study was to compare the effects of BSCs and fibroblasts on angiogenesis *in vivo*.

Preparation of Polyethylene Discs

The porous polyethylene structure used in the experiments was Medpor (Porex, Newnan, GA). The blocks of polyethylene were cut using a 5mm punch into 5mm diameter and 3mm thick discs. Four discs were soaked in a 50ml centrifuge tube containing $4x10^6$ cells and 1ml thrombin (Baxter AG, Vienna, Austria) for 10 minutes, allowing the cells to infiltrate into the pores of the polyethylene structure. After the porous polyethylene discs had been loaded with the cell-thrombin composite, they were moved to a 96-well culture plate. Two hundred microliters of fibrinogen (Baxter AG, Vienna, Austria) were added to each well to form a fibrin cuff that kept the infiltrated cells in the polyethylene pores. After 10 minutes, the polyethylene discs were taken from the plate and the surplus fibrin was removed from the exterior of the disc.

The polyethylene discs were divided into three groups according to the cells with which they had been mixed. In groups A and B, the discs were loaded with the BSCs and fibroblasts, respectively. In group C, the discs were filled with fibrin only. Sixteen discs per group (total 48 discs) were made for the angiogenesis study.

Implanting and Harvesting Discs

Eight Sprague-Dawley white rats, weighing 150 to 200g, were used. Three separate midline incisions approximately 5mm in length were made to create six pockets (two pockets per one incision) in the back of each rat, into which six discs (two discs per group) were implanted. Each pocket was closed by subcutaneous sutures not to influence each other.

At three time intervals from 1 to 3 weeks, the implanted discs were harvested and the connective tissues around the discs were carefully removed for the histological study.

Histological Evaluation of Angiogenesis

After the discs had been fixed with formalin in order to establish the paraffin block, a longitudinal section was cut to a thickness of 6 μm in the middle of each disc. HandE staining was performed after dissolving the paraffin and polyethylene substances by processing with xylene (Duksan Chem.) for one minute. The extent of angiogenesis was quantified by two histologists, who measured the microvascular density (MVD) as the number of microvessels in the middle of a biopsy specimen under a 100x magnification field.

Statistical comparisons were performed using the Kruskal-Wallis test and Bonferroni correction was employed. A p value < 0.05 was considered significant.

All the polyethylene discs examined after implantation were firmly linked with the adjacent tissues by the growth of connective tissue into the pores (Figure 15). There was little difference in the MVD among the three groups by the second week. Significant differences between all groups were noted at week three. The average MVD of the BSC group was 52.88, which is much higher than that of the fibroblast group (26.12), and of the control group (17.50) (P=0.00, Figures 16 and 17).

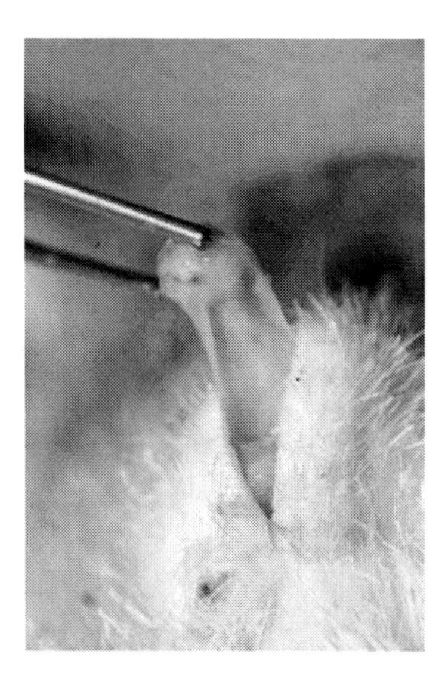

Figure 15. One week after implantation, a disc was harvested for histologic study. Growth of connective tissue into the pores of the polyethylene implant was observed.

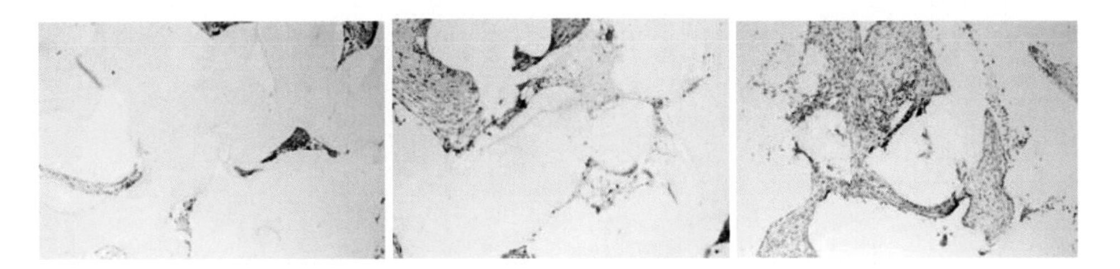

Figure 16. Microvascular density at 3 weeks after implantation under a 100-magnification field. Control (left), fibroblast (center), and bone marrow stromal cell (right) groups show 14, 26, and 62 microvessels, respectively.

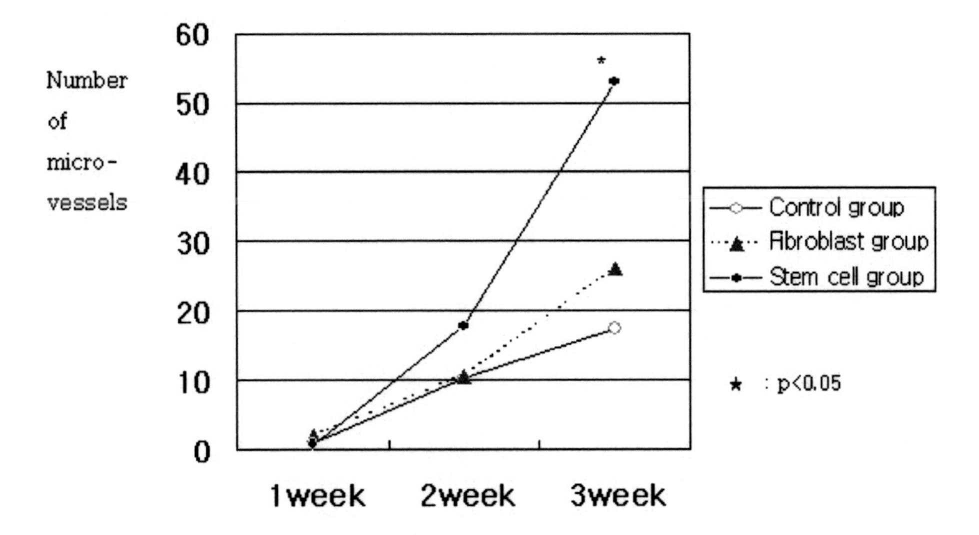

Figure 17. Microvascular density under a 100-magnification field (* $p<0.05$).

The results of this experiment allow a comparison of the effect of BSCs with that of fibroblast products on angiogenesis. Angiogenesis is one of the most important fields in plastic surgery because it has a significant clinical importance not only in the wound healing process but also in tissue grafts, flaps, and replantations.

Regarding the results of our *in vivo* rat model, however, further studies will be needed. There is a possibility that immune-mediated inflammation may be influencing or confounding the results. The increased MVD noted for cell-containing discs compared with the vehicle controls could be related to an inflammatory reaction to the heterogenic cellular implants. The difference in MVD between the fibroblast and BSC containing discs could be due differences in the intensity of this inflammatory reaction to the different cell types. To measure the MVD more clearly, it would be better to conduct immunohistochemical staining with the CD34 or CD14 endothelial markers. However, the reagent in the market was only for frozen specimens. The reagent required to conduct the staining of the specimen in the paraffin block could not be acquired. Nevertheless, it can be presumed that such staining would not cause any problems in comparing the MVD since two histologists measured it without difficulty.

Despite the study's weak points, we would like to say that cell therapy for angiogenesis using BSCs may have more potential benefit than that using fibroblasts. Furthermore, BSCs are capable of expanding profoundly while still showing a stable phenotype and remaining immunologically tolerable. BSCs open the door to transplantation with the aim of angiogenesis.

CONCLUSION

This chapter compared the wound healing activity of human BSCs with that of dermal fibroblasts both *in vitro* and *in vivo*. In the *in vitro* study, there were not great differences for cell proliferation and TGF-β production. However, BSCs produced much higher amounts of collagen, bFGF and VEGF. Especially for the level of VEGF, the difference was twelve fold.

In vivo study by xenotransplantations of such cells also showed that the application of BSCs had a more promising effect on collagen synthesis, epithelization, and angiogenesis. These results suggest a great capacity of BSCs in promoting wound healing and further suggest that they may possibly replace the current cell therapy method that uses fibroblasts.

REFERENCES

[1] Bennett, SP; Griffiths, GD; Schor, AM; et al. Growth factors in the treatment of diabetic foot ulcers. *Br. J. Surg.* 2003;90, 133-146.

[2] Embil, JM; Papp, K; Sibbald, G; et al. Recombinant human platelet–derived growth factor-BB (becaplermin) for healing chronic lower extremity diabetic ulcers: an open-label clinical evaluation of efficacy. *Wound Rep. Reg.* 2000;8, 162.

[3] Kano, M ; Masuda, Y ; Tominaga, T ; et al. Collagen synthesis and collagenase activity of cryopreserved heart valves. *J. Thorac. and Cardiovasc. Surg.* 2001;122, 706-711.

[4] Tsang, MW; Wong, WKR; Hung, CS; et al. Human epidermal growth factor enhances healing of diabetic foot ulcers. *Diabetes Care.* 2003;26, 1856.

[5] Kuroyanagi, Y ; Yamada, N ; Yamashita, R ; et al. Tissue-engineered product: allogenic cultured dermal substitute composed of spongy collagen with fibroblasts. *Artif. organs.* 2001;25, 180.

[6] Gohari, S; Gambla, C; Healey, M; et al. Evaluation of tissue-engineered skin (human skin substitute) and secondary intention healing in the treatment of full thickness wounds after Mohs micrographic or excisional surgery. *Derm. Surg.* 2002;28, 1107-1114.

[7] Mansbridge, JN; Liu, K; Patch, R; et al. Three-dimensional fibroblast culture implant for the treatment of diabetic foot ulcers: metabolic activity and therapeutic range. *Tissue Engineering.* 1998;4, 403.

[8] Han, SK; Choi, KJ; Kim, WK. Clinical application of fresh fibroblast allografts for the treatment of diabetic foot ulcers: A pilot study. *Plast. Reconstr. Surg.* 2004;114, 1783-1789.

[9] Prockop, DJ; Marrow stromal cells as stem cells for nonhematopoietic tissues. *Science.* 1997;276, 71–74.

[10] Galmiche, MC ; Koteliansky, VE ; Brie`re, J ; et al. Stromal cells from human long-term marrow cultures are mesenchymal cells that differentiate following a vascular smooth muscle differentiation pathway. *Blood.* 1993;82, 66–76.

[11] Pereira, RF; Halford, KW; O'Hara, MD; et al. Cultured adherent cells from marrow can serve as long-lasting precursor cells for bone, cartilage, and lung in irradiated mice. *Proc. Nat. Acad. Sci. USA.* 1995;92, 4857–4861.

[12] Watanabe, N; Woo, SLY; Papageorgiou, C; at al. Fate of donor bone marrow cells in medial collateral ligament after simulated autologous transplantation. *Microsc. Res. Tech.* 2002;58, 39-44.

[13] Quirici, N; Soligo, D; Bossolasco, P; et al. Isolation of bone marrow mesenchymal stem cells by anti-nerve growth factor receptor antibodies. *Exp. Hematol.* 2002;30, 783 – 791.

[14] Conget, PA; Minguell, JJ. Phenotypical and functional properties of human bone marrow mesenchymal progenitor cells. *J. Cell. Physiol.* 1999;181, 67-73.

[15] Naumann, A ; Dennis, J ; Staudenmaier, R ; et al. Mesenchymal stem cells: a new pathway for tissue engineering in reconstructive surgery. *Laryngorhinootologie.* 2002;81, 521-527.

[16] Kittler, ELW; McGrath, H; Temeles, D; et al. Biologic significance of constitutive and subliminal growth factor production by bone marrow stroma. *Blood.* 1992;79, 3168–3178.

[17] Chichester, CO; Ferna´ndez, M; Minguell, JJ. Extracellular matrix gene expression by human bone marrow stroma and by marrow fibroblasts. *Cell Adhesion Commun.* 1993;1, 93–99.

[18] Deans, RJ; Moseley, AB. Mesenchymal stem cells: biology and potential clinical uses. *Exp. Hematol.* 2000;28, 875.

[19] Hansen, SL; Young, DM; Boudreau, NJ. HoxD3 expression and collagen synthesis in diabetic fibroblasts. *Wound Rep. Reg.* 2003;11, 474-480.

[20] Goodson, WH; Hunt, TK. Deficient collagen formation by obese mice in a standard wound model. *Am. J. Surg.* 1979;138, 692-694.

[21] Frank, S; Hubner, G; Breier, G; et al. Regulation of vascular endothelial growth factor expression in cultured keratinocytes. Implication for normal and impaired wound healing. *J. Biol. Chem.* 1995;270, 12607-12613.

[22] Pierce, GF. Inflammation in nonhealing diabetic wounds: the space-time continuum does matter. *Am. J. Pathol.* 2001;159, 399-403.

[23] Schultz, GS ; Sibbald, RG ; Falanga,V ; et al. Wound bed preparation: a systematic approach to wound management. *Wound Rep. Reg.* 2003;11, 1-28.

[24] Belgore, F; Lip, GYH; Blann, AD; et al. Basic fibrobrast growth factor induces the secretion of vascular endothelial growth factor by human aortic smooth muscle cells but not by endothelial cells. *Europ. J. Cl. Investigation.* 2003;33, 833-839.

[25] Zhang, F; Oswald, T; Lin, S; et al. Vascular endothelial growth factor (VEGF) expression and the effect of exogenous VEGF on survival of a random flap in the rat. *Br. J. Plast. Surg.* 2003;56, 653-659.

[26] Yang, R; Thomas, G.R; Bunting, S; et al. Effects of vascular endothelial growth factor on hemodynamics and cardiac performance. *J. Cardiovasc. Pharmacol.* 1996;27, 838-844.

[27] Alon, T; Hemo, I; Itin, A; et al. Vascular endothelial growth factor acts as a survival factor for newly formed retinal vessels and has implications for retinopathy of prematurity. *Nat. Med.* 1995;1, 1024-1028.

[28] Wilkins, BS; Jones, DB. Immunohistochemical characterization of intact stromal layers in long-term cultures of human bone marrow. *Br. J. Haematol.* 1995;90, 757–766.

[29] Lerman, OZ ; Galiano, RD ; Armour, M ; et al. Cellular dysfunction in the diabetic fibroblast. Impairment in migration, vascular endothelial growth factor production, and response to hypoxia. *Am. J. Pathol.* 2003;162, 303-312.

[30] Seifter, E ; Rettura, G ; Padawer, J ; et al. Impaired wound healing in streptozotocin diabetes. Prevention by supplemental vitamin A. *Ann. Surg.* 1981;194, 42-50.

[31] Simmons, PJ; Torok-Storb, B. CD34 expression by stromal precursors in normal human adult bone marrow. *Blood.* 1991;78, 2848.

[32] Dennis, JE; Caplan, AI. Differentiation potential of conditionally immortalized mesenchymal progenitor cells from adult marrow of a H-2Kb-tsA58 transgenic mouse. *J. Cell Physiol.* 1996;167, 523-538.

[33] Kadiyala, S; Young, RG; Thiede, MA.et al. Culture expanded canine mesenchymal stem cells possess osteochondrogenic potential in vivo and in vitro. *Cell Transplant. 1997;* 6, 125-134.

[34] Chesney, J; Metz, CN; Stavitsky, AB. et al. Regulated production of type I collagen and inflammatory cytokines by peripheral blood fibrocytes. *J. Immunol.* 1998;160, 419-425.

[35] Mast, B-A; Cohen, I-K. *Plastic Surgery: Normal wound healing.* In: Achauer BM, Eriksson E, Guyuron B, et al., St. Louis: Mosby; 2000; 37-52.

In: Stem Cell Research Trends
Editor: Josse R. Braggina, pp. 277-292
ISBN: 978-1-60021-622-0

Chapter X

ANTIANGIOGENESIS IN ARTHRITIS THERAPY: ENDOSTATIN AND ITS MECHANISM OF ACTION

Daitaro Kurosaka
Jikei University School of Medicine, Japan

ABSTRACT

Rheumatoid arthritis is a chronic inflammatory disease, and angiogenesis observed in synovial tissue involves various factors. Based on the hypothesis that angiogenesis inhibition suppresses arthritis, studies have shown anti-arthritic effects in animal models of arthritis after the administration of angiogenesis inhibitors. Endostatin is a fragment of the C-terminal non-collagenous domain of type XVIII collagen, and has a potent angiogenesis-inhibiting effect. The administration of endostatin to a mouse model of arthritis inhibited angiogenesis, resulting in the suppression of arthritis. No side effects were observed during the period of drug administration. As a therapeutic agent for rheumatoid arthritis, endostatin includes the advantages of having few side effects, an action mechanism different from those of conventional therapeutic agents for rheumatoid arthritis, an angiogenesis-inhibiting effect through a broad-spectrum mechanism, and little potential for inducing tolerance; therefore, it may be a promising therapeutic agent for rheumatoid arthritis.

INTRODUCTION

Rheumatoid arthritis (RA) is a chronic autoimmune disease of unknown etiology, mainly affecting synovial tissue. Inflammatory synovial tissue in RA contains many new blood vessels, which serve as pathways for the supply of nutrients and oxygen, migration of inflammatory cells, and transport of various cytokines and disease-aggravating factors to the synovium [1-5]. Many angiogenesis promoters are known to perform important functions in the formation of new blood vessels; indeed, an increase in various angiogenesis promoters is observed in the inflammatory synovial tissue of RA patients. In addition, although angiogenesis is generally controlled by a balance between proangiogenic and antiangiogenic factors, RA patients have been reported to show a marked imbalance between them [6].

Interestingly, it has been noted that many antirheumatic drugs in clinical use partially inhibit angiogenesis [7-9]. For example, methotrexate [7], D-penicillamine8), and gold preparations [9] are known to inhibit endothelial cell growth factor (ECGF)-induced rabbit corneal angiogenesis, and clinically suppress inflammation in RA, thereby inhibiting bone destruction to some extent, suggesting that such antirheumatic activity is partially mediated through the inhibition of angiogenesis. In addition, a new treatment targeting TNF-α, a proangiogenic molecule, has recently been developed, and shown to markedly inhibit the progression of bone destruction in RA.

Thus, angiogenesis is considered to play a crucial role in the pathogenesis of RA, and its control is considered to constitute a therapeutic strategy for RA5, [10-12]. Therefore, various attempts have recently been made to develop therapies targeting angiogenesis.

Angiogenesis in arthritis involves angiogenesis promoters such as vascular endothelial growth factor (VEGF) [13], fibroblast growth factor (FGF) [14], interleukin-18 (IL-18) [15], and $\alpha V\beta 3$ integrin [16]. It has been reported that blocking these angiogenesis promoters suppresses arthritis in an animal model of RA.

Antiangiogenesis therapy using angiogenesis inhibitors has also been studied. Known endogenous angiogenesis inhibitors include angiostatin [17], and endostatin [18-21]. The administration of these endogenous as well as non-endogenous angiogenesis inhibitors to animal models of RA has been reported to suppress the arthritis [22-24].

To date, we have examined the effects of administration of the potent angiogenesis inhibitor endostatin to mouse models of arthritis. In this paper, we introduce our findings, and discuss the potential of antiangiogenesis therapy for the treatment of arthritis.

TREATMENT OF ARTHRITIS WITH ENDOSTATIN

Several studies have explored the possibility of endostatin as a therapeutic agent for arthritis. Matsuno et al. reported that human RA synovial tissue implanted in SCID mice underwent marked regression after the local injection of endostatin [18]. Yin et al. reported that the direct injection of endostatin gene-expressing lentiviral vectors into the joint of arthritic mice improved the arthritis [19]. Yue et al. [20] and our group [21] reported the arthritis-inhibiting effect of systemically administered endostatin. We evaluated the arthritis-inhibiting effect of recombinant endostatin that had been administered to mice with antibody-induced arthritis before its development. To investigate the arthritis-inhibiting effect of recombinant endostatin, Yue et al. administered it to mice with adjuvant arthritis at its onset.

Changes in Endostatin Levels in RA Patients

Nagashima et al. [6] reported that the endostatin levels in the peripheral blood and synovial fluid of RA patients did not differ from those of non-RA patients including osteoarthritis patients, that the levels of the angiogenesis promoter VEGF in the peripheral blood and synovial fluid were higher in RA than in non-RA patients, and that, after therapy, VEGF and endostatin levels tended to decrease and increase, respectively. Thus, they considered that a balance between angiogenesis promoters such as VEGF and angiogenesis inhibitors such as endostatin was important in RA, and that this balance was in favor of angiogenesis promoters in the active phase of RA.

Structure of Endostatin

Endostatin was isolated from the culture supernatant of mouse endothelioma cells during the screening for tumor cell-produced endogenous angiogenesis inhibitors using endothelial cell proliferation-inhibiting activity as an index25). Amino acid sequencing identified endostatin as a fragment of the C-terminal noncollagenous domain of type XVIII collagen [25] (Figure 1). Type XVIII collagen is widely distributed in vascular and subepithelial basement membrane layers, although its function has not been fully elucidated [26]. Endostatin has a molecular weight of 20 kDa, and consists of 184 or 183 amino acid residues in mice and humans, respectively.

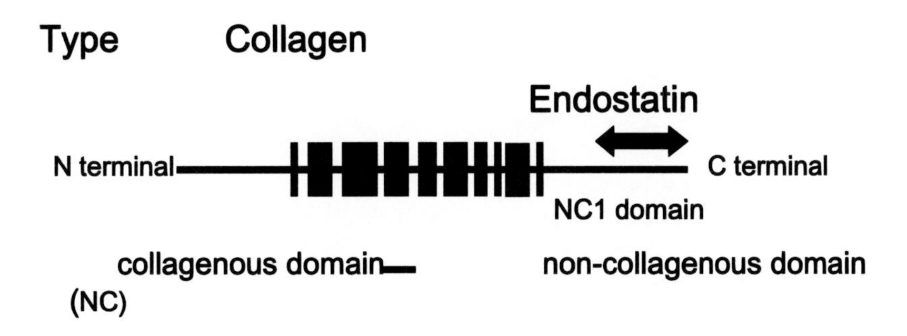

Figure 1. Structure of endostatin.

Antiangiogenic Action Mechanism of Endostatin

Endostatin is considered to act specifically on vascular endothelial cells, and inhibits VEGF-induced endothelial cell proliferation and migration *in vitro* (Figure 2) [27]. *In vivo*, endostatin inhibits angiogenesis; for example, *E. coli*-derived recombinant endostatin inhibits tumor angiogenesis in a mouse tumor model [25].

It has been reported that endostatin inhibits vascular endothelial cell migration by binding to integrin on endothelial cells [28, 29]. Involvement of E-selectin has also been noted30). Endostatin has been reported to inhibit not only membrane type 1-matrix metalloproteinase(MT1-MMP) -activating metalloproteinase-2(MMP-2) but also activated MMP-[2, 31] and to block the activity of MMP-9 and MMP-[13.32]

Recently, an interesting paper in the context of the mechanism of angiogenesis inhibition by endostatin has been published [33].

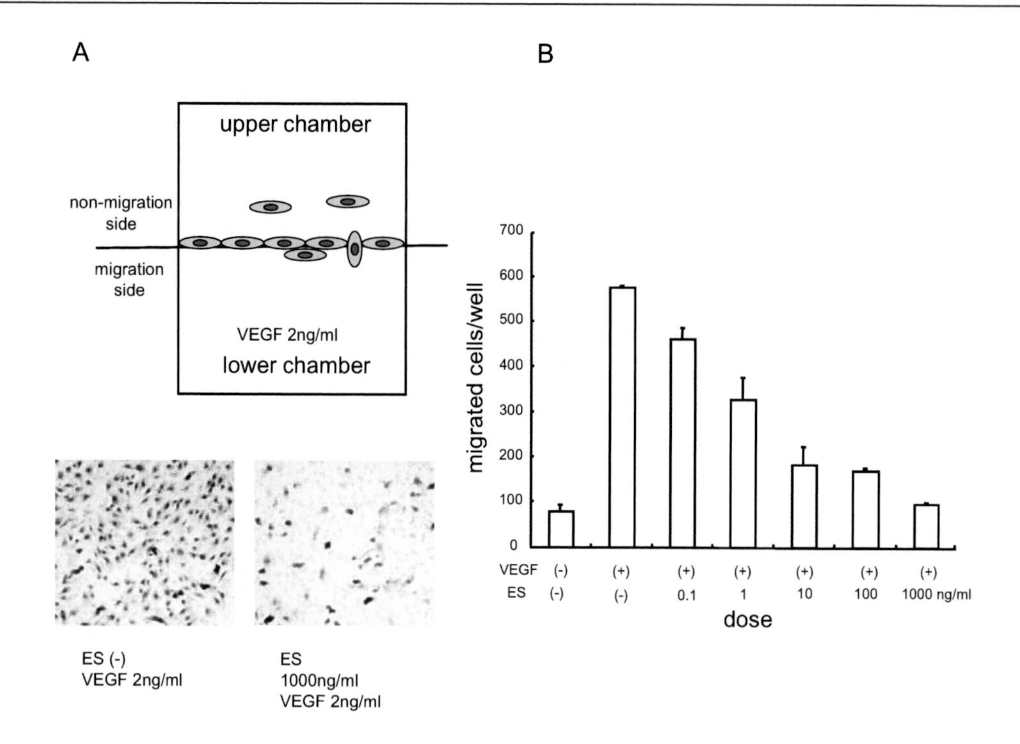

Figure 2. Effects of endostatin evaluated by the human umbilical vein endothelial cell migration assay. A, A type 1 collagen-coated polycarbonate filter was sandwiched between the upper and lower chemotaxis chambers. A suspension of human umbilical vein endothelial cells was placed in the upper chamber, and VEGF was added to the culture medium in the lower chamber. After 6 hours of culture, cells that had migrated through the filter and were adherent to the other side were observed. The addition of endostatin to the culture medium reduced the number of cells that had migrated. B, The longitudinal axis shows the number of endothelial cells that migrated per well. Endostatin at concentrations of 0.1-10 ng/ml inhibited the VEGF-promoted migration of endothelial cells. HUVEC stands for human umbilical vein endothelial cells, VEGF for vascular endothelial growth factor, and ES for endostatin.

Abdollahi et al. administered endostatin to endothelial cells, examined them for changes in gene expression, and found that endostatin influenced the expression of many genes. Combining these findings with the results of RT-PCR and protein phosphorylation experiments, they concluded that mainly the factors shown in Table 1 were negatively regulated by endostatin. Among these factors, activator-1 (AP-1), E26 transformation specificity-1 (ETS-1), hypoxia-inducible factor 1α (HIF-1α), inhibitor of DNA binding (Id), nuclear factor of kappa light chain gene enhancer in B cells (NF-κB), and signal transducer and activator of transcription factor (STAT) are transcription factors, and AP-1, NF-κB, and STAT are involved in the expression of many genes including those for cytokines. Angiogenesis promoters such as VEGF induce the expression of ETS-1, which induces the expression of protease genes and angiogenesis-related genes such as integrin β334). HIF-1α induces VEGF gene expression in hypoxia35). Id protein is also considered to play an important role in angiogenesis. It has been reported that Id-1(+/-)Id-3(-/-) mice are born, but when a tumor is implanted in them, the tumor is poorly vascularized and shows delayed growth [36]. Ephrin is a membrane protein bound to Eph, a receptor-type tyrosine kinase.

Eph and Ephrin are well known to regulate the formation of metamerically arranged cell masses and the extension of axons.

Table 1. The main factors negatively controlled by endostatin

AP-1 (activator protein 1)
Jun, Jun-β, FoS
ETS-1 (E26 transformation specific-1)
HIF1-α (hypoxia inducible factor 1-α)
Id (inhibitor of DNA binding)
Id1, Id3
NF−κB (nuclear factor of kappa light chain gene enhancer in B cells)
p50, p65
STAT (signal transducer and activator of transcription factor)
STAT-1, STAT-3
Ephrin
EphrinA1
MAP kinase
P-JNK, P38
Trombin receptor
PAR-1, PAR-2

Recently, their function in the formation of the blood vessel wall and in angiogenesis has attracted attention [37]. MAP kinase is an intracellular signal transduction molecule that controls cell proliferation and differentiation. Thrombin receptors are also known to be involved in angiogenesis [38]. Thus, endostatin influences the expression of many genes. In the context of these observations, Folkman referred to treatment with endostatin as broad-spectrum antiangiogenic therapy [39].

On the other hand, endostatin has little effect on wound healing and reproduction [40], and mainly inhibits pathological angiogenesis. Although its mechanism has not been fully elucidated, it has been suggested that endostatin may block the involvement of integrin [28, 29, 41] and E-selectin [30] in pathological angiogenesis.

Arthritis-Inhibiting Effect of Endostatin in a Model of Arthritis

Arthritis was induced in 6-week-old male BALB/c mice by administering four kinds of monoclonal anti-type II collagen antibodies, followed 3 days later by LPS. Groups of mice were injected with 10 mg/kg/day of endostatin in two divided doses subcutaneously in the back for 13 days prior to the development of arthritis, and were evaluated for the severity of arthritis. A control group received PBS. It is of note that during the period of endostatin administration, the mice displayed no grossly observable side effects. The severity of arthritis was evaluated by the arthritis score and hind paw thickness. The arthritis scores (Figure 3A) as well as hind paw thicknesses (Figure 3B) during the time-course of arthritis were significantly lower in the endostatin group than in the control group [20]. Analysis of X-rays taken on the last day of the experiment also revealed that bone destruction was less severe in the endostatin group than in the control group (Figure 4) [42].

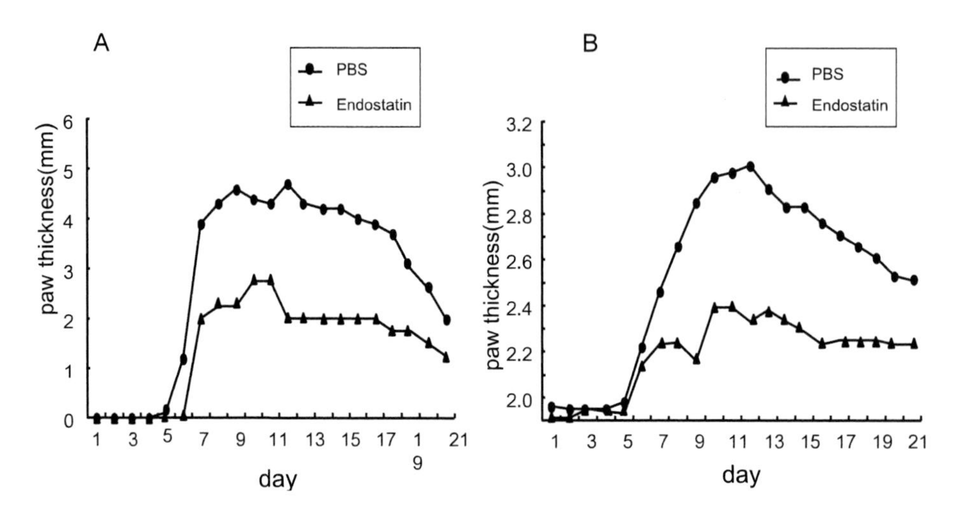

Figure 3. Arthritis scores and the time-course of hind paw swelling. A, Time-course of arthritis scores. The severity of arthritis was scored on a scale of 0 to 3 for each of the four paws, and evaluated by the sum of scores for the four paw joints: 0 = no swelling, 1 = mild swelling and redness, 2 = marked swelling (edema), and 3 = ankylosis. The arthritis scores during the time-course of arthritis were significantly lower in the endostatin 10 mg/kg/day group than in the control group. B, Hind paw thicknesses also remained significantly smaller in the endostatin than in the control group.

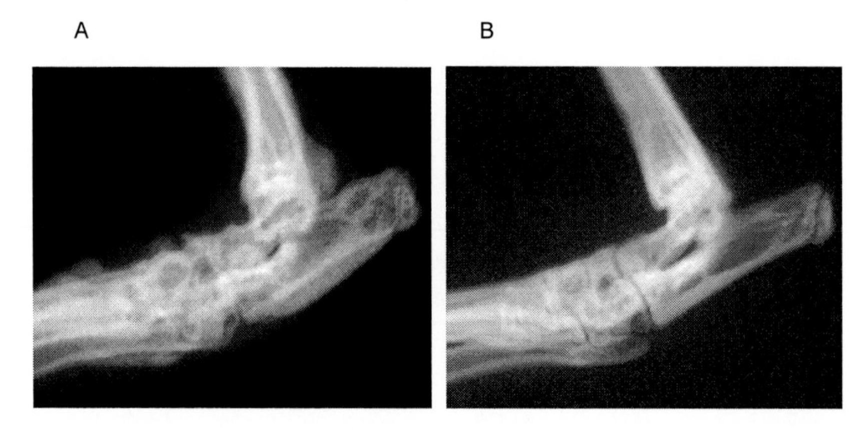

Figure 4. Comparison of the endostatin and control groups for X-ray findings in the hind paws. A, X-ray image of the hind paw in the control group. B, X-ray image of the hind paw in the endostatin 10 mg/kg/day group. Bone destruction was milder in the endostatin than in the control group.

Histopathological examination of the paw joint on the last day of the experiment revealed bone and cartilage erosion and the formation of inflammatory synovial granulation tissue (pannus) in the control group, whereas these changes were significantly suppressed in the endostatin group (Figure 5) [20, 42].

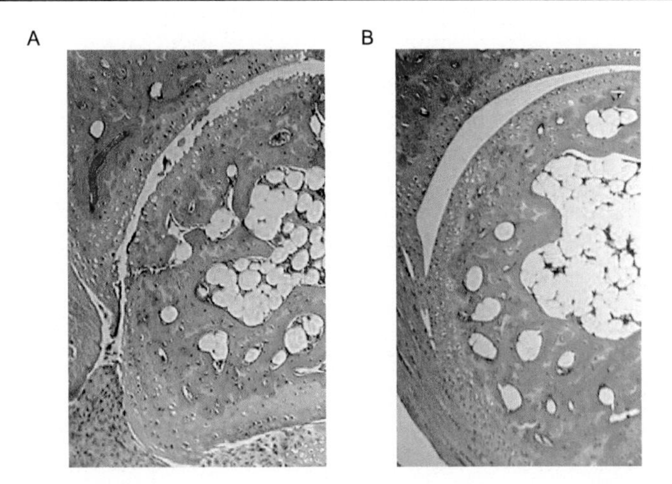

Figure 5. Comparison of the endostatin and control groups for histopathological findings in the hind paw joints. A, Histopathological findings in the hind paw joint in the control group. B, Histopathological findings in the hind paw joint in the endostatin 10 mg/kg/day group. Bone and cartilage erosion and pannus formation, as observed in the control group, were significantly suppressed in the endostatin group.

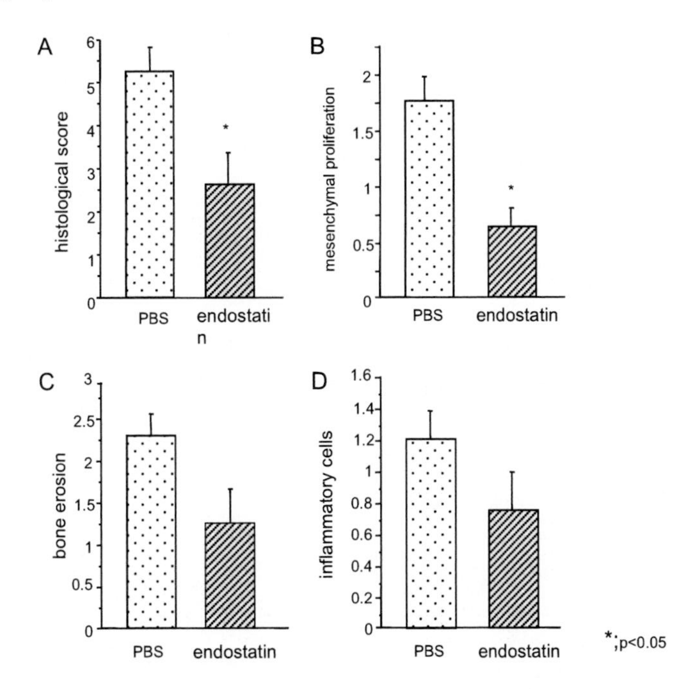

Figure 6. Comparison of the endostatin and control groups for histopathological scores. In the final experiment, synovial thickening (pannus formation with mesenchymal proliferation), subchondral bone erosion, and periarticular inflammation (extent of inflammatory cell infiltration) were scored on a scale of 0 to 3: 0= normal, 1 = mild, 2 = moderate, and 3 = severe. The histological score was defined as the sum of the scores for the three parameters, with a maximum of 9. All scorings were performed by the pathologist in a blind manner. A, Histological score. A significant difference in the histological score was noted between the control and endostatin 10 mg/kg/day groups. B, Synovial thickening score. C, Subchondral bone erosion score. D, Periarticular inflammation score.

In addition, statistical analysis of histopathological arthritis scores showed a significant difference between the 10 mg/kg/day and control groups (Figure 6) [20].

EFFECTS OF ENDOSTATIN ON THE EXPRESSION OF VARIOUS GENES AT THE ARTHRITIC SITE

An endostatin administration experiment was also performed in mice with type II collagen-induced arthritis. The results were similar to those obtained in the mice with monoclonal [43, 44] antibody-induced arthritis. In this experiment, expressions of mRNA for angiogenesis-related factors (Figure 7) and inflammatory cytokines such as IL-1, IL-6, and TNF-α (Figure 8) were analyzed [43, 44].

In a mouse model of arthritis, VEGFR-1 mRNA expression has been reported to increase with the progression of arthritis [13], and the blockage of the VEGF-VEGFR-1 signaling pathway has been reported to suppress arthritis [45, 46].

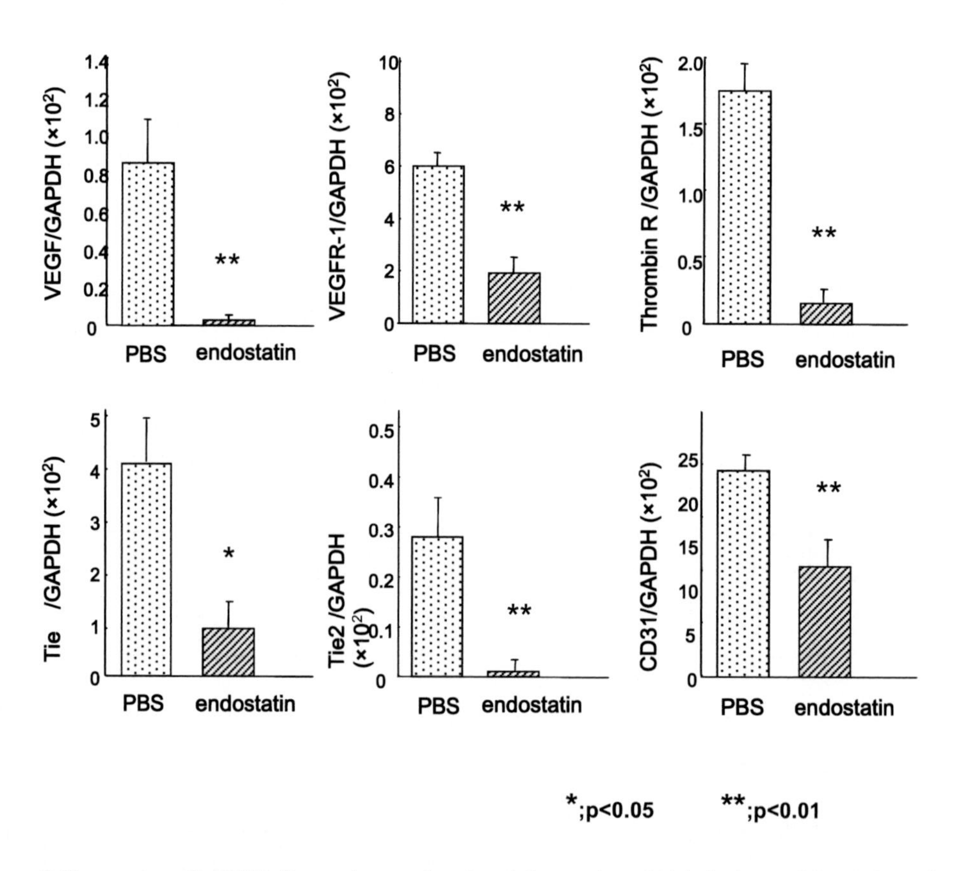

Figure 7. Expression of mRNA for angiogenesis-related factors in arthritic lesions of the endostatin and control groups. RNA was extracted from arthritic lesions of the endostatin 10mg/Kg/day and control groups, mRNA for angiogenesis-related factors was detected by the RNase protection assay, and digitized for analysis. The expression of mRNA for angiogenesis-related factors was lower in the endostatin 10mg/Kg/day than in the control group.

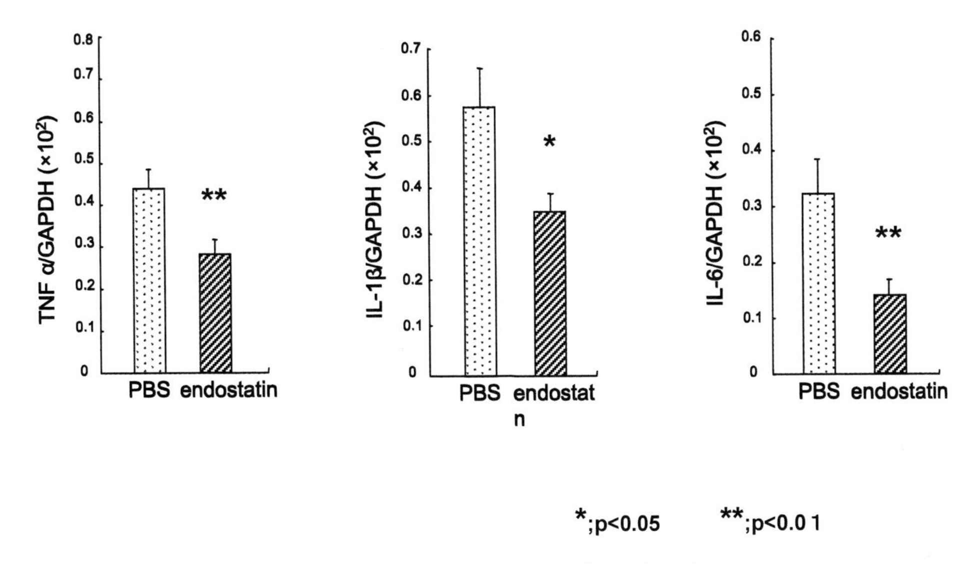

Figure 8. The expression of mRNA for inflammatory cytokines in arthritic lesions of the endostatin and control groups. RNA was extracted from arthritic lesions of the endostatin 10 mg/kg/day and control groups, and mRNA for IL-β, IL-6, and TNF-α was detected by RT-PCR. The expression of mRNA for these inflammatory cytokines was lower in the endostatin than in the control group.

Among the angiogenesis-related factors examined in this study, VEGF is considered the most important, and its expression is increased in rheumatoid arthritic synovial tissue [47]. This has also been reported in a mouse model of arthritis [13]. Therefore, antiangiogenesis therapy targeting VEGF has been attempted [13, 48]. The expression of these factors is increased in arthritic lesions, and is considered to be important in the pathogenesis of arthritis.

Thus, inhibition of the expression of these factors by endostatin administration was an important finding in considering the mechanism of the antiarthritic action of endostatin. The expression of thrombin receptors is increased in the synovial tissue of RA patients [49]. Similar findings have been reported in a mouse model of arthritis. The administration of thrombin receptor antagonists suppresses arthritis in a mouse model [50]. Expressions of Tie-1 and Tie-2 are increased in the synovial tissue of RA patients [51]. In particular, Tie-2, a ligand of Ang-1 and Ang-2, performs an important function in angiogenesis in arthritis [52]. CD31 expression is also increased in the synovial tissue of RA patients [53], and increases with the progression of arthritis in a mouse model [54].

The expressions of IL-1, IL-6, and TNF-α mRNA were lower in the endostatin group than in the control group. Considering that endostatin mainly targets vascular endothelial cells, we speculate that the mechanisms of the inhibition of inflammatory cytokine mRNA expression by endostatin are as follows. The first mechanism is mediated by vascular endothelial cells, that is, endostatin directly inhibits the expression of these cytokines in vascular endothelial cells. Since endothelial vascular cells express IL-1 and IL-6 mRNA, endostatin may directly inhibit their expression. This mechanism is very likely to operate in the case of IL-6, which was reported by Adollahi et al. to be negatively controlled by endostatin, as cited above. The second mechanism is mediated by non-endothelial cells. In this mechanism, endostatin inhibits angiogenesis, resulting in the suppression of inflammation, thereby reducing the expression of mRNA for these cytokines in non-endothelial cells such as macrophages. This mechanism may be operative in the expression of any cytokine. Next, the possibility should be considered of a mechanism in which endostatin directly acts on cytokine-expressing non-endothelial cells. To date, no studies have reported that endostatin acts directly on macrophages and lymphocytes; however, this possibility requires further study. These findings do not clarify the mechanisms of the inhibition of inflammatory cytokine mRNA expression by endostatin, but suggest that endostatin has an anti-inflammatory action in addition to an antiangiogenic action.

Advantages of Endostatin as a Therapeutic Drug for Arthritis

Folkman has reported that anticancer drugs are classified into broad-spectrum drugs and drugs that have a narrow target cell specificity [39]. The latter drugs have the advantage of a maximal pharmacological activity against tumor cells, with minimal toxicity to normal tissue, but have the disadvantage of easily producing drug resistance. In contrast, broad-spectrum drugs have the disadvantage of causing severe side-effects. This view holds true for cytotoxic anticancer therapy, but not necessarily for angiogenesis inhibitors. Endostatin is a broad-spectrum drug, but has few side-effects.

In the present study, the mice displayed no observable severe side effects after the administration of recombinant endostatin. This is important for a therapeutic agent for arthritis. Endostatin is present in the body, and can be measured in the blood to a certain extent. This may be related to the fact that endostatin has few side-effects.

However, since endostatin suppresses the migration of vascular endothelial cells to inhibit angiogenesis, it may aggravate atherosclerotic lesions. Studies have reported increased plasma levels of endostatin in diseases such as systemic scleroderma, suggesting its involvement in vascular lesions in such diseases [55]. Since there are considerable numbers of elderly people with atherosclerotic lesions and RA patients with peripheral blood flow disturbance, future studies of the angiotoxicity of endostatin are needed to enable its systemic administration.

A second advantage of endostatin relates to its action mechanism. Anti-cytokine therapy for RA has recently attracted attention; however, the main action mechanism of this therapy is not angiogenesis inhibition. In contrast, endostatin therapy mainly targets angiogenesis; therefore, it may be applicable to pathological conditions that cannot be treated by anti-cytokine therapy.

The third advantage of endostatin is that it is a broad-spectrum drug. Many factors are involved in angiogenesis in arthritis as in tumor angiogenesis; for example, even if VEGF alone is blocked, other factors may well promote angiogenesis.

The fourth advantage of endostatin is that it has little potential for inducing tolerance. From experience with the administration of endostatin to humans, Folkman reported that the drug was less likely to induce tolerance [39]. This is related to it being a broad-spectrum drug, and seems to be an advantage, because RA is a chronic disease.

Future Possibilities of Endostatin as a Drug

Unfortunately, clinical trials of endostatin as an anticancer drug in the United States were terminated, probably because of problems with the supply of stably active endostatin and the maintenance of its blood levels. On the other hand, Folkman considered that the advantages of endostatin, that is, having a broad-spectrum activity, little potential for inducing tolerance, and few side-effects, favored the combined use of endostatin and other drugs, and proposed combination therapy with endostatin and other anticancer drugs [39]. Currently, clinical trials of combination therapy with recombinant endostatin and anticancer drugs for non-small cell lung cancer are in progress in China, reportedly producing good results, although no final conclusions have been reached yet [39]. At the experimental level, endostatin has shown antitumor activity against many tumors; therefore, various improvements are very likely to lead to it becoming an anticancer drug [56-59]. Studies of endostatin itself have also advanced; for example, the biological activity of endostatin was reported to reside in 27 amino acid residues in the NH2-terminal domain [60-65].

Endostatin has the potential for use as a unique therapeutic agent for arthritis with an action mechanism different from that of conventional drugs. However, the clinical application of endostatin to the treatment of arthritis requires a number of issues to be resolved. Moreover, the pathogenesis of angiogenesis in inflammatory diseases has not been fully elucidated. It is hoped that further advances will be made in research in this area.

References

[1] Colville-Nash P.R. and Scott D.L. (1992) Angiogenesis and rheumatoid arthritis: pathogenic and therapeutic implications. *Ann. Rheum. Dis.,* 51(7):919-925.

[2] Szekanecz Z., Szegedi G.and Koch A.E. (1998) Angiogenesis in rheumatoid arthritis: pathogenic and clinical significance. *J. Investig. Med.,* 46(2):27-41.

[3] Koch A.E. (2000) The role of angiogenesis in rheumatoid arthritis: recent developments. *Ann. Rheum. Dis.,* 59 Suppl 1:i65-71. Walsh D.A.and Pearson C.I. (2001).

[4] Angiogenesis in the pathogenesis of inflammatory joint and lung diseases. *Arthritis Res.,* 3(3):147-153.

[5] Koch A.E. (2003) Angiogenesis as a target in rheumatoid arthritis. *Ann. Rheum. Dis.,* 62 Suppl 2:ii60-ii67.

[6] Nagashima M., Asano G. and Yoshino S. (2000) Imbalance in production between vascular endothelial growth factor and endostatin in patients with rheumatoid arthritis. *J. Rheumatol.,* 27(10): 2339-2342.

[7] Hirata S., Matsubara .T, Saura R., Tateishi H. and Hirohata K. (1989) Inhibition of in vitro vascular endothelial cell proliferation and in vivo neovascularization by low-dose methotrexate. *Arthritis Rheum.,* 32(9):1065-1073.

[8] Matsubara T., Saura R., Hirohata K. and Ziff M. (1989) Inhibition of human endothelial cell proliferation in vitro and neovascularization in vivo by D-penicillamine. *J. Clin. Invest.,* 83(1):158-167.

[9] Matsubara T. and Ziff M. Inhibition of human endothelial cell proliferation by gold compounds. *J. Clin. Invest.*, 79(5): 1440-1446.

[10] Paleolog E.M.and Miotla J.M. (1998) Angiogenesis in arthritis: role in disease pathogenesis and as a potential therapeutic target. *Angiogenesis*, 2(4):295-307.

[11] Brenchley P.E. (2000) Angiogenesis in inflammatory joint disease: a target for therapeutic intervention. *Clin. Exp. Immunol.,* 121(3):426-429.

[12] Walsh D.A and Haywood L. (2001) Angiogenesis: a therapeutic target in arthritis. *Curr. Opin. Investig. Drugs,* 2(8):1054-1063.

[13] Lu J, Kasama T., Kobayashi K., Yoda Y., Shiozawa F., Hanyuda M., Negishi M., Ide H. and Adachi M. (2000) Vascular endothelial growth factor expression and regulation of murine collagen-induced arthritis. *J. Immunol.,* 164:5922-5927.

[14] Yamashita A., Yonemitsu Y., Okano S., Nakagawa K., Nakashima Y., Irisa T., Iwamoto Y., Nagai Y., Hasegawa M and Sueishi K. (2002) Fibroblast growth factor-2 determines severity of joint disease in adjuvant-induced arthritis in rats. *J. Immunol.,* 168(1):450-457.

[15] Canetti C.A., Leung B.P., Culshaw S., McInnes I.B., Cunha F.Q. and Liew F.Y. (2003) IL-18 enhances collagen-induced arthritis by recruiting neutrophils via TNF-alpha and leukotriene B4. *J. Immunol.,* 171(2):1009-1015.

[16] Storgard C. M, Stupack D.G., Jonczyk A., Goodman S.L., Fox R.I. and Cheresh D.A. (1999) Decreased angiogenesis and arthritic disease in rabbits treated with an alphavbeta3 antagonist. *J. Clin. Invest.,* 103(1):47-54.

[17] Kim J.M., Ho S.H., Park E.J., Hahn W., Cho H., Jeong J.G., Lee Y.W.and Kim S. (2002) Angiostatin gene transfer as an effective treatment strategy in murine collagen-induced arthritis. *Arthritis Rheum.,* 46(3):793-801.

[18] Matsuno H., Yudoh K., Uzuki M., Nakazawa F., Sawai T., Yamaguchi N., Olsen B.R. and Kimura T. (2002) Treatment with the angiogenesis inhibitor endostatin: a novel therapy in rheumatoid arthritis. *J. Rheumatol.,* 29(5):890-895.

[19] Yin G., Liu W., An P., Li P., Ding I., Planelles V., Schwarz E.M.and Min W. (2002) Endostatin gene transfer inhibits joint angiogenesis and pannus formation in inflammatory arthritis. *Mol. Ther.,* 5(5 Pt 1):547-554.

[20] 20.Yue L., Wang H., Liu L.H., Shen Y.X.and Wei W. (2004) Anti-adjuvant arthritis of recombinant human endostatin in rats via inhibition of angiogenesis and proinflammatory factors. *Acta Pharmacol. Sin.,* 25(9):1182-1185.

[21] Kurosaka D., Yoshida K., Yasuda J., Yokoyama T., Kingetsu I., Yamaguchi N., Joh K., Matsushima M., Saito S.and Yamada A. (2003) Inhibition of arthritis by systemic administration of endostatin in passive murine collagen induced arthritis. *Ann. Rheum. Dis.,* 62(7): 677-679.

[22] Peacock D.J., Banquerigo M.L.and Brahn E.(1992) Angiogenesis inhibition suppresses collagen arthritis. *J. Exp. Med.,* 175(4): 1135-1138.

[23] Oliver S.J., Banquerigo M.L.and Brahn E. (1994) Suppression of collagen-induced arthritis using an angiogenesis inhibitor, AGM-1470, and a microtubule stabilizer, taxol. *Cell Immunol.,* 157(1):291-299.

[24] Bandt M.D, Grossin M., Weber A.J., Chopin M., Elbim C., Pla M., Gougerot-Pocidalo M.A.and Gaudry M. (2000) Suppression of arthritis and protection from bone

destruction by treatment with TNP-470/AGM-1470 in a transgenic mouse model of rheumatoid arthritis. *Arthritis Rheum.*, 43(9):2056-2063.

[25] O'Reilly M.S., Boehm T., Shing Y., Fukai N., Vasios G., Lane W.S, Flynn E.,Birkhead J.R, Olsen B.R.and Folkman J. (1997) Endostatin: an endogenous inhibitor of angiogenesis and tumor growth. *Cell*, 88(2):277-285.

[26] Halfter W., Dong S., Schurer B.and Cole G. J. (1998) Collagen XVIII is a basement membrane heparan sulfate proteoglycan. *J. Biol. Chem.*, 273(39):25404-25412.

[27] Yamaguchi N., Anand-Apte B., Lee M., Sasaki T., Fukai N., Shapiro R. R., Que I., Lowik C., Timpl R.and Olsen B.R. (1999) Endostatin inhibits VEGF-induced endothelial cell migration and tumor growth independently of zinc binding. *EMBO J.*, 18(16):4414-4423.

[28] Sudhakar A., Sugimoto H., Yang C., Lively J., Zeisberg M.and Kalluri R. (2003) Human tumstatin and human endostatin exhibit distinct antiangiogenic activities mediated by alpha v beta 3 and alpha 5 beta 1 integrins. *Proc. Natl. Acad. Sci. USA*, 100(8):4766-4771.

[29] Wickstrom S.A., Alitalo K.and Keski-Oja J. (2002) Endostatin associates with integrin alpha 5 beta 1 and caveolin-1, and activates Src via a tyrosyl phosphatase-dependent pathway in human endothelial cells. *Cancer Res.*, 62(19):5580-5589.

[30] Yu Y., Moulton K.S., Khan M.K., Vineberg S., Boye E., Davis V.M., O'Donnell P.E., Bischoff J.and Milstone D.S. (2004) E-selectin is required for the antiangiogenic activity of endostatin. *Proc. Natl. Acad. Sci. USA*, 101(21):8005-8010.

[31] Kim Y.M., Jang J.W., Lee O.H., Yeon J., Choi E.Y., Kim K.W., Lee S.T.and Kwon Y.G. (2000) Endostatin inhibits endothelial and tumor cellular invasion by blocking the activation and catalytic activity of matrix metalloproteinase. *Cancer Res.*, 60:5410-5413.

[32] Nyberg P, Heikkila P, Sorsa T, Luostarinen J, Heljasvaara R, Stenman UH, Pihlajaniemi T.and Salo T. (2003) Endostatin inhibits human tongue carcinoma cell invasion and intravasation and blocks the activation of matrix metalloprotease-2, -9, and.-13. *J. Biol. Chem.*, 278(25):22404-22411.

[33] Abdollahi A., Hahnfeldt P., Maercker C., Grone H.J., Debus J., Ansorge W, Folkman J., Hlatky L.and Huber P.E. (2004) Endostatin's antiangiogenic signaling network. *Mol. Cell*, 13(5):649-663.

[34] Sato Y. (2001) Role of ETS family transcription factors in vascular development and angiogenesis. *Cell Struct. Funct.*, 26(1):19-24.

[35] Damert A., Ikeda E.and Risau W. (1997) Activator-protein-1 binding potentiates the hypoxia-induciblefactor-1-mediated hypoxia- induced transcriptional activation of vascular-endothelial growth factor expression in C6 glioma cells. *Biochem. J.*, 327 (Pt 2):419-423.

[36] 36.Lyden D., Young A.Z., Zagzag D., Yan W., Gerald W., O'Reilly R., Bader B.L., Hynes R.O., Zhuang Y., Manova K.and Benezra R. (1999) Id1 and Id3 are required for neurogenesis, angiogenesis and vascularization of tumour xenografts. Nature, 401(6754): 670-677.

[37] Pandey A., Shao H., Marks R.M., Polverini P.J.and Dixit V.M. (1995) Role of B61, the ligand for the Eck receptor tyrosine kinase, in TNF-alpha-induced angiogenesis. *Science,* 268(5210):567-569.

[38] Tsopanoglou N.E.and Maragoudakis M.E. (1999) On the mechanism of thrombin-induced angiogenesis. Potentiation of vascular endothelial growth factor activity on endothelial cells by up-regulation of its receptors. *J. Biol. Chem.*, 274(34):23969-23976.

[39] Folkman J. (2006) Antiangiogenesis in cancer therapy--endostatin and its mechanisms of action. *Exp. Cell Res.*, 312(5):594-607.

[40] Berger A.C., Feldman A.L., Grant M.F., Kruger E.A., Sim B.K., Hewitt S., Figg W.D., Alexander H.R.and Libutti S.K. (2000) The angiogenesis inhibitor, endostatin, does not affect murine cutaneous would hearing. *J. Surg. Res.,* 91(1):26-31.

[41] Sund M., Hamano Y., Sugimoto H., Sudhakar A., Soubasakos M., Yerramalla U., Benjamin L.E., Lawler J., Kieran M., Shah A.and Kalluri R. (2005) Function of endogenous inhibitors of angiogenesis as endothelium- specific tumor suppressors. *Proc. Natl. Acad. Sci. USA*, 102(8):2934-2939.

[42] Kurosaka D, Yoshida K. (2004) Arthritis inhibiting effect of endostatin. *Inflammation and Regeneration*, 24(2):107-112.

[43] Kurosaka D., Yoshida K., Yasuda C., Yasuda J., Toyokawa Y., Yokoyama T., Kingetu I., Yamaguchi N., Joh K., Saito S.and Yamada A. (2005) The rple of inflammatory cytokines in arthritis inhibiting effect of endosutatin. *Modern Rheum.,* 15:S182.

[44] Kurosaka D., Yoshida K., Yasuda J. ,Yasuda C., Noda K., Toyokawa Y.and Yamada A. (2006) Arthritis-inhibiting effecy of endostatin on collagen-induced arthritis. *Modern Rheum.*, 16:178.

[45] Miotla J., Maciewicz R., Kendrew J., Feldmann M.and Paleolog E.(2000) Treatment with soluble VEGF receptor reduces disease severity in murine collagen-induced arthritis. *Lab. Invest.,* 80:1195-1205.

[46] Luttun A., Tjwa M., Moons L.,Wu Y., Angelillo-Scherrer A., Liao F., Nagy J.A., Hopper A., Priller J., Klerck B.D., Compernolle V., Daci E., Bohlen P., Dewerchin M., Herbert J.M., Fava R., Matthys P., Carmeliet G., Collen D., Dvorak H.F, Hickin D.J.and Carmeliet P. (2002) Revascularization of is chemic tissues by PIGF treatment, and inhibition of tumor angiogenesis, arthritis and atherosclerosis by anti-Flt1. *Nature* Med, 8(8):831-840.

[47] Fava R.A., Olsen N.J., Spencer-Green G, Yeo K.T., Berse B., Jackman R.W., Senger D.R., Dvorak H.F.and Brown L.F. (1994) Vascular permeability factor/endothelial growth factor (VPF/VEGF): accumulation and expression in human synovial fluids and rheumatoid synovial tissue. *J. Exp. Med.*, 180(1):341-346.

[48] Mould A.W., Tonks I.D., Cahill M.M., Pettit A.R., Thomas R., Hayward N.K.and Kay G.F.(2003) Vegfb gene knockout mice display reduced pathlogy and synovial angiogenesis in both antigen-induced and collagen-induced models of arthritis. *Arthritis Rheum.*, 48(9):2660-2669.

[49] Morris R., Winyard P.G., Brass L.F., Blake D.R.and Morris C.J. (1996) Thrombin receptor expression in rheumatoid and osteoarthritic synovial tissue. *Ann. Rheum. Dis.,* 55(11):841-843.

[50] 50.Ferrell W.R., Lockhart J.C., Kelso E.B., Dunning L., Plevin R., Meek S.E., Smith A.J., Hunter G.D., McLean J.S., McGarry F., Ramage R., Jiang L., Kanke T.and Kawagoe J.(2003) Essential role for proteinase-activated receptor-2 in arthritis. *J. Clin. Invest.,* 111(1):35-41.

[51] Shahrara S., Volin M.V., Connors M.A., Haines G.K.and Koch A.E. (2002) Differential expression of the angiogenic t ie receptor family in arthritic and normal synovial tissue. *Arthritis Res.,* 4(3):201-208.

[52] Debusk L.M., Chen Y., Nishishita T., Chen J., Thomas J.W.and Lin P.C. (2003) Tie2 receptor tyrosine kinase, a major mediator of tumor necrosis factorα-induced angiogenesis in rheumatoid arthritis. *Arthritis Rheum.,* 48(9):2461-2471.

[53] Szekanecz Z., Haines G.K., Harlow L.A., Shah M.R., Fong T.W., Fu R., Lin S.J.and Koch A.E. (1995) Increased synovial expression of the adhension molecules CD66a, CD66b, and CD31 in rheumatoid and osteoarthritis. *Clin. Immunol. Immunopathol.,* 76(2):180-186.

[54] Volin M.V., Szekanecz Z., Harloran M.M., Woods J.M, Magua J., Damergis J.A. Jr, Haines K.G. 3rd, Crocker P.R.and Koch A.E. (1999). PECAM-1 and leukosialin(CD43) expression correlate with heightened inflammation in rat adjuvant-induced arthritis. *Exp. Mol. Pathol.,* 66(3):211-219.

[55] Hebbar M., Peyrat J.P., Hornez L., Hatron P.Y., Hachulla E.and Devulder B. (2000) Increased concentrations of the circulating angiogenesis inhibitor endostatin in patients with systemic sclerosis. *Arthritis Rheum.,* 43(4):889-893.

[56] Wu X., Huang J., Chang G.and Luo Y. (2004) Detection and characterization of an acid-induced folding intermediate of endostatin. *Biochem. Biophys. Res. Commun.,* 320(3):973-978.

[57] Li B., Wu X., Zhou H., Chen Q.and Luo Y. (2004) Acid-induced unfolding mechanism of recombinant human endostatin. *Biochem*, 43(9):2550-2557.

[58] Zhou H., Wang W.and Luo Y. (2005) Contributions of disulfide bonds in a nested pattern to the structure, stability, and biological functions of endostatin. *J. Biol. Chem.,* 280(12):11303-11312.

[59] He Y., Zhou H., Tang H.and Luo Y. (2006) Deficiency of disulfide bonds facilitating fibrillogenesis of edndostatin. *J. Biol. Chem*., 28(2):1048-1057.

[60] Tjin Tham Sjin R.M., Satchi-Fainaro R., Birsner A.E., Ramanujam V.M., Folkman J.and Javaherian K. (2005) A 27-amino-acid synthetic peptide corresponding to the NH2-terminal zinc-binding domain of endostatin is responsible for its antitumor activity. *Cancer Res.,* 65(9):3656-3663.

[61] Cattaneo M.G., Pola S., Francescato P., Chillemi F.and Vicentini L.M. (2003) Human endostatin-derived synthetic peptides possess potent antiangiogenic properties in vitro and in vivo. *Exp. Cell Res.,* 283(2):230-236.

[62] Morbidelli L., Donnini S., Chillemi F., Giachetti A.and Ziche M. (2003) Angiosuppressive and angiostimulatory effects exerted by synthetic partial sequences of endostatin. *Clin. Cancer Res.,* 9(14):5358-5369.

[63] Chillemi F., Francescato P., Ragg E., Cattaneo M.G., Pola S.and Vicentini L. (2003) Studies on the structure-activity relationship of endostatin:synthesis of human

endostatin peptides exhibiting potent antiangiogenic activities. *J. Med. Chem.*, 46(19):4165-4172.

[64] Olsson A.K., Johansson I., Akerud H., Einarsson B., Christofferson R., Sasaki T., Timpl R.and Claesson-Welsh L. (2004) The minimal active domain of endostatin is a heparin-binding motif that mediates inhibition of tumor vascularization. Cancer Res, 64(24):9012- 9017.

[65] Wickstorm S.A., Alitalo K.and Keski-Oja J. (2004) An endostatin- derived peptide interacts with integrins and regulates actin cytoskeleton and migration of endothelial cells. *J. Biol. Chem.*, 279(19):20178-20185.

In: Stem Cell Research Trends
Editor: Josse R. Braggina, pp. 293-303
ISBN: 978-1-60021-622-0

Chapter XI

Body Building Exercise: Maintenance and Loss of Stemness in Spinal Cord Neural Stem Cells

Isabelle Roszko and Luc Mathis*
Unité de Biologie Moléculaire du Développement,
CNRS URA 2578, Institut Pasteur, 25, rue du Docteur Roux,
75724 Paris Cedex 15, France

Abstract

Stem cells are a major stake in biology, biotechnology and therapy. The ability of certain cells to regenerate tissues and functions has a great theoretical importance in Developmental Biology and constitutes a great hope to treat debiliting conditions resulting from aging or injuries. The development of efficient methods to amplify, differentiate and graft stem cells is a major challenge for today's biotechnology industry. In the recent past, major progresses have arisen from the field of developmental biology. A clear example is the use of the sequence of signals in the embryo to drive the differentiation of stem cells into preselected cell types in vitro. Recent results indicate the importance of a tissue-level organization in the maintenance and differentiation of the pools of stem cells in vivo. In the present article, we describe how spatially organized stem cells are responsible for the elongation of the spinal cord. We also show that a limited number of constraints are responsible for the orientation of cell divisions of neuroepithelial stem cell, a parameter critical for the maintenance of stemness in these cells. Taken together, these results suggest a lineage of stem cells from ES cells to neural differentiation that may be significant for the embryo. In addition, these results suggest that future strategies for the manipulation of stem cell in vitro will require to change growth methods into three-dimensional matrices and reconstitution of apico-basal organization of the cells. This should help fine tune the properties of the stem cells and provide a better spatial control of the sequence of events that take place in the embryo.

* Author for correspondence: lmathis@pasteur.fr, Phone: 00-33-676548264. Present address: Cellectis SA, Biocitech, 102 rue de Noisy, 93235 Romainville, France.

INTRODUCTION

At the moment, there are two major sources of stem cells: embryonic and adult stem cells. Embryonic stem cell lines have been derived from human blastocysts and a number are available for research purposes. The major limitation of the use of embryonic stem cells in cellular therapy is that they are heterologous, and may be rejected by the immune system when grafted. Autologous adult stem cells may resolve the rejection issue, but they require a personalized and expensive manipulation. In fact, totipotent adult stem cells are found in the special places formed body, such as the bone marrow and the umbilical cord blood. More specialized adult stem cells are found in dedicated niches, the satellite muscular stem cells or the cells that self renew the intestine, the liver or the spermatozoids. The biotechnology field has to resolve a number of issues before getting into the industrial age of therapeutic use of stem cells.

Recent results indicate the importance of a tissue-level organization in the maintenance and differentiation of the pools of stem cells. This could pave the way for future improved strategies for the manipulation of stem cell in vitro. In addition, these recent findings on neural stem cells in spinal cord formation indicate a lineage of stem cells from early stages of neural morphogenesis to neurogenesis stages.

STEM CELLS FOR AXIS ELONGATION

Stem cells are self-renewing cells whose undifferentiated properties are maintained by signals and cytokines such as FGF. Stem cells have been identified early in embryonic development (the ES cells) as well as later on, during organogenesis (e.g. the neural or hematopoietic stem cells). The possibility of a direct link between these undifferentiated stem cells has been explored, and lead to the identification of an intermediate stem cell population, dependent on FGF, that generates the body axis [1, 2]. Analysis of the constraints acting on this cell population provides indications as to the mechanisms involved in and show the importance of the three- (and even four-) dimensional organization of growth.

Axis elongation is the process that transforms the initial polarity of the embryo into the formed body. Two main processes have been proposed to lead to trunk elongation, the convergence extension movements and the generation of descendants from a pool of selfrenewing stem cells. Convergence-extension is a general cellular mechanism and has been also described in the elongation of the germ band in Drosophila [3] or the formation of the notochord in ascidians [4]. In zebrafish and Xenopus embryos, time-lapse analyses of gastrulation movements have shown that axis elongation results from the convergence of cells toward the midline and intercalation of cells, resulting in narrowing and lengthening of the embryo (Figure .1A) [5]. During gastrulation, the neural plate cells proliferate and the orientation of cell divisions are also involved in axis elongation (Figure 1B). The elongation of the axis is based on a posterior growth zone in many invertebrate species including crustaceans, arachneids, leeches and long germ band insects [6]. In mouse and chick embryos, elongation appears to result from the proliferation of cells resident in the node region, in the primitive streak and later in the tail bud (Figure 1C) [7].

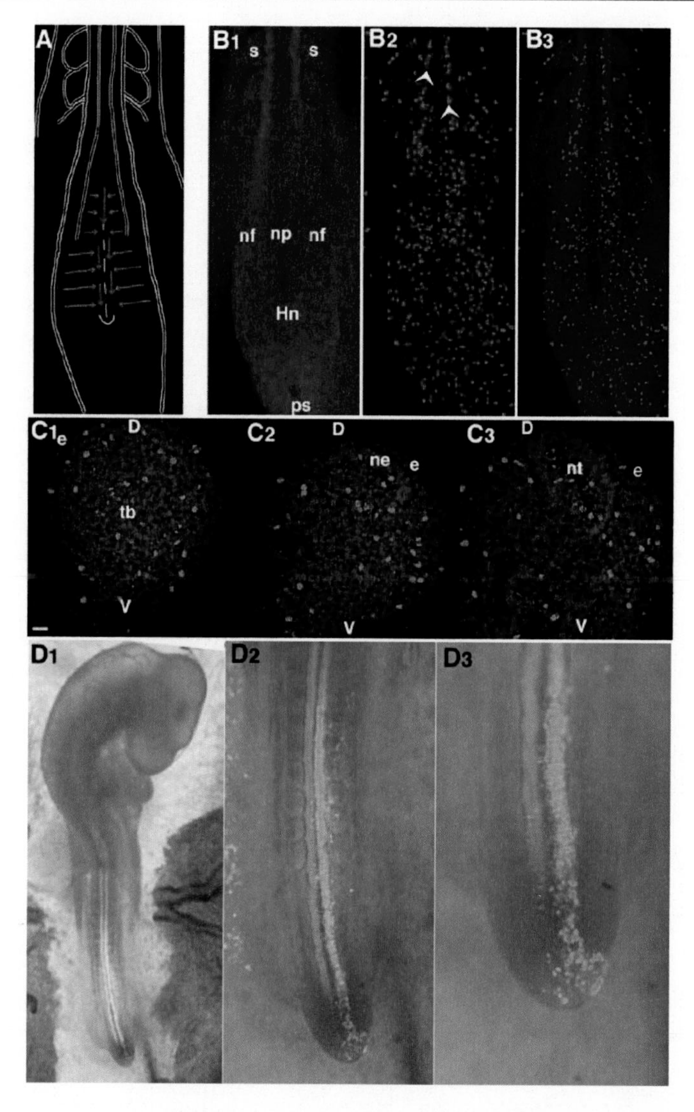

Figure 1. Neural stem cells during spinal cord elongation. A) Convergence (red) and extension (blue) movements in the neural plate of the 6-somitestage chick embryo. B) Morphology of the neural plate at 6-somite-stage chick embryo and localizations of the divisions occurring in the neural plate. B1) To-Pro nuclear staining, B2) H3P staining, B3) Merge. C1-C3) Transversal sections of chick embryo tail bud stained with To-Pro, a DNA marker in blue and with a mitotic marker, H3P in red. D1) is the most posterior section and shows the randomly organized cell nuclei in the tail bud. D2) is the section of the median tail bud region and shows the neuroepithelium formation by cell (and nuclei) reorganization in the dorsal part of the tail bud. D2) is the most anterior tail bud section, which shows the completely formed neural tube. Scale bar 20 microns. D1-D3) Thee pictures at different magnification of a chick embryo at tail bud stage, electroporated previously in the neural plate at 9-somite-stage with a plasmid coding for a membrane-GFP. The neural plate cells expressing the GFP protein contributed to the neural tube formation and relative extension of the antero-posterior axis. GFP expressing cell still located in the tail bud and contribute to the neural tube extension by producing cells, which enter in the neural tube tissue (potential stem cells). Note the presence of some neuroepithelial cells in the left side (non electroporated) of the most posterior neural tube (red arrowhead). e: surface ectoderm, h: head, Hn: Hensen's node, ne: neuroepithelium, nf: neural fold, np: neural plate, nt: neural tube, ps: primitive streak, psm: pre-somitic mesoderm, s: somite, tb, tail bud, D: Dorsal side of the embryo, V: Ventral.

Key patterning events for the regional organization of the vertebrate central nervous system (CNS) occur during the longitudinal development of the neural plate and neural tube. In the mouse embryo, early signals for forebrain development are provided by the anterior visceral endoderm [8], and the first regionalized expression of genetic markers in the neuroepithelium (e.g. otx-2) is established at the early primitive streak stage (E6.5) [9, 10]. Inductive signals emanating from the node [11-15] the pre-chordal plate (mesendoderm) [16] and the paraxial mesoderm [17, 18] are also involved in the early A-P patterning of the CNS.

Morphological regionalization of the neural tube becomes evident with the subdivision of the longitudinal axis into four major domains: the prosencephalon, the mesencephalon, the rhombencephalon and the spinal cord (SC). The elongation of the spinal cord is a key event in the formation of the central nervous system of vertebrates. After a phase of longitudinal dispersion of CNS founder cells located in the epiblast at gastrulation (around E6.5 in the mouse) [19] fate of brain versus spinal cord progenitors becomes rapidly restricted [20, 21]. Spinal cord progenitors undergo proliferation and cell movements that lead to elongation of the posterior CNS [22]. Fate maps of the chick or mouse neural primordium show that the precursors of the spinal cord, located in a very small region of the epiblast lateral to the node, give rise to descendants dispersed extensively along the rostrocaudal axis [1, 13, 23-30]. Time-lapse analyses in the chick following electroporation of the neural plate further showed that cells remain resident in the posterior neural plate and then in the tail bud to produce the elongating axis (Figure 1D) [2]. These data have suggested that spinal cord elongation results from a stem cell zone of growth that is progressively laid down in the neural tube from anterior to posterior [20, 31].

The antero-posterior dispersion of clonally related cells is a prominent feature of axis elongation in vertebrate embryos. Two major models have been proposed: (i) the intercalation of cells by convergent-extension and (ii) the sequential production of the forming axis by stem cells. The relative importance of these cell behaviors during the long period of elongation is poorly understood. We propose that a pre-existing stem zone of growth becomes predominant to form the posterior half of the axis while the neural plate reduces in size. This may provide an explanation to the apparent difference in elongation patterns between higher (birds and mammals) and lower Vertebrates (fishes and amphibians), as well as a formal explanation to the maintenance of the genetic organization during axis elongation.

The anterior and posterior neural system precursors become separated early during gastrulation. In principle, this separation could reflect a restriction of cell dispersion at the interface between the two clonal domains (due to a clonal boundary or to differences in adhesive properties of the precursors of adjacent domains) or to cell movements in opposite directions. The separation is most likely due to the regression of a pool of self-renewing cells and to a relatively coherent growth in the brain. Therefore, complementing other mechanisms such as cell sorting [32], slow cell intermixing [33], an arrest of cell dispersion [34] or the restriction of cell movements at boundaries [35], a differential mode of growth is a possible mechanism for the separation of pools of progenitor cells in the embryo.

What controls this mode of growth? A possibility is that posterior developmental signals such as bFGF [36] and retinoic acid [37] may endow posterior CNS founder cells with this mode of growth. In a polarized self-renewing system, the clonal history (i.e. the number of cell generations preceding the arrest of cell dispersion) is longer for the caudal part of the trunk than for the rostral part. This correlates with the fact that the number of expressed *Hox* genes increases from rostral to caudal [38, 39]. A possibility [40, 41] is that some A-P positional information could be delivered by this temporal system, as it is in simpler organisms [42].

Maintenance of Stem Cells in the Neural Tube

The neural stem cells are contained during embryogenesis in a tubular structure called the neural tube (Figure 2A). The neural tube is formed by a pseudostratified epithelium, the neuroepithelium (Figure 2E). This tissue is therefore polarised (express epithelial polarity markers) with the apical pole near the lumen of the neural tube and the basal pole is near the somite. (Figure 2B-D) During the neural development of the embryo the neuroepithelial cell divisions occur at the apical pole (Figure F-H). The first step in differentiation of neural stem cells occurs in the neuroepithelium, when the progenitor cell divides asymmetrically. The neuroblast expresses early neuronal markers quickly after the division (Figure 2I).

The orientation of cell divisions is likely to be an important parameter for the generation of cellular diversity from multipotent self-renewing stem cells. Current models suggest that different fates can be adopted by sister cells that have asymmetrically inherited cytoplasmic or plasma-membrane-associated determinants during the division of neural stem cells [43-45]. In the vertebrate central nervous system, though a significant fraction of the cells undergo apico-basal cell divisions at late stages of neurogenesis (ABcd) [46] the vast majority of neuroepithelial (NE) cells divide within the plane of the neuroepithelium (planar cell divisions, Pcd, [47]. Huttner and colleagues have suggested that, because the apical surface of NE cells is small, even a small deviation in division plane can result in one cell acquiring all of the apical domain. A partial shift from planar to apico-basal oriented divisions in NE cells is associated with the period of neuronal differentiation [47-49]. Apico-basal divisions are usually observed in the neural tissue during the neurogenic phase resulting in the production of one differentiated cell and one cell that maintains stem cell characteristics.

In *Drosophila* and *C.elegans*, the orientation of cell divisions is critically dependent on mechanisms that position the spindle, including early rotation events and maintenance of spindle orientation before anaphase [50-52]. A number of studies have identified molecular regulators of spindle formation and positioning including, amongst others, small G-protein family members [53-56]. In the cerebral cortex, impaired signaling through the G□ □subunits of the heterotrimeric G proteins increases the frequency of planar cell divisions and favors neural differentiation [55]. The constraints acting on the spindle to orient most cell divisions within the plane of the neuroepithelium, and its shift to an apico-basal orientation during neurogenesis, have been adressed. NE cells divide in the plane of the neuroepithelium as a combined result of an early rotation that aligns the spindle in the neuroepithelial plane within the first 15 minutes of metaphase and limited oscillations of the spindle that maintain this orientation until anaphase [57]. This conclusion is somewhat reminiscent of the finding that

NE cells undergoing apico-basal cell divisions in the mammalian cerebral cortex display extensive metaphase plate oscillations [48].

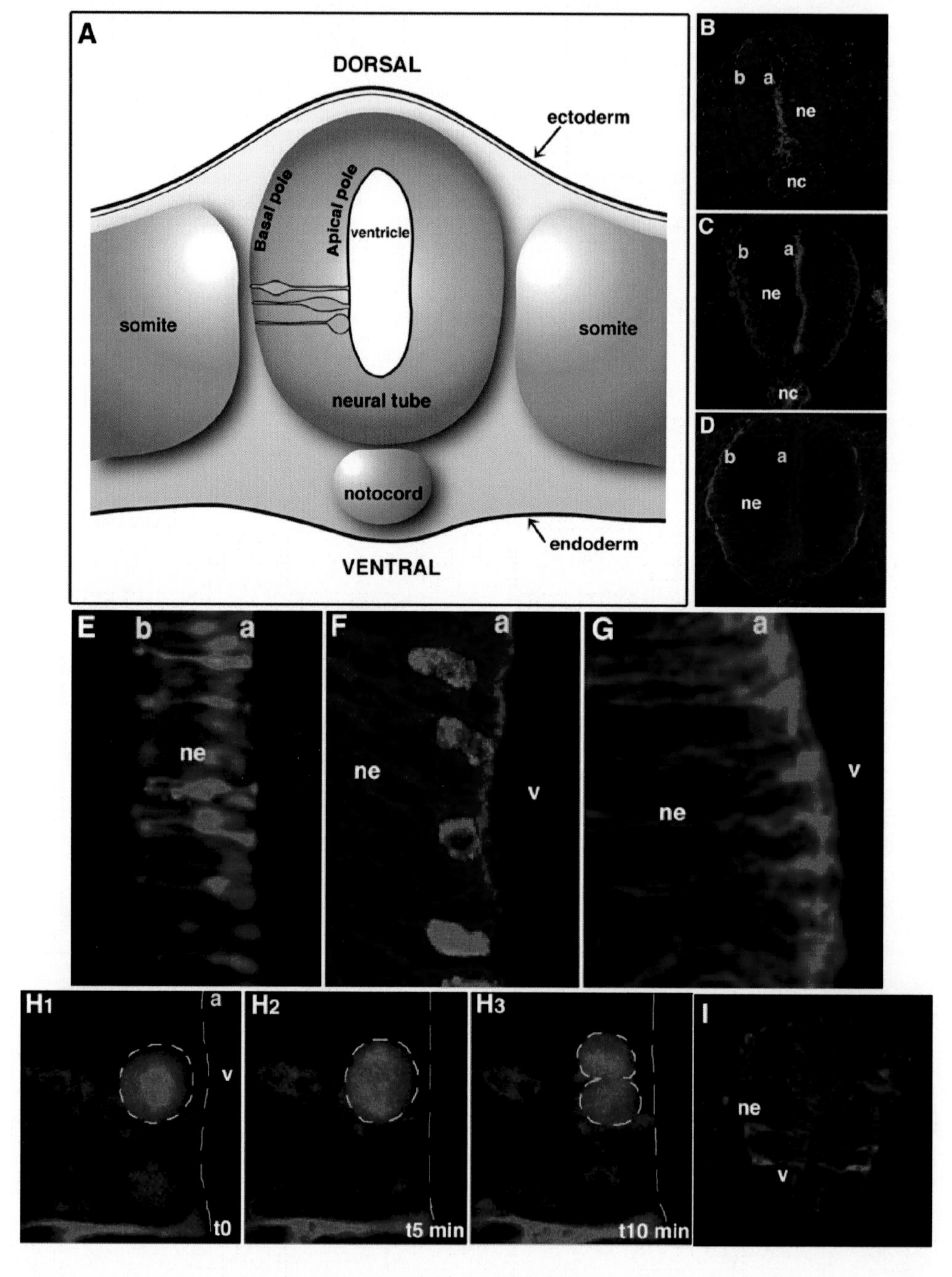

Figure 2. Organization of the neural stem cells in the neural tube. A) Schematic representation of trunk transversal section of the embryo. B) Neural tube section stained with phalloidin, polymerized actin marker (red) and To-Pro staining, DNA marker (blue). C) N-Cadherin staining of neural tube section showing the accumulation of NCadherins at the apical pole of the neuroepithelium (adherent junctions). D) β1-inetgrin staining of neural tube section showing the accumulation of N-Cadherins at the basal pole of the neuroepithelium. E) Dorsal view of the neuroepithelium showing the characteristic

neuroepithelial cells morphology (the chick embryo was electroporated with a plasmid coding for a cytoplasmic GFP). F) Magnification of the apical pole of the neuroepithelium stained with the apical marker (b-catenin, localized mainly at the adherant junctions, in blue) which shows the cells divisions (H3P marker (red)) at this pole. G) Magnification of the apical pole of the neuroepithelium stained with phalloidin (polymerized actin marker) which shows the polinerized actin accumulation at the apical pole of the neuroepithelial cells. H1-H3) Time-lapse experiment showing a cell division occurring in the plane of the neuroepithelium. The apical pole of the epithelium is represented by the yellow dash line. The neuroepithelial cells were transfected by electroporation and express EB1-GFP pfusion protein. EB1 is a microtubules binding protein. The strong signal visible in the cell is the mitotic spindle. I) Picture showing on one transversal section of the neural tube at the beginning of the neurogenesis a Tuj1 staining (marker of the βIII-tubulin, expressed in cells that adopt a neuroblastic fate). a: apical pole, b: basal pole, ne: neuroepithelium, nc notocord, v: ventricle.

Since they might both lead to asymmetry in apical plasma membrane inheritance), the spindle oscillations and the constraints that limit them might be significant in terms of cell fate determination.

An interesting candidate for the control of such constraints is the small RhoGTPase, RhoA. Global patterning defects of expressing a dominant negative RhoA have been previously described in Xenopus and Zebrafish. The function of RhoA is central for CNS morphogenesis, and is implicated in early convergent-extension motion in the neural plate [5] and later events of axons formation [58], Expressing a dominant-negative RhoA leads to apico-basal cell divisions after a correct initial rotation of the spindle [57]. Our data reveal a specific role for RhoA in the maintenance of spindle orientation, prior to anaphase. One possibility is that the defects in spindle orientation seen on dnRhoA expression are the consequence of other effects of RhoA, for example changes in cell shape. However, there might be no simple answer to this question: cell biologists, including Michel Bornens for instance, now argue that cell shape and the orientation of cell divisions might be deeply linked. Obviously, the molecular and cellular cascade linking RhoA and the orientation of division in the neuroepithelium is a new field of investigation. A number of approaches will be required to evaluate the relative importance of cell shape, the binding to the actin cortex, the actin cortex rigidity, the function of myosinII and/or other players, which are all potential key parameters involved. The phenotype obtained by impairing G®© subunits of the heterotrimeric G proteins is opposite to that reported for impaired RhoA, at least regarding the orientation of cleavage plane. This suggests a general control of cleavage plane orientation in the neuroepithelium through different GTPase and antagonist requirements for different players. Furthermore, given the interplay between RhoA and heterotrimeric G proteins [59], this could suggest the existence of feedback loop control mechanisms based on G proteins that fine-tunes the orientation of neuroepithelial progenitors cell divisions. Thus, RhoA appears as a key player potentially regulated by the neurogenic program or by the neural stem cell environment to control the balance between planar and apico-basal divisions, during normal or pathological development.

Conclusion

The development of efficient methods to amplify, differentiate and graft stem cells is a major challenge for today's biotechnology industry. An important work of characterization of stem cells at the genetic level has already taken place [60, 61] leading to the identification of a new markers and genetic networks involved in the maintenance or loss of stemness. In the recent past, major progresses have also arisen from the field of developmental biology. A clear example is the use of the sequence of signals in the embryo to drive the differentiation of stem cells into preselected cell types in vitro. A major route ahead will be to change growth methods into three-dimensional matrices and reconstitution of apico-basal organization of the cells. Taken together, these results suggest a lineage of stem cells from ES cells to neural differentiation that may be significant for the embryo. In particular, the constraints acting on cells and cell lineages might help or be required for the realization of the neural differenciation program. In addition, these results suggest that future strategies for the manipulation of stem cell in vitro will require to change growth methods into threedimensional matrices and reconstitution of apico-basal organization of the cells. This should help fine tune the properties of the stem cells and provide a better spatial control of the sequence of events that take place in the embryo. Together with the knowledge acquired on the cell biology of stem cells in the embryo, this approach is likely to open new routes that may ultimately lead to an industrial use of stem cells for therapeutic uses.

References

[1] Akai, J., P.A. Halley, and K.G. Storey, FGF-dependent Notch signaling maintains the spinal cord stem zone. *Genes Dev.,* 2005.

[2] Mathis, L., P.M. Kulesa, and S.E. Fraser, FGF receptor signalling is required to maintain neural progenitors during Hensen's node progression. *Nat. Cell Biol.,* 2001. 3(6): p. 559-66.

[3] Bertet, C., L. Sulak, and T. Lecuit, Myosin-dependent junction remodelling controls planar cell intercalation and axis elongation. *Nature,* 2004. 429(6992): p. 667-71.

[4] Munro, E.M. and G. Odell, Morphogenetic pattern formation during ascidian notochord formation is regulative and highly robust. *Development*, 2002. 129(1): p. 1-12.

[5] Keller, R., Shaping the vertebrate body plan by polarized embryonic cell movements. *Science*, 2002. 298(5600): p. 1950-4.

[6] Patel, N.H., The ancestry of segmentation. *Dev. Cell,* 2003. 5(1): p. 2-4.

[7] Mathis, L. and J.F. Nicolas, Cellular patterning of the vertebrate embryo. *Trends Genet*, 2002. 18(12): p. 627-35.

[8] Beddington, R.S. and E.J. Robertson, Axis development and early asymmetry in mammals. *Cell*, 1999. 96(2): p. 195-209.

[9] Rhinn, M., et al., Sequential roles for Otx2 in visceral endoderm and neuroectoderm for forebrain and midbrain induction and specification. *Development,* 1998. 125(5): p. 845-56.

[10] Varlet, I., J. Collignon, and E.J. Robertson, nodal expression in the primitive endoderm is required for specification of the anterior axis during mouse gastrulation. *Development*, 1997. 124(5): p. 1033-44.

[11] Bouwmeester, T., et al., Cerberus is a head-inducing secreted factor expressed in the anterior endoderm of Spemann's organizer. *Nature,* 1996. 382(6592): p. 595-601.

[12] Ruiz i Altaba, A., Pattern formation in the vertebrate neural plate. *Tr. Neurosci.*, 1994. 17(6): p. 233-43.

[13] Selleck, M.A. and C.D. *Stern, Evidence for stem cells in the mesoderm of Hensen's node and their role in embryonic pattern formation, in Formation and differentiation of early embryonic mesoderm*, R.B.e. al., Editor. 1992, Plenum Press: New York. p. 23-31.

[14] Stern, H.M., A.M. Brown, and S.D. Hauschka, Myogenesis in paraxial mesoderm: preferential induction by dorsal neural tube and by cells expressing Wnt-1. *Development,* 1995. 121(11): p. 3675-86.

[15] Storey, K.G., et al., Early posterior neural tissue is induced by FGF in the chick embryo. *Development*, 1998. 125(3): p. 473-84.

[16] Pera, E.M. and M. Kessel, Patterning of the chick forebrain anlage by the prechordal plate. *Development*, 1997. 124(20): p. 4153-62.

[17] Muhr, J., et al., Convergent inductive signals specify midbrain, hindbrain and spinal cord identity in gastrula stage chick embryos. *Neuron*, 1999. 23: p. 689-702.

[18] Woo, K. and S.E. Fraser, Specification of the zebrafish nervous system by nonaxial signals. *Science*, 1997. 277(5323): p. 254-7.

[19] Gardner, R.L. and D.L. Cockroft, Complete dissipation of coherent clonal growth occurs before gastrulation in mouse epiblast. *Development*, 1998. 125(13): p. 2397-402.

[20] Mathis, L. and J.F. Nicolas, Different clonal dispersion in the rostral and caudal mouse central nervous system. *Development*, 2000. 127: p. 1277-90.

[21] Quinlan, G.A., et al., Neuroectodermal fate of epiblast cells in the distal region of the mouse egg cylinder: implication for body plan organization during early embryogenesis. *Development,* 1995. 121: p. 87-98.

[22] Lawson, K.A., J.J. Meneses, and R.A. Pedersen, Clonal analysis of epiblast fate during germ layer formation in the mouse embryo. *Development*, 1991. 113: p. 891-911.

[23] Alvarez, I.S. and G.C. Schoenwolf, Patterns of neurepithelial cell rearrangement during avian neurulation are determined prior to notochordal inductive interactions. *Dev. Biol.,* 1991. 143(1): p. 78-92.

[24] Brown, J.M. and K.G. Storey, A region of the vertebrate neural plate in which neighbouring cells can adopt neural or epidermal fates. *Curr. Biol.,* 2000. 10(14): p. 869-72.

[25] Cambray, N. and V. Wilson, Axial progenitors with extensive potency are localised to the mouse chordoneural hinge. *Development*, 2002. 129(20): p. 4855-66.

[26] Catala, M., et al., A spinal cord fate map in the avian embryo: while regressing, Hensen's node lays down the notochord and floor plate thus joining the spinal cord lateral walls. *Development*, 1996. 122(9): p. 2599-610.

[27] Fernandez-Garre, P., et al., Fate map of the chicken neural plate at stage 4. *Development*, 2002. 129(12): p. 2807-22.
[28] Hatada, Y. and C.D. Stern, A fate map of the epiblast of the early chick embryo. *Development*, 1994. 120: p. 2879-89.
[29] Henrique, D., et al., cash4, a novel achaete-scute homolog induced by Hensen's node during generation of the posterior nervous system. *Genes Dev.,* 1997. 11(5): p. 603-15.
[30] Rodriguez-Gallardo, L., et al., Agreement and disagreement among fate maps of the chick neural plate. *Brain Res. Brain Res. Rev*., 2005. 49(2): p. 191-201.
[31] Vasiliauskas, D. and C.D. Stern, Patterning the embryonic axis: FGF signaling and how vertebrate embryos measure time. *Cell*, 2001. 106(2): p. 133-6.
[32] Mellitzer, G., Q. Xu, and D.G. Wilkinson, Eph receptors and ephrins restrict cell intermingling and communication. *Nature*, 1999. 400(6739): p. 77-81.
[33] Wetts, R. and S.E. Fraser, Slow intermixing of cells during xenopus embryogenesis contributes to the consistency of the blastomere fate map. *Development,* 1989. 105: p. 9-15.
[34] Mathis, L., et al., Successive patterns of clonal cell dispersion in relation to neuromeric subdivision in the mouse neuroepithelium. *Development*, 1999. 126: p. 4095-106.
[35] Fraser, S., R. Keynes, and A. Lumsden, Segmentation in the chick embryo hindbrain is defined by cell lineage restrictions. *Nature*, 1990. 344: p. 431-435.
[36] Lamb, T.M. and R.M. Harland, Fibroblast growth factor is a direct neural inducer, which combined with noggin generates anterior-posterior neural pattern. *Development,* 1995. 121(11): p. 3627-36.
[37] Blumberg, B., et al., An essential role for retinoid signaling in anteroposterior neural patterning. *Development*, 1997. 124(2): p. 373-9.
[38] Duboule, D. and P. Dollé, The structural and functional organization of the murine HOX gene family resembles that of Drosophila homeotic genes. *EMBO J*., 1989. 8(5): p. 1497-505.
[39] Graham, A., N. Papalopulu, and R. Krumlauf, The murine and Drosophila homeobox gene complexes have common features of organization and expression. *Cell*, 1989. 57: p. 367-78.
[40] Nicolas, J.F., L. Mathis, and C. Bonnerot, Evidence in the mouse for self-renewing stem cells in the formation of a segmented longitudinal structure, the myotome. *Development*, 1996. 122(1996): p. 2933-46.
[41] Dubrulle, J., M.J. McGrew, and O. Pourquie, FGF signaling controls somite boundary position and regulates segmentation clock control of spatiotemporal Hox gene activation. *Cell,* 2001. 106(2): p. 219-32.
[42] Martindale, M.Q. and M. Shankland, Intrinsic segmental identity of segmental founder cells of the leech embryo. *Nature*, 1990. 347: p. 672-4.
[43] Fishell, G. and A.R. Kriegstein, Neurons from radial glia: the consequences of asymmetric inheritance. *Curr. Opin. Neurobiol*., 2003. 13(1): p. 34-41.
[44] Roegiers, F. and Y.N. Jan, Asymmetric cell division. *Curr. Opin. Cell Biol.,* 2004. 16(2): p. 195-205.
[45] Wodarz, A. and W.B. Huttner, Asymmetric cell division during neurogenesis in Drosophila and vertebrates. *Mech. Dev.,* 2003. 120(11): p. 1297-309.

[46] Chenn, A. and S.K. McConnell, Cleavage orientation and the asymmetric inheritance of Notch1 immunoreactivity in mammalian neurogenesis. *Cell*, 1995. 82(4): p. 631-41.
[47] Das, T., et al., In vivo time-lapse imaging of cell divisions during neurogenesis in the developing zebrafish retina. *Neuron*, 2003. 37(4): p. 597-609.
[48] Haydar, T.F., E. Ang, Jr., and P. Rakic, Mitotic spindle rotation and mode of cell division in the developing telencephalon. *Proc. Natl. Acad. Sci. USA*, 2003. 100(5): p. 2890-5.
[49] Kosodo, Y., et al., Asymmetric distribution of the apical plasma membrane during neurogenic divisions of mammalian neuroepithelial cells. *Embo J.*, 2004. 23(11): p. 2314-24.
[50] Cowan, C.R. and A.A. Hyman, Asymmetric cell division in C. elegans: cortical polarity and spindle positioning. *Annu. Rev. Cell Dev. Biol.*, 2004. 20: p. 427-53.
[51] Glotzer, M., Cytokinesis: progress on all fronts. *Curr. Opin. Cell Biol.*, 2003. 15(6): p. 684-90.
[52] Wang, H. and W. Chia, Drosophila neural progenitor polarity and asymmetric division. *Biol. Cell,* 2005. 97(1): p. 63-74.
[53] Barros, C.S., C.B. Phelps, and A.H. Brand, Drosophila nonmuscle myosin II promotes the asymmetric segregation of cell fate determinants by cortical exclusion rather than active transport. *Dev. Cell*, 2003. 5(6): p. 829-40.
[54] Kaltschmidt, J.A., et al., Rotation and asymmetry of the mitotic spindle direct asymmetric cell division in the developing central nervous system. *Nat. Cell Biol.*, 2000. 2(1): p. 7-12.
[55] Sanada, K. and L.H. Tsai, G protein betagamma subunits and AGS3 control spindle orientation and asymmetric cell fate of cerebral cortical progenitors. *Cell*, 2005. 122(1): p. 119-31.
[56] Zheng, Y., G protein control of microtubule assembly. *Annu. Rev. Cell Dev. Biol.*, 2004. 20: p. 867-94.
[57] Roszko, I., et al., Key role played by RhoA in the balance between planar and apicobasal cell divisions in the chick neuroepithelium. *Dev. Biol.*, 2006. 212(1): p. 212-24.
[58] Nikolic, M., The role of Rho GTPases and associated kinases in regulating neurite outgrowth. *Int. J. Biochem. Cell Biol.*, 2002. 34(7): p. 731-45.
[59] Bhattacharya, M., A.V. Babwah, and S.S. Ferguson, Small GTP-binding proteincoupled receptors. Biochem Soc Trans, 2004. 32(Pt 6): p. 1040-4.
[60] Livesey, F.J., T.L. Young, and C.L. Cepko, An analysis of the gene expression program of mammalian neural progenitor cells. *Proc. Natl. Acad. Sci. USA*, 2004. 101(5): p. 1374-9.
[61] Suarez-Farinas, M., et al., Comparing independent microarray studies: the case of human embryonic stem cells. *BMC Genomics*, 2005. 6: p. 99.

In: Stem Cell Research Trends
Editor: Josse R. Braggina, pp. 305-319
ISBN: 978-1-60021-622-0

Chapter XII

LENTIVIRUS AND NEURAL STEM CELL: AN EXPERIMENTAL STUDY FOR THE REPAIR OF SPINAL CORD INJURY

Guang-yun Sun *[*1], **Pei-Qiang Cai**[1], **Pei-shu Cai**[1], **M. Oudega**[2], **Xue-wen Wang**[1], **Yun-bing Shu**[1], **Cheng Cai**[1], **Wei Li**[1] **and Hai-hao Yang**[1]*

[1]Department of Orthopaedics, the First People's Hospital of Yibin, Sichuan province, China, 644000
[2]The Miami Project to Cure Paralysis, University of Miami School of Medicine, USA

ABSTRACT

Objective

Neural stem cells have been shown to participate in the repair of experimental CNS disorders due to their self-renewal and multi-potency. In our study, we attempt to explore the feasibility for the therapy of spinal cord injury (SCI) by combining neural stem cell (NSC) with lentivirus.

Methods

Following the construction of the genetic engineering NSC modified by Lentivirus to secrete both neurotrophic factor-3 (NT-3) and green fluorescence protein (GFP), hemisection of spinal cord at the level of T_{10} was produced in 48dult Wistar rats that were randomly divided into four groups (n=16), namely three treated groups and one control group. The treated groups were dealed with NSC, genetic engineering NSC

* Correspondence to : CAI Pei-qiang: E-mail: cpq20032002@yahoo.com.cn. This study was supported by the Foundation of YiBin, SiChuan, China.

respectively. Then used fluorescence microscope to detect the transgenic expression in vitro and in vivo, migration of the grafted cells in vivo and used the method of BBB to assess the function recovery.

Results

The transplanted cells could survive for long time in vivo and migrate for long distances; the stably transgenic expression could be detected in vivo; the hind-limb function of the injured rats, especially for the rats that had been dealed with genetic engineering NSC, had obviously improved.

Conclusion

The genetic engineering NSC modified by lentivirus to deliver NT-3, acting as a source of neurotrophic factors and function cell in vivo, have the potential to participate in spinal cord repair.

Keywords: lentivirus, spinal cord injury, neural stem cell, neurotrophic factor-3.

The study for the functional repair after spinal cord injury (SCI) is one of the focuses in neuroscience in recent years. Since the famous neuroanatomist Ramon y Cajal wrote in the early 20th century that the central nervous system (CNS) does not regenerate once it is injured, this theory has been popularly accepted. However, large numbers of studies have indicated that the introduction of an appropriate environment into the injured site could cause the injured axons to functionally regenerate the 1980s[1,2,3,4]. According to the modern views: The lack of regenerative properties of the mammalian CNS, especially in the spinal cord, could be attributable to a combination of factors: including the death of large numbers of the functional cells after SCI, due to the primary injury and the secondary pathophysiologic changes in vivo, such as ischemia, edema, inflammation and so on, which can result in the death of a mass of neurons, astrocytes and oligodendrocytes; the lack of sufficient trophic support [5]; the existence of the myelin-associated inhibitors, including Nogo, myelin-associated glycoprotein (MAG) and oligodendrocyte-myelin glycoprotein (OMgp) [4]; and injury-induced glial scars [6]. Just these factors cause the obstacle of the functional regeneration after SCI. Therefore given certain appropriate local microenvironment, the functional regeneration of the injured spinal cord should be really possible. This paper explored the feasibility for the therapy of SCI by combine NSC with lentivirus, attempted to establish certain bases and provide some valuable clues for the further study of SCI

METHODS

Cell Culture

The brain cortex tissues of E15 rat were dissected in chilled sterile phosphate buffered saline (PBS, pH 7.4). Identified pieces were incubated in 0.25% trypsin for 20 min at 37°C. Following three washes in PBS the tissues were triturated in DMEM/F12 (1:1,Sigma) using a fine polished Pasteur pipette. Cell counts showed greater than 90% viable cells in all cases. Cells were seeded at a concentration of 500000 per ml into 75cm^2 tissue culture flasks. The growth medium consisted of DMEM/F12 (1:1), penicillin G, streptomycin sulphate (1:100; Gibco), B27 (1:50; Gibco), human recombinant bFGF (20ng/ml; Sigma). Passage was carried out every 6-7 days and consisted of a gentle mechanical dissociation using a fine polished Pasteur pipette.

Extraction of the Lentiviral Particles

18–24 h prior to transfection, plated 3 million 293T cells per 10cm plate in 8mL of DMEM 10%fetal bovine serum (FBS). It would be 80% confluent by the following day. Next day, transfected the 293T cells with the lentiviral plasmids (all of these plasmids were constructed by Blits B and Oudega M) by the method of CaPO4-transfect, for two 10 cm plates, the plasmids were Lentivirus-NT-3 (which carries both NT-3 and GFP) 10ug, PCMV847 6.5ug, VSVG 3.5ug, respectively, 24 hours after transfection began to collect the viral supernatant and replace the fresh medium every 24h for 3d. Then centrifuged the medium at 50,000*g* for 2 h at 4°C to concentrate the viral supernatant, then store aliquots of the virus at –80°C.

The Construction of the Genetic Engineering NSC

Seeded 2×10^6 NSC in 5ml fresh growth medium in the presence of 8μg/ml polybrene in a 15-ml round-bottom polypropylene centrifuge tube, then added 5ul concentrated viral supernatant. Centrifuged cells and vector-conditioned medium at 2000*g* for 2h at room temperature. Resuspended the cell pellet in 5mL of fresh growth medium and incubated overnight at 37 °C, 5% CO2. Repeated the above procedures to enhance transduction once a day for 2d.

NT-3 Bioassay

Conditioned media from the genetic engineering NSCs and normal NSCs were tested for the production of bioactive NT-3 using a fetal rat dorsal root ganglion (DRG) neurite outgrowth assay. Briefly, E17 rat embryos were dissected and the DRG aseptically removed from the spinal cord. DRG were pooled in ice-cold L-15 Leibovitz medium with L-glutamine (Gibco BRL) and transferred in 96-well plate (*n*=one per well, together 24 wells), which was coated with poly-L-Lysine. Conditioned culture medium (50μl per well) was then applied onto wells containing the DRG explants and the 96-well plate were cultured for 48 h at 37 °C, 5% CO2. Neurite outgrowth from the DRG was examined after 48h using an inverted phase contrast microscope.

Western Blotting and Slot Blotting Analysis of Transgenic Expression

To verify NT-3 expression by the transgenic engineering NSCs, immunoblotting with polyclonal anti-NT-3 antibody (Promega) was carried out. Briefly, the mediums (including the genetic engineering NSCs and normal NSCs) were collected and then heated for 10 minutes in 100°C water. Then the mediums were loaded onto adjacent lanes and separated by 10% SDS–PAGE and then transferred onto nitrocellulose (NC) membranes and processed for Western blotting. The NC membranes were first incubated with 5% nonfat dry milk in TTBS buffer (0.1% Tween 20, 150mM NaCl, 50 mM Tris–HCl, pH 7.6) for 1 h, followed by another hour incubation with primary antibodies for NT-3 (Promege, dilution 1:2000). The membranes were then rinsed 30 minutes with TTBS buffer and incubated with HRP-conjugated goat anti-rabbit secondary antibody (diluted 1:4000) for 1 h. The immunoreactivity was visualized by super signal west Pico chemiluminescent substrate reagents (Pierce 34079ZZ). For slot blot analysis, drew three rings on an appropriate NC membrane with a pencil, then added the samples (including the transgenic engineering NSCs, normal NSCs and the medium of DMEM/F12) into the three rings with a microsyringe, 2ul per time. Repeated the same procedures 6-8 times after the NC membrane dried up. Then processed for immunostaining with the NT-3 antibodies as described for the Western blot, but the immunoreactivity was detected by diaminobenzidine solution (DAB).

Surgical Procedure and Cell Graft

Forty-eight female adult Wistar rats (250-300g body weight) were anesthetized with an intraperitoneal (ip) injection of pentobarbital (30mg per 1000g body weight). All rats were underwent laminectomy at the T10 spinal cord level to expose one spinal cord segment. Following the spinal cords were hemisected at the T10 level; the muscles and skin were sutured separately. During surgery, animals were kept on a heating pad to maintain body temperature at 37°C. Then the rats were allowed to recover in warmed cages with food and water readily available. Bladders were emptied manually twice a day until bladder function returned. All the injured rats were randomly divided into three groups (n=16), namely two therapeutic groups and one control group. After one or two weeks of the first surgery, the spinal cords were opened again and the rats of the therapeutic groups were injected 5μl of NSC suspension (10^5 cells/μl), 5μl of genetic engineering NSC suspension (10^5 cells/μl) that carried NT-3 and GFP with a 10μl microsyringe, respectively, the control group were injected with 5ul medium of DMEM/F12. All the rats received the same tendance as above.

Assessment of Locomotor Performance

Hind-limb function of the rats were assessed at the time of 1d, 1w, 2w, 4w, 6w, 8w and 10w after SCI, using the Basso, Beattie and Bresnahan (BBB) open-field locomotor test [7]. Two different observers blinded to evaluate rat performance independently. Statistical analysis was performed using one-way analysis of variance with repeated measures, followed by the posthoc Newman-Keuls test. Values were considered statistically different when $P<0.05$. Statistical analysis was performed using SPSS 11.0 software.

Tissue Preparations

Animals were anesthetized with an ip injection of sodium pentobarbital (100 mg/kg) and transcardially perfused with 200 ml of normal saline followed by 500 ml of ice-cold 4% paraformaldehyde in 0.1 M phosphate buffer (pH 7.4). A spinal cord tissue block (approximately 2 cm in length) containing the graft site was removed, then postfixed in the same fixative for 3 h, followed by cryoprotection in 30% sucrose (in 0.1-M phosphate buffer, pH 7.4) for 2 days. Tissues were then embedded in OCT compound (Leica Jung Frigo-Cut.2800E) and cut into 20- to 40-μm-thick longitudinal sections or transverse sections on a cryostat.

Immunohistochemistry

NSCs were identified using anti-Nestin (1:50, BOSTER) and their differentiated offspring were identified with anti-NF-200 (1:200, BOSTER), anti-GFAP (1:500, BOSTER). All the primary antibodies and the secondary antibodies (1:50,BOSTER) were diluted in PBS containing 10% normal goat serum and 0.01% Triton X-100. All of the staining procedures were carried out according to the specification. The sections were visualized with diaminobenzidine solution.

Results

Establishment of the Neural Stem Cell Line

The cells, which were separated from the E14 rats, formed the suspending spheres in the growth medium in response to the mitogens FGF-2 while growing. The numbers of cells in each sphere were from several cells to tens, even over one hundred cells. The majority of cells within the growing spheres were found to be positive for nestin, a marker for undifferentiated neuroepithelial stem cells (Figure 1A).

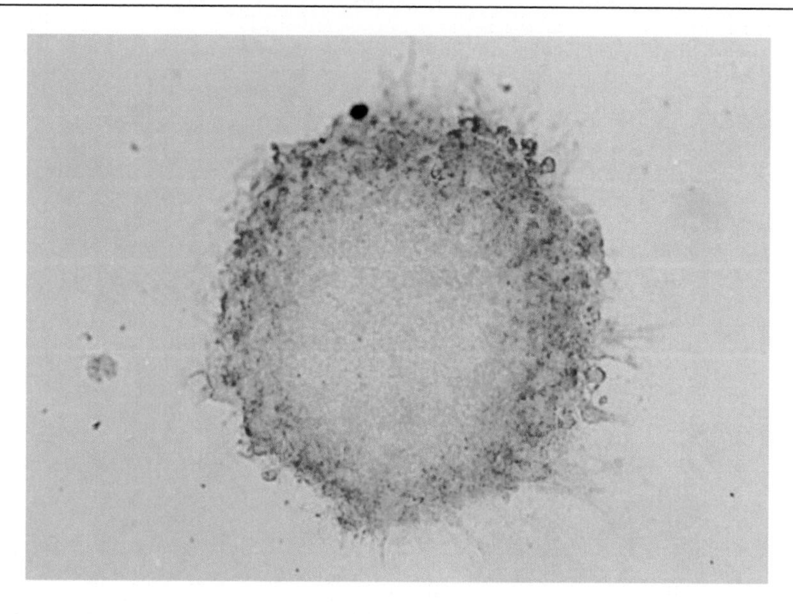

Figure 1A. The Nestin positive neural spheres (x 200)

In addition, when these neural spheres were plated onto poly-L-Lysine coated coverslips in DMEM/F12 and 1% FBS to induce differentiation, they rapidly attached and many cells could be seen migrating out from the spheres within hours. Between 2 and 7 days, radiating processes often developed which sketched from the edge of the sphere. Migrating cells eventually formed a monolayer around the plated sphere. Among these migrating cells, some were NF-200 positive (Figure 1B), the others were GFAP positive (Figure 1C). These results indicated that they were neurons and astrocytes respectively.

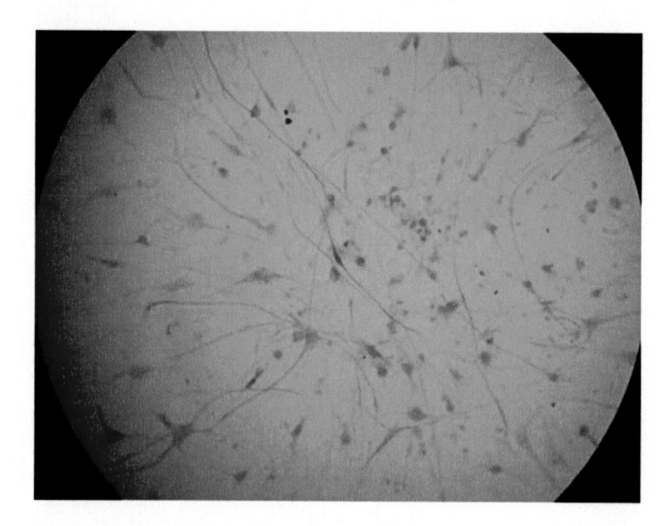

Figure 1B. The NF-200 positive neuron (x 100)

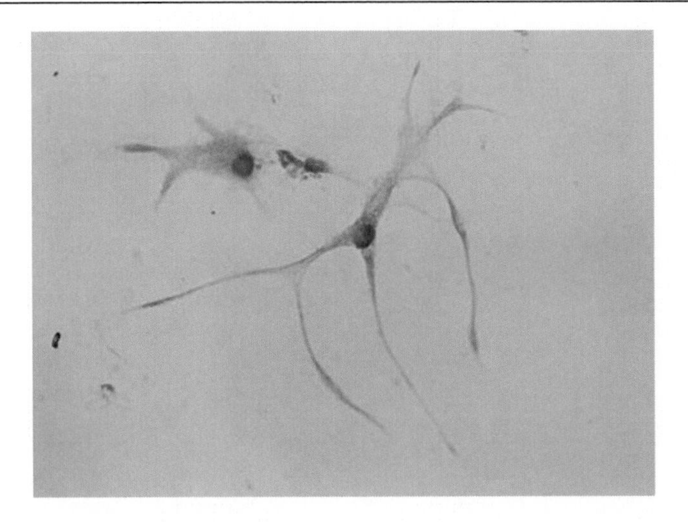

Figure 1C. The GFAP positive astrocyte (x 250)

Fluorescence Microscope of Genetic Engineering NSC

The genetic engineering NSCs exhibited a moderate green fluorescence under the fluorescence microscope. The fluorescence was weak in a single cell; however, neural spheres exhibited light green fluorescence (Figure 2).

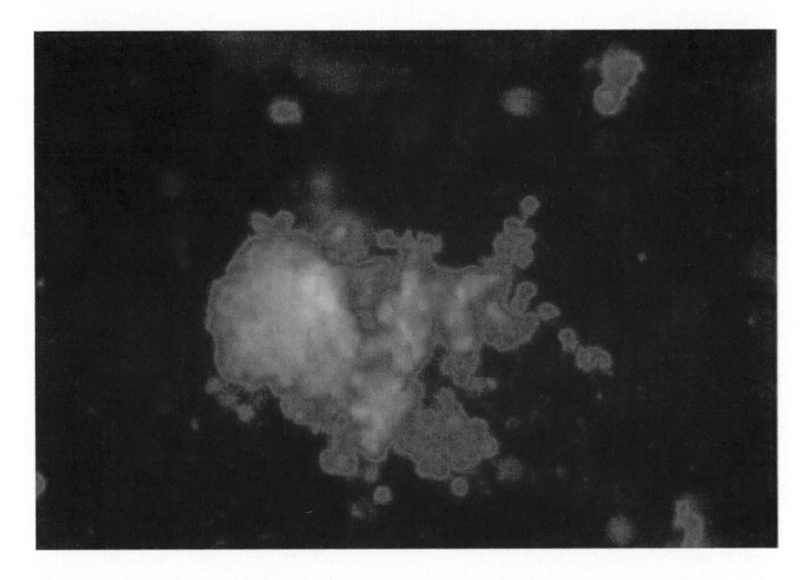

Figure 2. The genetic engineering NSC under the fluorescence microscope (x 200)

NT-3 Bioassay

Non-neuronal cells migrated from DRG explants after receiving the usual NSC medium and the neurite outgrowth was poor (Figure 3A). Non-neuronal cells also migrated from the DRG explants after receiving medium from the genetic engineering NSC.

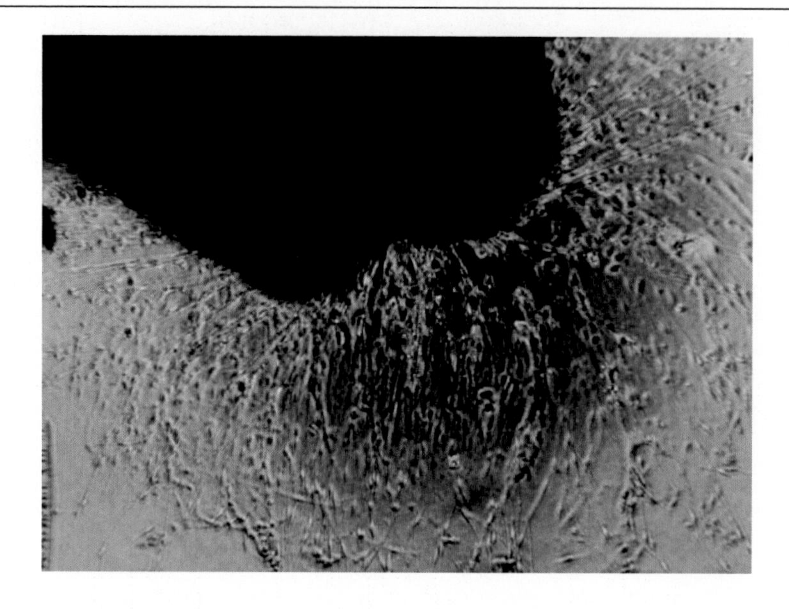

Figure 3A. The fetal rat DRG that were cultured by the medium of the genetic engineering NSC.(x 100))

However, in contrast to the controls, the neurite outgrowth in these cultures was robust (Figure 3B). These data indicated that the genetic engineering NSC actively synthesized and secreted biologically active NT-3 protein.

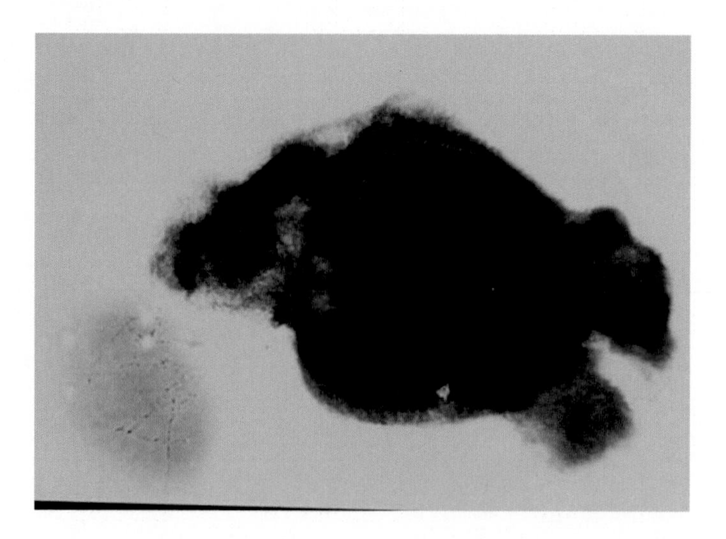

Figure 3B. The fetal rat DRG that were cultured by the medium of the normal NSC. (x 40)

Western Blotting

Conditioned media from the genetic engineering NSC was separated on 10% SDS–PAGE and immunoblotted with a polyclonal antibody against NT-3 protein. It left a few deeply dark strips, which indicated the genetic engineering NSC produced high levels of NT-

3 protein. However, the conditioned medium of the usual NSC showed only a little trace labeling (Figure 4).

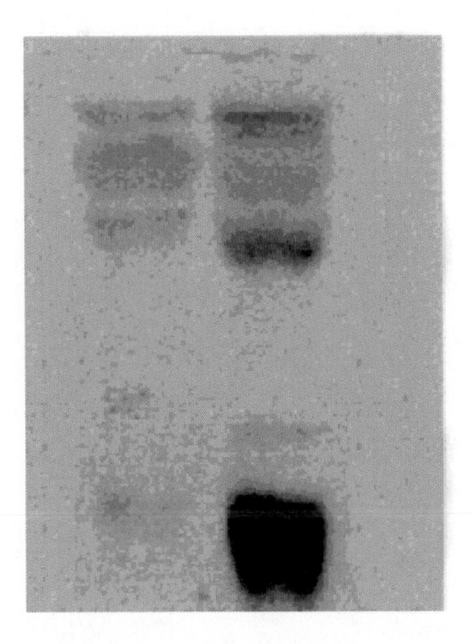

Figure 4. Western blot

Assessment of Locomotor Performance

Hind-limb performance was evaluated using the BBB locomotor test [7]. One day after surgery, the scores of all rats were approximately equal, about 1.2 point. Then hind-limb performance gradually improved in all groups. Following treatment with NSC graft and genetic engineering NSC graft, a modest yet significant improvement in hind limb performance was observed compared with control animals. At 4 weeks post-injury, the mean scores of treated groups were all higher than the control group, and they remained higher thereafter. The one-way analysis of variance revealed that both treated groups were significantly different from the control group. A statistically significant difference ($P<0.05$) between the animals of the experimental groups and the animals of the control group was present from 4 weeks to 10 weeks after implantation of cells.

At 10 weeks, animals treated with genetic engineering NSC exhibited an average score of 14.4, which were 5.4 higher than the scores of the control group that received DMEM/F12. Meanwhile, from the beginning of 4 weeks, the scores of rats treated with genetic engineering NSC were higher obviously than the other experimental group and there was a statistically significant difference ($P<0.05$) using the test of Student-Newman-Keuls. The above results revealed that both kinds of treatment were effective to improve the hind-limb function, however, transplant of the genetic engineering NSC was a more effective method.

The Detection of the Transgenic Expression *In Vivo*

The cryo-sections were observed under a fluorescence microscope. For the group of the genetic engineering NSC graft, at the 2w post-implant, the graft cells were distributed mainly around the injected site. At the 4w post-graft, the cells at the injected site decreased obviously because many transplanted cells migrated to the rostral and the caudal (Figure 5A.B). And these cells were decreasing with time. However, there were still a few cells that could be detected by the fluorescence microscope until 10w (Figure 5c).

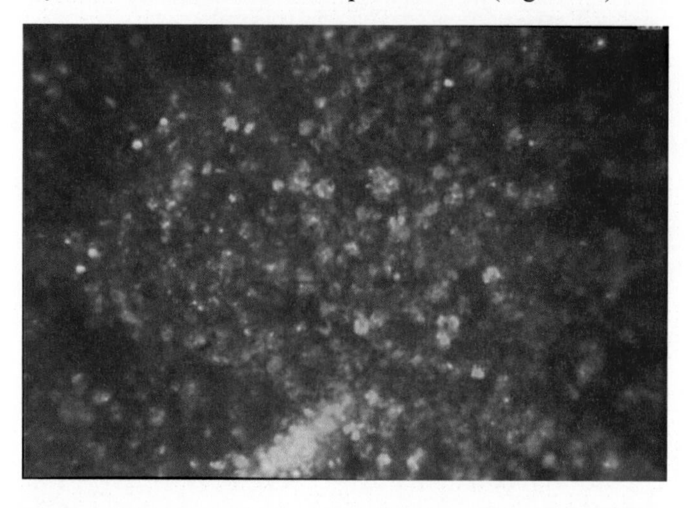

Figure 5A. The transgenic expression of the genetic engineering NSC in the injured spinal cord. At two weeks after SCI (x 100)

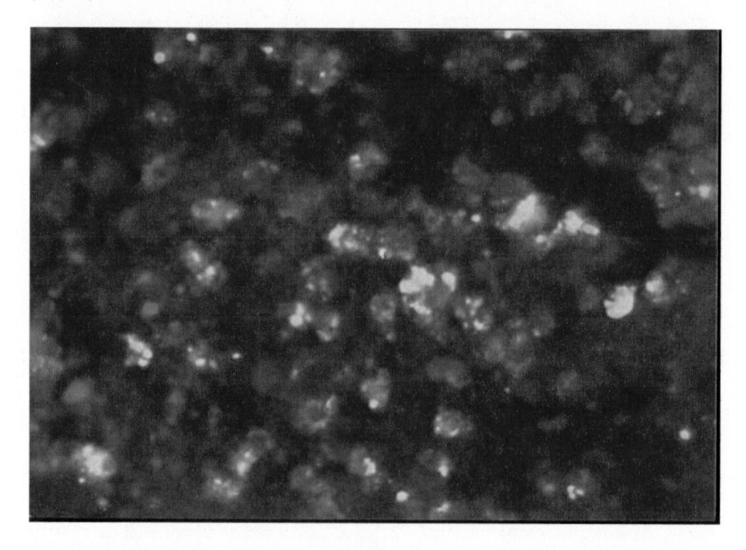

Figure 5B. At four weeks after SCI (x 200)

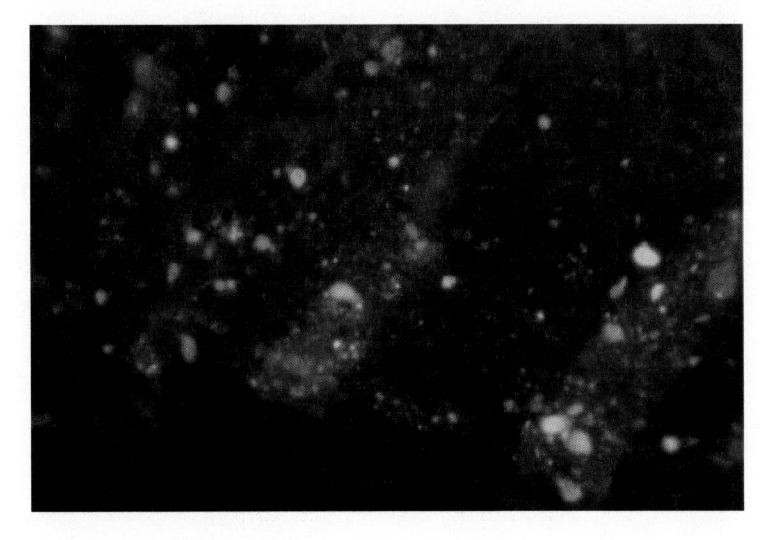

Figure 5C. At ten weeks after SCI(x 200)

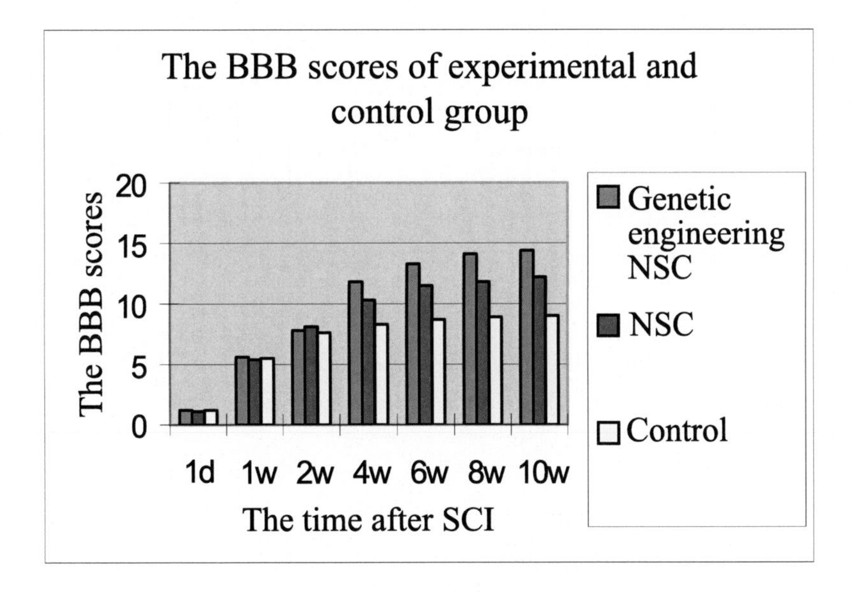

Figure 6. The BBB scores of the experimental groups and the control group at different time after SCI.

Discussion

NSCs are undifferentiated nervous system cells that are capable of proliferation, repeated subculture, and differentiation into all the major cell types of the adult central nervous system, including neuron, astrocyte and oligodendrocyte. And we have cultured the rat NSC for over four months and the cells have been passaged over 18 passages, which still kept their capable of self-renewing, meanwhile we could acquire neurons and astrocytes by inducing NSC of the different stages.

Lentiviral vectors are promising tools for gene transfer [8]. Like oncoretroviral vectors, they offer the unique advantage of stably integrating into the genome of the host cell, thus providing the basis for sustained gene expression. However, compared with other classical oncoretrovirus derived vectors, such as retrovirus (RV) and adeno-associated virus (AAV), Lentivirus has many advantages: In contrast to RV that can only transduce dividing cells, lentiviral vectors are highly efficient at infection of both dividing and nondividing cells because of the presence of nuclear localization signals on several virion associated proteins, which include matrix, viral protein R, and integrase. Another feature of the lentiviral vector system is that the virions can be constructed to express a surface protein of the vesicular stomatitis virus glycoprotein (VSV-G), which allows the virions to transduce a wide range of cell types from various species, including human and the mouse. In addition, the lentiviral particles with VSV-G are highly stable, and virus stocks containing this glycoprotein can be concentrated by ultracentrifugation, without apparent loss of infectivity. Lentivirus can carry and express multiple genes stably at the same time [9], which have also proved in our experiment. The ability to express multiple genes by a lentiviral vector may present important benefits. For example, if certain vector expresses both a therapeutic gene and a reported gene, such as GFP or EGPF, just as the lentivirus that was used in our experiment, we can easily discriminate transduced vs untransduced cells in the same experiment. Therefore multigene lentiviral vectors may be useful in basic research and even in the possible clinic application in future. Lentiviral vectors have the capacity to carry as much as 8 kb of genetic material into the target cell (approx twice the capacity of AAV vectors), facilitating the transfer of larger expression cassettes. This capacity should allow for the generation of a wide variety of useful experimental vectors. In order to prevent oncogene activation and to improve safety, the self-inactivating (SIN) vectors have been constructed by deleting the sequences in the U3 region of the 3′ long-terminal repeat (LTR). Therefore the lentiviral safety have been enhanced greatly and the safety have been demonstrated fully in vivo by transducting brain, eye, liver, and muscle tissues using mice, rats, and monkeys [10]. Thus in considering safety, lentiviral SIN vectors are the preferable choice for future clinical applications. Lentivirus is able to mediate long-term expression of therapeutic genes because the recombinant provirus is integrated into the host genome and can persist for the lifetime of the cell. Some of these properties of lentivirus have also been demonstrated in our study.

There exist two main factors in vivo which can affect the functional regeneration of the injured axons after SCI, one is inhibitory factors, namely all kinds of myelin-associated inhibitors, including Nogo, MAG and OMgp[11]. The other is stimulative factor, including all kinds of neurotrophic factors(NFs), such as nerve growth factor (NGF), neurotrophic factor-3, neurotrophic factor -4/5, brain derived neurotrophic factor (BDNF) and ciliary neurotrophic factor, and so on. To sum up the effect of these NFs: To rescue and survive the injured neurons. Paul Lu [12] had ever done a study: Adult Fischer 344 rats underwent subcortical lesions followed by grafts to the lesion cavity of syngenic fibroblasts genetically modified to secrete high amounts BDNF or, in control group, the reporter gene green fluorescent protein. In control subjects, only 36.2±7.0% of the retrogradely labeled corticospinal neurons survived the lesion, whereas 89.8±5.9% ($P < 0.01$) of the corticospinal neurons survived in animals that received BDNF-secreting grafts. To promote the growth of the injured axons and improve the functional recovery of hind-limb. For example, primary fibroblasts genetically modified to express NT-3 were transplanted into the rat spinal cord that had been underwent mid-thoracic spinal cord dorsal hemisection lesions three months ago. Three months later, compared to control-grafted animals, NT-3-grafted animals exhibited significant growth of corticospinal axons up to 15 mm distal to the lesion site and showed a modest but significant 1.5-point improvement in locomotor scores on the BBB scale [1]. To induce the definite differentiation of the exogenous and/or endogenous neural stem cell. Some recent studies showed that the soluble factors from neurons, including the BDNF and NT-3, strongly induced multipotent cortical stem cells to acquire neuronal identity, while the factors from astrocytes, such as ciliary neurotrophic factor, promoted astrocytic differentiation [13]. Though all NTs can promote the functional regeneration of the injured axons, yet different NF has its different effect for the functional regeneration of the injured axons after SCI. NGF is mainly elicits growth of sensory and noradrenergic axons. In addition, because the receptors for BDNF are mostly distributed at the cell soma, therefore BDNF can improve the functional recovery by rescuing and surviving injured cells. On the contrary, the receptors for NT-3 (trkC) are most present in the axons, thus NT-3 can promote the functional recovery of SCI by inducing the corticospinal growth [5].

In our study, we had used two different methods to cure the rats of spinal cord injury, namely to transplant the genetic engineering NSC modified by lentivirus to secrete high amounts NT-3 (group genetic engineering NSC), and to transplant the NSC (group NSC). Both methods could promote the functional recovery of SCI probably by adjusting the inappropriate local microenvironment of injured spinal cord.

For the group genetic engineering NSC, the transplanted cells could mainly acted several aspects of effects. Cells replacement [14]: According the properties of NSC, they are capable of proliferation, repeated subculture, and differentiation into the three types of cells composing the central nervous system. Furthermore, they could survive for a long time in vivo. Therefore they could constantly supply and replace the injured or even dead functional cells in CNS by proliferating and differentiating continuously in vivo. Therefore genetic engineering NSC is a potential expandable source of graft material for transplantation aimed at repairing the injured spinal cord. Neurotrophic supports: on the one hand, because the genetic engineering NSC was modified by lentivirus to express both NT-3 and GFP, it could directly deliver high level NT-3 in vivo continuously. On the other hand, NSC itself is a source of the multiple neurotrophic factors, just as a recent study showed that NSC had the ability to naturally continuously secrete significant quantities of several neurotrophic factors, including NGF, BDNF and glial cell line-derived neurotrophic factor both in vitro and in vivo, which could be detected quantificationally by specific ELISA [15]. Just as discussed above, these neurotrophic factors could promote the outgrowth of the injured axons after SCI. To induce the definite differentiation of the exogenous and/or endogenous NSC. A recent study demonstrated that the soluble factors from neurons, including the BDNF and NT-3, strongly induced multipotent cortical stem cells to acquire neuronal identity [12]. To rescue and survive the injured function cells in vivo [11].

As for the group NSC, it played two roles for the function repair of SCI, one is the cell replacement; the other is the neurotrophic supports just as discussed above. Of course, the effect of cell replacement is more important, though NSC itself could secrete multiple NFs, however, the amount of these NFs is limited, and whether it could act an important effect for the function repair of SCI is still unknown.

In conclusion, our experiment data showed that both methods have certain effect for the functional repair of SCI. However, compared to the other methods, the transplantation of genetic engineering NSC is a more efficient method for the function repair of SCI. These data indicate that it is feasible for the repair of SCI to combine NSC with lentivirus, especially for the genetic engineering NSCs that were modified by lentivirus to deliver NT-3, acting as a source of neurotrophic factors and function cell in vivo, have the potential to participate in spinal cord repair. We hope these data could provide some valuable clues for the further study of the therapy for SCI.

REFERENCES

[1] Tuszynski MH., Grill R, Leonard L, et al. NT-3 gene delivery elicits growth of chronically injured corticospinal axons and modestly improves functional deficits after chronic scar resection. *Exp. Neurol.,* 2003, 181: 47-56.
[2] Lu P, Jones L.L., Tuszynski M.H., et al. BDNF-expressing marrow stromal cells support extensive axonal growth at sites of spinal cord injury. *Exp. Neurol.*, 2005, 191: 344-360.
[3] Blits B, Oudega M, Boer GJ, et al. Adeno-associated viral vector-mediated neurotrophin gene transfer in the injured adult rat spinal cord improves hind-limb function. *Neurosci,* 2003, 118:271–281.
[4] GrandPre T, Li S, Strittmatter SM. Nogo-66 receptor antagonist peptide promotes axonal regeneration. *Nature,* 2002, 417, 547-551.
[5] Blesch A, Lu P, Tuszynski MH. Neurotrophic factors, gene therapy, and neural stem cells for spinal cord repair. *Brain Res. Bull.*, 2002, 57: 833-838.
[6] Fawcett J, Richard. A. The glial scar and central nervous system repair. *Brain Research Bulletin,* 1999, 49:377-391.
[7] Basso DM, Beattie MS, Bresnahan JC. A sensitive and reliable locomotor rating scale for open field testing in rats. *Neurotrauma*, 1995,12: 1–21.
[8] Vigna E Naldini, L. Lentiviral vectors: excellent tools for experimental gene transfer and promising candidates for gene therapy. *Gene Med.* 2000, 2: 308–316.
[9] Zhu Y, Feuer G, Day S L. et al. Multigene lentivirus vectors based on differential splicing and translational control. *Mol. Ther*. 2001,3: 375-382.
[10] Blomer U, Naldini L, Kafri T, et al. Highly effcient and sustained gene transfer in adult neurons with a lentivirus vector. *J. Virol.* 1997,71: 6641–6649.
[11] McGee AW, Strittmatter SM. The Nogo-66 receptor: focusing myelin inhibition of axon regeneration. *Trends Neurosci.* 2003,26: 193-198 .
[12] Lu P, Blesch A, Tuszynski MH, Neurotrophism without neurotropism: BDNF promotes survival but not growth of lesioned corticospinal neurons. *J. Comp. Neurol,* 2001,436: 456–470.

[13] Chang MY, Son H, Lee YS, et al. Neurons and astrocytes secrete factors that cause stem cells to differentiate into neurons and astrocytes, respectively. *Mol-Cell-Neurosci,* 2003, 23: 414-426.

[14] Liu Y, Himes BT, Solowska J, et al. Intraspinal Delivery of Neurotrophin-3 Using Neural Stem Cells Genetically Modified by Recombinant Retrovirus. *Exp. Neurol.,* 1999, 158,9-26.

[15] Lu P, Jones LL, Snyder EY, et al. Neural stem cells constitutively secrete neurotrophic factors and promote extensive host axonal growth after spinal cord injury. *Exp. Neurol.,* 2003, 181: 115–129.

In: Stem Cell Research Trends
Editor: Josse R. Braggina, pp. 321-328
ISBN: 978-1-60021-622-0

Chapter XIII

THERAPEUTIC POTENTIAL OF EMBRYONIC STEM CELLS

Philippe Taupin *

National Neuroscience Institute, Singapore
National University of Singapore
Nanyang Technological University, Singapore

ABSTRACT

Embryonic stem cells (ESCs) are self-renewing pluripotent cells that generate all the cell types of the body. ESCs hold the potential to cure a broad range of diseases and injuries, from diabetes, heart diseases, muscle damages, to neurological diseases and injuries. Because ESCs are derived from embryos, their use for clinical research and therapy is the source of debates and controversies. The stringent conditions require for maintaining them in culture, and their risk to form tumors upon grafting limit their therapeutic application. In this chapter, we will review the therapeutic potential of ESCs. We will discuss the ethical and political debates and controversies, and the limitations associated with the use of ESCs for therapy.

INTRODUCTION

ESCs are undifferentiated cells that proliferate indefinitely in vitro while maintaining the potential to differentiate into derivatives of all three embryonic germ layers, the endoderm, mesoderm and neurectoderm, and the germ cells. Because ESCs generate the different cell types of the body, they hold the potential to cure a broad range of diseases and injuries [1]. ESCs are derived from the inner cell mass (ICM) of blastocyst, at a stage before the blastocyst implants the uterine wall. ESCs were first isolated and characterized from mouse

* Correspondence: 11 Jalan Tan Tock Seng, Singapore 308433. Tel. (65) 6357 - 7533. Fax (65) 6256 - 9178. Email obgpjt@nus.edu.sg

blastocysts by Evans and Kaufman (1981) [2]. Their isolation, expansion and maintenance in vitro require stringent culture conditions, particularly the use of mouse fetal fibroblastic feeder layer. In 1994, Bongso et al. reported the first isolation and characterization in vitro of cells derived from the ICM of human blastocysts [3], and in 1995 and 1998, Thomson et al. derived ESCs from primates, non-human - rhesus monkey- and human [4, 5]. The derivation of ESCs from human (hESCs) provides a source of tissue for cellular therapy. However, the use of ESCs for clinical research and therapy is not without limitations [6]. Among them, the maintenance of ESCs in culture requires the use of mouse fibroblastic feeder layers, and ESCs carry the risk to form tumors upon grafting. The use of hESCs for therapy is subject to ethical and political debates, and controversies. To overcome these issues, investigators have proposed strategies to derive ESCs in defined medium and without the destruction of embryos, and to differentiate ESCs to minimize their tumorigenic risk. In this manuscript, we will review the potential of ESCs for therapy, and strategies being developed to bring ESCs to therapy. We will discuss the controversies, debates and limitations over their use for therapy.

POTENTIAL OF ESCs FOR THERAPY

ESCs generate the different cell types of the body; as such they hold the potential to cure a broad range of diseases and injuries. The therapeutic potential of ESCs lies in the ability to provide a source of tissue for transplantation suitable for therapy.

Derivation of hESCs for Therapy

Protocols originally established to isolate and maintain ESCs in culture involve the use of mouse fetal fibroblastic feeder layer, and other animal derivatives, like coating substrates, trophic factors, serum [2]. The first isolations and characterizations of cells from the ICM and ESCs derived from primate blastocysts reportedly used protocols similar to those established for the derivation of mouse ESCs [3-5, 7]. Particularly, they used mouse fetal fibroblastic feeder layer, coating substrates and other reagents, some of which originated from animals. The use of animal derived products in cell culture is not without limitations for therapy, as it could transmit diseases or trigger immune reaction upon transplantation into the host [8]. Particularly, hESCs derived in these conditions have been shown to express N-glycolyl-neuraminic acid residues [9]. N-glycolyl-neuraminic acid is a sugar present on the surface of most mammal and rodent cells, but not in humans, and against which most humans have circulating antibodies [10]. The incorporation of N-glycolyl-neuraminic acid residues on hESCs would result in the rejection of the graft. Thereby, limiting the use of existing hESC lines generated using animal derivatives, like mouse feeder layer.

To overcome this issue, investigators have aimed at isolating and establishing hESCs in vitro, free of animal contaminants. New cell lines have been derived with human feeder layer [11, 12], and without feeder layer [13-15]. More recently, Ludwig et al. successfully reported the isolation and characterization of hESCs in culture, in defined medium, in the absence of animal derived product [16]. The derivation of new hESCs free of animal contaminants provides a source of tissue for therapy.

Maintenance of ESCs for Therapy

The isolation and maintenance of ESCs in vitro require strict culture conditions. As the therapeutic potential of ESCs lies in their ability to maintain their potential to differentiate into the various lineages of the body, the importance of maintaining strict culture condition is crucial. Long-term culturing is known to have adverse effect on cells, like aneuploidie. Some investigators have reported that hESCs do not maintain their normal karyotypes [17-19], while others have confirmed that some established cell lines remain stable overtime [18, 19]. Therefore, established cell lines must be maintained under stringent culture standard, and checked overtime for normal chromosomal content.

Generation of Isogeneic Lines of hESCs for Therapy

One of the main concerns in cellular therapy involving tissue or cell graft is the risk of immune rejection in heterologous transplantation. Autologous transplantation, the transplantation of cells from the patient himself, is considered the best option for cellular therapy, as it avoids the risk of immune rejection and increases the chance of successful graft integration into the host [20]. ESCs derived from human blastocysts are used in allogeneic transplantation, and therefore, carry the risk of immune rejection upon grafting. Limiting the risk of immune rejection requires genetically matching the donor to the recipient, and the patient to follow an immunosuppressive treatment, that may not be well tolerated and may have secondary effects [20].

There is the potential to generate isogeneic ESCs from the patients by somatic cell nuclear transfer (SCNT) [21, 22]. Isogeneic cells have the patient's genetic make up, and would not be rejected upon grafting. SCNT consists in isolating nucleus of a somatic cell type (fibroblast for example) harvested from the future recipient into an enucleated oocyte [23]. By mechanisms still unknown, the cytoplasm of the oocyte reprograms the chromosomes of the somatic cell's nucleus. The cloned cell develops into a blastocyst from which ESCs can be derived. These ESCs carry a set of chromosomes identical to that of the donor, and therefore are unlikely to be rejected by that donor/future recipient upon grafting [24-26]. Though SCNT has such potential, the isolation and generation of ESCs by SCNT for therapy is not without limitations and controversies. Among them, there are ethical and political issues over SCNT and therapeutic cloning [27, 28], the cloning of individuals to get matching cells, tissues, or organs for therapy [29]. There are also unknown regarding the behavior of the generated cell lines and tissues by SCNT, in term of viability, development [30-32]. Finally, the generation of isogeneic hESCs by SCNT remains to be achieved [33].

In all, these data show strategies to establish lines of ESCs for cellular therapy have been devised, and would provide suitable source of tissue for therapy. However, the use of ESCs for therapy is not without controversies, debates, and limitations.

Controversies, Debates and Limitations over the Use of ESCs for Therapy

Ethical and Political Controversies and Debates

The use of human embryos and fetuses for research and therapy is under strict regulation, and associated with ethical issues. Because it involves the destruction of embryos, the derivation of ESCs from human blastocysts and their use in clinical research and therapy is subject to intense debates and controversies [34]. In the US, only research project working on a list of 78 hESC lines established prior August 9, 2001 can be funded by federal organisms. As these cell lines were established using animals derivatives, they are unsuitable for therapy.

In an attempt to overcome the ethical and political gridlock over the use of ESCs for cellular therapy, investigators are aiming at developing strategies to derive ESCs without the destruction of embryos. Chung et al. (2005) used single-cell embryo biopsy, a technique similar to pre-implantation genetic diagnosis (PGD) of genetic defects used in fertility clinics, to generate ESCs [35]. PGD consists in extracting a cell -to be used for genetic testing- from an 8 cells stage embryo (blastomere), a procedure that does not interfere with the developmental potential of embryos. The investigators extracted and culture single cells from eight-cell mouse blastomeres (2 days old). The isolated cells behaved like ES cells. The embryos went on to produce mice. In another study, Meissner and Jaenisch (2005) used a variation of SCNT, called altered nuclear transfer (ANT), to derive ESCs [36]. ANT has been proposed as a variation of nuclear transfer [37]. In ANT, a gene crucial for trophectoderm development, like the gene CDX2, is inactivated in the donor. CDX2 encodes the earliest-known trophectoderm-specific transcription factor that is activated in the 8-cell embryo and is essential for establishment and function of the trophectoderm lineage. Inactivating the gene CDX2 would eliminate the potential to form the fetal-maternal interface, but will spare the ICM. The eggs created by nuclear transfer from the nucleus inactivated for the gene crucial for trophectoderm development produce embryos that are unable to implant into the uterus and so do not pursue their developments, but ESCs could be generated from the eggs. Meissner and Jaenisch (2005) reported that inactivation of CDX2 using a lentivirus produces cloned blastocysts morphologically abnormal that are lacking functional trophoblast and fails to implant into the uterus. Yet, the eggs divided and grew enough to yield stem cells [36].

In both studies, the authors were able to generate ESCs without the destruction of embryos. However, the acceptance of such strategies to derive ESCs for therapy is not without problems, controversies and debates [38-40]. The derivation of ESCs from single blastomere suggests that when clinics perform PGDs, the isolated cells could be grown. Resulting cultures would then be used for genetic testing, and to establish stem cell lines. This strategy has been recently applied to derive hESCs [41]. However, the practice of such strategy is not without limitations and limited to case where PGD is performed, in most case due to genetic malformation, limiting the use of the established cell lines for therapy. Regarding the derivation of ESCs by ANT, one of the limitations in the reported procedure is the use of virus (lentil virus) to inactivate the gene CDX2. Such genetic manipulation may affect adversely the ES cells, and may present some risks for the recipient. Further, in the case of ANT, it has been argued that finding acceptable to destroy a CDX2 mutant embryo but not a normal embryo is "a flawed proposal", as there is no basis for concluding that the action of CDX2, or any other gene, represents a transition point at which a human embryo acquires moral status [42]. Thus, whether ANT solves the ethical dilemma of whether the mutant embryo is equivalent to normal embryo remain the source of debates and controversies [43]. In all, although these reports proposed alternative protocols to derive ES cells, their use for therapy remains the source of controversies and debates [38-40].

Tumorigenic Risk

ES cells have the potential to form tumor upon grafting [44]. The formation of teratomas would be associated with the undifferentiated state of ESCs. To reduce the risk of tumor formation, it is proposed to pre-differentiate ESCs in vitro to the desired lineage, and to remove the cells that have not differentiated from the cellular graft prior transplantation. Protocol leading to 100% differentiation, or purification by positive or negative selection -by isolating the differentiated cells from the bulk culture- would provide strategies to this aim. Experimental studies in which high yield differentiated ESCs into oligodendrocytes and cell sorting of pre-differentiated ESCs were transplanted into animal models, reported successful integration of ESCs without tumor formation [45, 46]. This highlights the potential of ESCs for therapy.

Conclusion

The derivation of hESCs has been a major step toward bringing ESC research to therapy. However, there are major obstacles to overcome before ESC research can be applied in therapy. Particularly, the generation of isogeneic lines of ESCs would represent a model of choice for stem cell therapy, and the risk of tumor formation associated with ECSs must be circumvented. The ethical and political concerns remain a major limitation to the progress of ESC and SCNT research. Alternative procedures to derive ESCs without the destruction of embryos and controversies remain to be established and accepted. Nonetheless, ESCs have a tremendous potential to cure a broad range of diseases and injuries, but also for our understanding of developmental biology. Future research will aim at unraveling the mechanisms of differentiation of ESCs to generate various lineages, tissues and organs for therapy.

Acknowledgments

P.T. is supported by grants from the NMRC, BMRC, and the Juvenile Diabetes Research Foundation.

REFERENCES

[1] Wobus AM, Boheler KR. (2005) Embryonic stem cells: prospects for developmental biology and cell therapy. *Physiol. Rev.* 85, 635-78.

[2] Evans MJ, Kaufman MH. (1981) Establishment in culture of pluripotential cells from mouse embryos. *Nature*. 292, 154–56.

[3] Bongso A, Fong CY, Ng SC, Ratnam S. (1994) Isolation and culture of inner cell mass cells from human blastocysts. *Hum. Reprod.* 9, 2110-7.

[4] Thomson JA, Kalishman J, Golos TG, Durning M, Harris CP, Becker RA, Hearn JP. (1995) Isolation of a primate embryonic stem cell line. *Proc. Natl. Acad. Sci. USA*. 92, 7844-8.

[5] Thomson JA, Itskovitz-Eldor J, Shapiro SS, Waknitz MA, Swiergiel JJ, Marshall VS, Jones JM. (1998) Embryonic stem cell lines derived from human blastocysts. Science. 282, 1145-7. Erratum in: (1998) *Science*. 282, 1827.

[6] Taupin. (2006) Derivation of embryonic stem cells for cellular therapy: challenges and new strategies. *Med. Sci. Monit.* 12, RA75-8.

[7] Reubinoff BE, Pera MF, Fong CY, Trounson A, Bongso A. (2000) Embryonic stem cell lines from human blastocysts: somatic differentiation in vitro. Nat Biotechnol. 18, 399-404. Erratum in: (2000) *Nat. Biotechnol.* 18, 559.

[8] Pedersen EB, Widner H. (2000) Xenotransplantation. *Prog. Brain Res.* 127, 157-88.

[9] Martin MJ, Muotri A, Gage F, Varki A. (2005) Human embryonic stem cells express an immunogenic nonhuman sialic acid. *Nat. Med.* 11, 228-32.

[10] Higashi H, Naiki M, Matuo S, Okouchi K. (1977) Antigen of "serum sickness" type of heterophile antibodies in human sera: indentification as gangliosides with N-glycolylneuraminic acid. *Biochem. Biophys. Res. Commun.* 79, 388-95.

[11] Richards M, Fong CY, Chan WK, Wong PC, Bongso A. (2002) Human feeders support prolonged undifferentiated growth of human inner cell masses and embryonic stem cells. *Nat. Biotechnol.* 20, 933-6.

[12] Stojkovic P, Lako M, Stewart R, Przyborski S, Armstrong L, Evans J, Murdoch A, Strachan T, Stojkovic M. (2005) An autogeneic feeder cell system that efficiently supports growth of undifferentiated human embryonic stem cells. *Stem Cells*. 23, 306-14.

[13] Xu C, Inokuma MS, Denham J, Golds K, Kundu P, Gold JD, Carpenter MK. (2001) Feeder-free growth of undifferentiated human embryonic stem cells. *Nat. Biotechnol.* 19, 971-4.

[14] Klimanskaya I, Chung Y, Meisner L, Johnson J, West MD, Lanza R. (2005) Human embryonic stem cells derived without feeder cells. *Lancet.* 365, 1636-41.

[15] Xu RH, Peck RM, Li DS, Feng X, Ludwig T, Thomson JA. (2005) Basic FGF and suppression of BMP signaling sustain undifferentiated proliferation of human ES cells. *Nat. Methods.* 2, 185-90.

[16] Ludwig TE, Levenstein ME, Jones JM, Berggren WT, Mitchen ER, Frane JL, Crandall LJ, Daigh CA, Conard KR, Piekarczyk MS, Llanas RA, Thomson JA. (2006) Derivation of human embryonic stem cells in defined conditions. *Nat. Biotechnol.* 24, 185-7.

[17] Draper JS, Smith K, Gokhale P, Moore HD, Maltby E, Johnson J, Meisner L, Zwaka TP, Thomson JA, Andrews PW. (2004) Recurrent gain of chromosomes 17q and 12 in cultured human embryonic stem cells. *Nat. Biotechnol.* 22, 53-4.

[18] Buzzard JJ, Gough NM, Crook JM, Colman A. (2004) Karyotype of human ES cells during extended culture. *Nat. Biotechnol.* 22, 381-2.

[19] Maitra A, Arking DE, Shivapurkar N, Ikeda M, Stastny V, Kassauei K, Sui G, Cutler DJ, Liu Y, Brimble SN, Noaksson K, Hyllner J, Schulz TC, Zeng X, Freed WJ, Crook J, Abraham S, Colman A, Sartipy P, Matsui SI, Carpenter M, Gazdar AF, Rao M, Chakravarti A. (2005) Genomic alterations in cultured human embryonic stem cells. *Nat. Genet.* 37, 1099-103.

[20] Barker RA, Widner H. (2004) Immune problems in central nervous system cell therapy. *NeuroRx.* 1, 472-81.

[21] Trounson A, Pera M. (1998) Potential benefits of cell cloning for human medicine. *Reprod. Fertil. Dev.* 10, 121-5.

[22] Lanza RP, Cibelli JB, West MD. (1999) Human therapeutic cloning. *Nat. Med.* 5, 975-7.

[23] Campbell KH, McWhir J, Ritchie WA, Wilmut I. (1996) Sheep cloned by nuclear transfer from a cultured cell line. *Nature.* 380, 64-6.

[24] Kawase E, Yamazaki Y, Yagi T, Yanagimachi R, Pedersen RA. (2000) Mouse embryonic stem (ES) cell lines established from neuronal cell-derived cloned blastocysts. *Genesis.* 28, 156-63.

[25] Wakayama T. (2003) Cloned mice and embryonic stem cell lines generated from adult somatic cells by nuclear transfer. *Oncol. Res.* 13, 309-14.

[26] Wakayama T. (2006) Establishment of nuclear transfer embryonic stem cell lines from adult somatic cells by nuclear transfer and its application. *Ernst Schering Res. Found Workshop.* 60, 111-23.

[27] Jaenisch R. (2004) Human cloning - the science and ethics of nuclear transplantation. *N. Engl. J. Med.* 351, 2787-91.

[28] Holden C. (2005) Stem cell research. Korean cloner admits lying about oocyte donations. *Science*. 310, 1402-3.

[29] Lanza RP, Cibelli JB, West MD. (1999) Human therapeutic cloning. *Nat. Med.* 5, 975-7.

[30] Shiels PG, Kind AJ, Campbell KH, Waddington D, Wilmut I, Colman A, Schnieke AE. (1999) Analysis of telomere lengths in cloned sheep. *Nature*. 399, 316-7.

[31] Rhind SM, Taylor JE, De Sousa PA, King TJ, McGarry M, Wilmut I. (2003) Human cloning: can it be made safe? *Nat. Rev Genet*. 4, 855-64.

[32] Wakayama S, Jakt ML, Suzuki M, Araki R, Hikichi T, Kishigami S, Ohta H, Van Thuan N, Mizutani E, Sakaide Y, Senda S, Tanaka S, Okada M, Miyake M, Abe M, Nishikawa S, Shiota K, Wakayama T. (2006) Equivalency of nuclear transfer-derived embryonic stem cells to those derived from fertilized mouse blastocysts. *Stem Cells*. 24, 2023-33.

[33] Kennedy D. (2006) Editorial retraction *Science.* 311, 335.

[34] Childress JF. (2004) Sources of stem cells: ethical controversies and policy developments in the United States. *Fetal. Diagn. Ther*. 19, 119-23.

[35] Chung Y, Klimanskaya I, Becker S, Marh J, Lu SJ, Johnson J, Meisner L, Lanza R. (2006) Embryonic and extraembryonic stem cell lines derived from single mouse blastomeres. *Nature*. 439, 216-9.

[36] Meissner A, Jaenisch R. (2006) Generation of nuclear transfer-derived pluripotent ES cells from cloned Cdx2-deficient blastocysts. *Nature*. 439, 212-5.

[37] Hurlbut WB. (2005) Altered nuclear transfer. *N. Engl. J. Med.* 352, 1153-4.

[38] Weissman IL. (2006) Medicine: politic stem cells. *Nature*. 439, 145-7.

[39] Solter D. (2005) Politically correct human embryonic stem cells? *N. Engl. J. Med.* 353, 2321-3.

[40] Jaenisch R, Meissner A. (2006) Politically correct human embryonic stem cells? *N. Engl. J. Med.* 354, 1208-9.

[41] Klimanskaya I, Chung Y, Becker S, Lu SJ, Lanza R. (2006) Human embryonic stem cell lines derived from single blastomeres. *Nature.* 444, 481-5. Erratum in: (2006) *Nature*. 444, 512.

[42] Melton DA, Daley GQ, Jennings CG. (2004) Altered nuclear transfer in stem-cell research - a flawed proposal. *N. Engl. J. Med.* 351, 2791-2.

[43] Hurlbut WB. (2005) Altered nuclear transfer as a morally acceptable means for the procurement of human embryonic stem cells. *Perspect. Biol. Med.* 48, 211-28.

[44] Wakitani S, Takaoka K, Hattori T, Miyazawa N, Iwanaga T, Takeda S, Watanabe TK, Tanigami A. (2003) Embryonic stem cells injected into the mouse knee joint form teratomas and subsequently destroy the joint. *Rheumatology* (Oxford). 42, 162-5.

[45] Menard C, Hagege AA, Agbulut O, Barro M, Morichetti MC, Brasselet C, Bel A, Messas E, Bissery A, Bruneval P, Desnos M, Puceat M, Menasche P. (2005) Transplantation of cardiac-committed mouse embryonic stem cells to infarcted sheep myocardium: a preclinical study. *Lancet*. 366, 1005-12.

[46] Keirstead HS, Nistor G, Bernal G, Totoiu M, Cloutier F, Sharp K, Steward O. (2005) Human embryonic stem cell-derived oligodendrocyte progenitor cell transplants remyelinate and restore locomotion after spinal cord injury. *J. Neurosci.* 25, 4694-705.

INDEX

A

B

C

D

E

F

G

H

I

J

K

L

M

N

O

P

Q

R

S

T